VOLUME ONE HUNDRED AND ONE

ADVANCES IN FOOD AND NUTRITION RESEARCH

Emerging Sources and Applications of Alternative Proteins

Edited by

JIANPING WU

Department of Agricultural, Food and Nutritional Science, University of Alberta, Edmonton, AB, Canada

ELSEVIER

ACADEMIC PRESS

An imprint of Elsevier

Academic Press is an imprint of Elsevier
50 Hampshire Street, 5th Floor, Cambridge, MA 02139, United States
525 B Street, Suite 1650, San Diego, CA 92101, United States
The Boulevard, Langford Lane, Kidlington, Oxford OX5 1GB, United Kingdom
125 London Wall, London, EC2Y 5AS, United Kingdom

First edition 2022

ISBN: 978-0-323-98380-8
ISSN: 1043-4526

For information on all Academic Press publications
visit our website at https://www.elsevier.com/books-and-journals

Publisher: Zoe Kruze
Acquisitions Editor: Sam Mahfoudh
Developmental Editor: Jhon Michael Peñano
Production Project Manager: Vijayaraj Purushothaman
Cover Designer: Vicky Pearson Esser

Typeset by STRAIVE, India

Contents

Contributors

Fatma Boukid
Food Safety and Functionality Programme, Institute of Agriculture and Food Research and Technology (IRTA), Monells, Spain

Xinyi Cao
Key Laboratory of Environmental Medicine and Engineering of Ministry of Education, and Department of Nutrition and Food Hygiene, School of Public Health, Southeast University, Nanjing, Jiangsu, China

Lingyun Chen
Department of Agricultural, Food & Nutritional Science, University of Alberta, Edmonton, AB, Canada

Sghaier Chriki
ISARA – Agro School for Life, Agroecology and Environment Unit, Lyon, France

Marie-Pierre Ellies-Oury
Bordeaux Sciences Agro, Gradignan; INRAE, University of Clermont-Ferrand, VetAgro Sup, Saint Genès Champanelle, France

Mohammed Gagaoua
Food Quality and Sensory Science Department, Teagasc Food Research Centre, Ashtown, Dublin, Ireland

Jean-François Hocquette
INRAE, University of Clermont-Ferrand, VetAgro Sup, Saint Genès Champanelle, France

Shan Hong
Department of Grain Science and Industry, Kansas State University, Manhattan, KS, United States

Weijuan Huang
College of Food Science, South China Agricultural University, Guangzhou, China; Department of Agricultural, Food & Nutritional Science, University of Alberta, Edmonton, AB, Canada

Yonghui Li
Department of Grain Science and Industry, Kansas State University, Manhattan, KS, United States

Wang Liao
Key Laboratory of Environmental Medicine and Engineering of Ministry of Education, and Department of Nutrition and Food Hygiene, School of Public Health, Southeast University, Nanjing, Jiangsu, China

Andrea M. Liceaga
Protein Chemistry and Bioactive Peptides Laboratory, Department of Food Science, Purdue University, West Lafayette, IN, United States

Yanting Shen
Department of Grain Science and Industry, Kansas State University, Manhattan, KS, United States

Pedro F. Souza Filho
Laboratory of Biochemical Engineering, Chemical Engineering Department, Federal University of Rio Grande do Norte (UFRN), Natal, RN, Brazil

Amir Vahedifar
Department of Agricultural, Food and Nutritional Science, University of Alberta, Edmonton, AB, Canada

Shaokang Wang
Key Laboratory of Environmental Medicine and Engineering of Ministry of Education, and Department of Nutrition and Food Hygiene, School of Public Health, Southeast University, Nanjing, Jiangsu, China

Jianping Wu
Department of Agricultural, Food and Nutritional Science, University of Alberta, Edmonton, AB, Canada

Preface

This volume of *Advances in Food and Nutrition Research* comprises nine chapters written by energetic authors who are actively leading research and innovation in alternative proteins. Alternative proteins include proteins from plants (i.e., grains, legumes, pulses, and nuts), fungi (i.e., mushrooms), algae, insects, and cultured (lab-grown) meat, which can be used as alternatives to conventional animal proteins (livestock and fish). Except for cultured meat, most alternative proteins have been an integral part of human diet for centuries. Fueled by population growth as well as increasing environmental and sustainable concerns, alternative proteins are expected to provide a solution to meet the growing protein demand within environmental limits. The major concerns for human consumption of alternative proteins are their inferior organoleptic properties, consumer acceptability, affordability, and sustainability. This volume aims to provide an overview of the recent advances in researching and developing alternative proteins in food and nutritional products.

Chapter 1 provides a high-level overview of alternative proteins based on three fundamental questions: Why do we need alternative proteins? What are alternative proteins? What are the applications of alternative proteins? The chapter also highlights the need to develop culturally diversified alternative protein foods. Chapter 2 focuses on the extraction, nutrition, functionality, and commercial applications of canola proteins in food formulations. Despite not being considered traditional food proteins, the recent GRAS status awarded to canola proteins has paved the way for their use in food applications. Chapter 3 presents a comprehensive review of pea protein composition and structures, functional properties, modification methods, and specific food applications. In particular, it provides fundamental knowledge of pea protein structures and functional improvements for potential food applications. Chapter 4 provides a historic review of human consumption of edible insects (human entomophagy), the roles of traditional (e.g., sun-drying, roasting, boiling, roasting, baking, and frying) and commercial (e.g., pasteurization, enzymatic proteolysis, and high-pressure processing) processing technologies, applications of insect proteins in food and beverage formulations, and the challenges and future prospects of insect proteins. Chapter 5 reviews the production and use of single-cell proteins in new formulations and their associated regulatory, safety, nutritional, ethical, social, and environmental considerations. Chapter 6 describes the current knowledge on the production, research, and commercial activities, challenges, and issues related to the production, marketing, legislation, ethics, religious status, consumer perception, health and nutritional qualities, and environmental impact of cultured meat. Chapter 7 summarizes the main alternative proteins to muscle foods and discusses their advantages and disadvantages, including the current challenges and limitations in research and development of such new food products and meat alternatives. Chapter 8 introduces the major techniques for fabrication of nanoparticles and nanoemulsions from various food proteins and the approaches to modulate particle size and bioactive molecule encapsulation and release. The advantages and disadvantages of different techniques are also discussed. Because of their high nutritional value, good functional properties, and health-benefiting effects, nanoparticles and nanoemulsions developed from food proteins show great potential as novel delivery systems for poorly absorbed and/or labile bioactive molecules to improve their bioavailability. Chapter 9 provides an overview of the production, bioactivities, bioavailability, safety, and, more importantly, clinical evidence of bioactive peptides in the management of nutrition-related chronic diseases.

In this volume, soy proteins are not included as they have well-established uses in many food products. Regretfully, cereal proteins should have been included given their significance in protein nutrition. This volume has been made possible owing to the excellent contributions of committed authors and high-quality editorial assistance provided by Mr. Jhon Michael Peñano, Developmental Editor at Elsevier. Last but not least, the kind encouragement from Professor Fidel Toldrá, Editor of the *Advances in Food and Nutrition Research* book series, is greatly acknowledged.

Jianping Wu
Department of Agricultural, Food and Nutritional Science
University of Alberta, Edmonton, Alberta, Canada

CHAPTER ONE

Emerging sources and applications of alternative proteins: An introduction

Jianping Wu*
Department of Agricultural, Food and Nutritional Science, University of Alberta, Edmonton, AB, Canada
*Corresponding author: e-mail address: jwu3@ualberta.ca

Contents

Abstract

Food protein is an essential macronutrient. Even though daily per capita supply of protein has increased globally from 61 g in 1961 to 81 g in 2013, and most people in the developed world have sufficient protein intake from their diets, however, protein deficiencies continue to be pervasive globally. Protein deficiency is the single major factor responsible for impaired growth and suboptimal health worldwide. Animal proteins are high quality and contain adequate and balanced amino acids, animal protein production however is inefficient and resource intensive. Alternative proteins are expected to provide the solution to meet the growing protein demand within the environmental limits. Alternative proteins include proteins from plants (i.e., grains, legumes, pulse, and nuts), fungus (i.e., mushrooms), algae, insects and cultured (lab-grown) meat that can be used to replace conventional animal proteins. Major concerns for human consumption of alternative proteins are inferior organoleptic properties, consumer acceptability, affordability, and sustainability. There is a need to develop culturally diversified alternative proteins to mitigate global protein malnutrition. Food proteins are also found applications in biomaterials and as a source of bioactive peptides.

1. Why do we need alternative proteins?

Protein is an essential macronutrient that human cannot synthesize and must be obtained from foods, and adequate food protein intake is indispensable for optimal growth and health (WHO, 2007; Wu et al., 2014).

Advances in Food and Nutrition Research, Volume 101
ISSN 1043-4526
https://doi.org/10.1016/bs.afnr.2022.06.001

Food protein is the ultimate source of the amino acids used to make protein in humans. WHO recommends an intake level of 0.83 g/kg body weight per day for adults with a protein digestibility-corrected amino acid score value of 1.0 (WHO, 2007). Animal proteins are high quality and contain adequate and balanced amino acids, while most plant proteins are inferior in quality and are "incomplete" in that they are deficient in one or more of the essential amino acids, and contain several antinutritional factors such as protease inhibitors, phytates and phytochemicals that may inhibit protein digestion and absorption (Wu et al., 2014). Thus, adjustments need to be made for protein quality to assess protein adequacy (WHO, 2007). Plant- and animal-based foods contribute to ~60% and ~40%, respectively (FAO, 2020 via Our World in Data; Roser & Ritchie, 2013), of protein intake in human diets worldwide, compared to ~35% and ~65% in United States. Daily per capita supply of protein globally has increased from 61 g in 1961 to 81 g in 2013 (FAO, 2020 via Our World in Data; Roser & Ritchie, 2013), and most people in the developed world have sufficient protein intake from their diets, however, protein deficiencies continue to be pervasive globally. There is approximately 1 billion people worldwide have inadequate protein intake (Wu et al., 2014); especially for infants and children, protein deficiency is the single major factor responsible for impaired growth and suboptimal health worldwide (Ghosh, Suri, & Uauy, 2012).

Although animal protein is superior to plant proteins in terms of quality, animal protein production however is inefficient and resource intensive; for example, livestock production uses 77% agricultural land and one-third fresh water, and generates ~58% greenhouse gas emissions of all food production while providing only ~18% of calories and ~37% of total protein (FAO, 2020 via Our World in Data; Ritchie & Roser, 2021a). Despite this, a 76% increase in global meat consumption is projected by 2050 (Alexandratos & Bruinsma, 2012), which is driven by population growth and increase in average individual incomes. Meeting the growing global protein demand while being sustainable is a daunting challenge. Protein sustainability is further complicated by recent economic downturns caused mainly by the global pandemic, and the global climate change, such as the 2022 Canada's heat dome, deadly floods in Europe, deluge in China, and California's wildfires (FAO, IFAD, UNICEF, WFP, & WHO, 2021; Laborde, Martin, Swinnen, & Vos, 2020; Markandya, Salcone, Hussain, Mueller, & Thambi, 2021; Rzymski et al., 2021; Van Vliet, Kronberg, & Provenza, 2020).

In addition to environmental and sustainable concerns over animal protein production, a high animal protein consumption is associated with an increased risk of obesity, type 2 diabetes, cardiovascular disease, and some cancers like colorectal cancer (Bouvard et al., 2015; Hu, Otis, & McCarthy, 2019). Consumption of plant-based proteins, on the other hand, has been reported to exert health benefits such as reduced low-density lipoprotein cholesterol and body weight (Crimarco et al., 2020), reduced risk of type 2 diabetes (De la Cruz-Ares et al., 2021; Sluik et al., 2019) and cardiovascular disease (Song et al., 2016), and a reduction in all-cause mortality (Huang et al., 2020). Canada's Food Guide (2022) recommends "eat a variety of protein foods as part of a healthy eating pattern"; similarly, Dietary Guidelines for American 2021–2025 (USDA & HHS, 2020) also recommends "select a wide variety of protein foods to get more of the nutrients your body needs and for health benefits," suggesting the significance of plant proteins and alternatives on body needs and health benefits. Alternative proteins are generally considered to be healthier and more environmentally friendly than traditional meats, and the market share is expected to grow continuously (Gravely & Fraser, 2018; Onwezen, Bouwman, Reinders, & Dagevos, 2021). A transition to alternative proteins, such as plant, insect, fungal and microbial proteins, provides a promise to feed an estimated 10 billion people in 2050 (Willett et al., 2019).

2. What are the alternative proteins?

Broadly speaking, alternative proteins are alternatives to conventional animal proteins. They include plants (i.e., grains, legumes, pulse, and nuts), fungus (i.e., mushrooms), algae, insects and cultured (lab-grown) meat (Bekhit, Riley, & Hussain, 2022; Hartmann & Siegrist, 2017; van der Weele, Feindt, van der Goot, van Mierlo, & van Boekel, 2019). Most alternative proteins have been an integrated part of human diet in regions for centuries, but are not necessarily acceptable to other regions of the world where consumption of these alternative proteins is perceived as novel. For example, insects are consumed by many Asian and African countries but considered new to most Western consumers. Familiarity is a key factor for consumer acceptance of alternative proteins (Onwezen et al., 2021); among alternative proteins, plant-based meat substitutes are among the most accepted ones, followed by cultured meat and insects (Circus & Robison, 2018; Iannuzzi, Sisto, & Nigro, 2019). Soy proteins are by far the dominate plant protein in the food industry. Soy is traditionally processed in China and other Asian countries into soy milk, tofu, and among others as the major source of protein. In comparison, soy is fractionated for edible oil initially, and then proteins (soy flour, soy concentrate, and soy isolate) in US. Worldwide soy production has been increasing from 20 to 30 million tonnes per year in the 1960s to >350 million tonnes per year now (Food and Agriculture Organization (FAO) of the United Nations, 2020 via Our World in Data; Ritchie & Roser, 2021b), of which 19.2% is used for direct human consumption and 77% for animal feed uses (USDA, 2022). Soy proteins have the best functionalities compared to other plant proteins; however, soy proteins are associated with concerns such as a major allergic source and the prevalence of genetically modified varieties (Vogelsang-O'Dwyer, Zannini, & Arendt, 2021). Pulse proteins especially pea proteins are considered strong contender to soy proteins, while canola proteins are an emerging one.

- With a global production of 75 million tonnes in recent years, canola/rapeseed is second only to soy in terms of production (FAO, 2020). Canola proteins are not traditionally used for human consumption, probably due to the presence of antinutritional factors and the difficulty of protein extraction. Canola proteins have now received a status of Generally Approved as Safe (GRAS, 2010, 2011, 2016) by US Food and Drug Administration (FDA), and several companies, such as Merit Functional Foods and DSM's collaboration with Avril Group, have been actively pursuing their commercialization. Praised as an emerging contender to soy proteins, it is expected to see more food applications of canola proteins in the next 5–10 years. Chapter 2 (Vahedifar & Wu, 2022) focuses on methods of canola protein extraction, nutritional, biological and functional properties of canola proteins, as well as their applications in food formulations. Other important protein sources of oilseeds include sunflower, peanut, mustard, coconut, sesamum, castor seeds, cotton seeds, linseed, and hempseed.
- Pulse proteins are a real and strong contender for soy proteins. Unlike soy proteins, pulse proteins are positioned as MGO-free, and not a major source of allergen (which is debatable). Pulses refer to dry edible seeds of legumes that can include green and yellow peas, faba beans, lentils, chickpeas, adzuki beans and mung beans (Derbyshire & Delange, 2021), and are by the far the fastest growing legume, doubled to 70 million tonnes in 2010, and double again in 2050 (Global Pulse Confederation, 2022). The protein content vary among pulses and their cultivars; for example, the crude protein content of dry pea, chickpea, and lentils range from 23–31%, 16–26%, 21–30%, respectively (Chibbar, Ambigaipalan, & Hoover, 2010), whereas faba bean cultivars

typically range from 26% to 34% (Ivarsson & Neil, 2018). Not considered as oilseeds, most pulses also contain high amount of starch and other carbohydrates (Boye, Zare, & Pletch, 2010), which make them suitable for dry fractionation into a protein-rich fraction and a starch-rich fraction (Saldanha do Carmo et al., 2020). Dry fractionation can maintain the native protein structure with lower energy use compared to wet fractionation. At present, pea proteins are the only widely available pulse proteins for food application. Chapter 3 (Shen, Hong, & Li, 2022) provides a comprehensive review on pea protein composition and structures, functional properties, modification methods, and specific food applications. In particular, it provides the fundamental knowledge of pea protein structures and functional improvements for potential food applications.

- Insects are the most diverse animals. According to Royal Entomological Society, it is estimated that there may be as many as 10 million species of insects that make up to 90% of all species of animals on the planet, while only one-tenth has been discovered (Royal Entomological Society, 2022). Insects have been eaten as a food throughout the history of humanity, and continue to play a role in the diet of at least two billion people worldwide (van Huis, van Itterbeeck, Klunder, et al., 2013). Edible insects are nutritious and contain high quality proteins. The protein content of insects varies greatly, ranging from 13% to 77% on a dry matter basis (Akhtar & Isman, 2018). Touted as "minilivestock," with an estimated more than 2,000 edible insect species worldwide (Jongema, 2017), they can be produced anywhere, including in the zero gravity environment of outer space as the only sustainable protein source of space foods (Dunkel & Payne, 2016). There is a growing interest in edible insects worldwide as they provide an ideal solution to address protein sustainability and environmental concerns, due to their high fecundity, high fecundity, short reproductive cycle, high efficiency of feed conversion, and low level of greenhouse gases emission (Dunkel & Payne, 2016). Chapter 4 (Liceaga, 2022) provides a historic review of human consumption of insects (human entomophagy), the roles of traditional (e.g., sun-drying, roasting, boiling, roasting, baking, frying) and commercial processing technologies (e.g., pasteurization, enzymatic proteolysis, high pressure processing), and applications of insect proteins in food and beverage formulations, challenges and future prospects of insect proteins.
- Single-cell proteins (SCP) refer to microbial proteins that was coined by Carol. Wilson in 1966, as most microorganism grows as single or

filamentous individuals (Hosseininezhad & Hussain, 2022). The first industrial production of SCP was in Germany during the first World War using the yeast *Saccharomyces cerevisiae*. Besides yeast, fungi, algae and bacteria are also used for the production of SCP. SCP production has advantages in terms of high efficiency in substrate conversion, high rate of multiplication, independence of seasonal factors, and the ability of use a variety of raw materials (mostly agricultural, forestry and food byproducts and waste). However, commercial activities are still limited due to high capital and operating costs along with approval costs. The most well-known mycoprotein is the Quorn™ brand launched by Marlow Foods (UK) in 1985 using the filamentous fungus *Fusarium venenatum* (Ritala, Häkkinen, Toivari, & Wiebe, 2017). Chapter 5 (Souza Filho, 2022) provides a review on the production and use of SCP in new formulations and their associated regulatory, safety, nutritional, ethics, social, and environmental considerations.

- Cultured meat, also called in vitro meat, is produced by growing stem cells collected from animals (livestock animals and fish) in a bioreactor. Cultured meat has caught extensive media attention since the launch of the first cell-based burger by Mosa Meat that costs almost around $400,000, and its ability to attract investments from major multinational companies such as Cargill, Mitsubishi, JBS, ADM, Tyson, Upside Foods, and celebrity investors such as Bill Gates, Richard Branson, Ashton Kutcher, Leonardo DiCaprio (Jones, 2022). Although most are at lab and pilot-scale stages, the Singapore Food Agency (SFA) has granted Eat Just (start-up) approval to sell its lab grown, cultured chicken in Singapore, and many companies claimed they are ready to market their products in a couple of years (Ellies-Oury, Chriki, & Hocquette, 2022). The market of cultured meat is estimated to be worth USD $25 billions by 2030 according to McKinsey and Company (2021). The same technology can be used to produce other types of animal products such as milk, egg white, seafood, and even honey. Cellular agriculture is a term widely used to describe the process of farming animal products from cells instead of animals (Cellular Agriculture Society, 2021). Chapter 6 (Ellies-Oury, Chriki, & Hocquette, 2022) describes current knowledge on the production, overview research and commercial activities around the world, challenges and issues related to production, market, legislation, ethics, religious status, consumer perception, health and nutritional qualities, and environmental impact.

With a protein content ranging from ~8% (rice) to ~12% (wheat), cereals have been overlooked as a significant source of proteins (Poutanen et al., 2022). As the staple foods, cereals provide more than half of the total protein production in the world, especially in developing countries where cereal proteins can account for as much as 70% to 90% of the total protein consumption (Lasztity, 2017). Gluten is probably the most extensively studied cereal proteins, due to its unique dough formation and baking properties. Gluten is also a major co-products of starch production, totaling ~1 million tonnes annually (IndexBox, 2022). Gluten has been traditionally processed into seitan as one of the oldest meat analogs; gluten remains an important ingredient in Beyond meat, while its major use is in the baking industry. In addition to one of the big eight food allergens, gluten is also the offending protein for celiac disease, or gluten-sensitive enteropathy, characterized by malabsorption of nutrition as a consequence of damage to the absorptive epithelial cells in the small intestine (Chaudhry, Jacobs, Green, & Rampertab, 2021). Driven by the gluten-free trend, the use of other cereals, such as rice, maize, sorghum, millet, buckwheat, is expected to rise. Cereal processing generates a great deal of by products that are promising source of cereal proteins. Secondly only to wheat, with a worldwide production over 780 million tonnes (Ritchie & Roser, 2020), rice processing can yield ~ 78 million tonnes of rice bran (based on average 10% yield) and ~62 million tonnes of broken rice (estimated on 8% yield) (Bodie, Micciche, Atungulu, Rothrock, & Ricke, 2019); the brewing industry generates about 40 million tonnes of brewers' spent grain worldwide which contains 19–30% protein (Mussatto, 2014); bio-ethanol industry generates millions of tonnes of DDGS that is suitable for protein extraction (Gupta, Wilson, & Vadlani, 2016); Chinese Baijiu sector produces roughly 100 million tons of vinasse every year (Liu et al., 2022).

3. What are the applications of alternative proteins?

Most alternative proteins, except for canola protein and cultured meat, have been a part of human diet for centuries. Humankind has developed methods to include them into their diet: they are minimally processed, i.e., sun-drying, roasting, boiling, steaming, frying, and consumed as whole without fractionation, such as algae, edible insects, mushroom; some plant materials are milled and further processed to improve the nutritional value and taste, such as cereals and the most well-known example of traditional soy products such as soy milk and tofu; they are also incorporated into different foodstuffs, often after appropriate processing, such as edible insect flour and pulse flour. The use of whole insect or pulse flour in food is preferred as this fulfills the trend of whole foods; for example, gluten-free bread made with cricket flour, rice flour and cornstarch showed acceptable hardness and chewiness of crumbs and a high protein content (40% and 100% increase in protein content at 10% and 20% addition of cricket flour, respectively, compared to the control) (da Rosa Machado & Thys, 2019); adding pulse flour into bread boosts both the levels of protein and fiber (Bresciani & Marti, 2019). Direct use in food formulation however may be restricted due to some sensorial and technological constraints, and the presence of anti-nutrient compounds; traditional processing including germination and fermentation are widely used to improve nutritional and sensorial properties.

Unlike the traditional uses, modern uses of alternative proteins are first to fractionate protein, and then to be incorporated in a wide range of food products such as cereal products, beverage, meat products, powder, and supplements. Even though the protein content has been significantly increased in comparison to the starting material, the structure of protein is likely affected during processing, and the functionality is impaired; the functionality challenge of protein extracts suggests the necessity of protein modifications (Akharume, Aluko, & Adedeji, 2021; Nasrabadi, Doost, & Mezzenga, 2021).

- The major application of alternative proteins is to replace animal proteins, in particular for use as meat alternatives. Alternative proteins are subjected to texturization to form meat-like fiber texture, but not the hierarchical organization of muscle tissue; products are mostly sold as restructured or formed meat mimetics (Sha & Xiong, 2020). According to Research and Markets, the global meat substitute market is expected to increase from USD 12.9 billion in 2022 to USD 234.7 billion by 2030, at a CAGR of 42.1% from 2022 to 2030 (Research and Markets, 2022); in comparison, the global meat sector is forecasted to increase from USD 838.3 billion in 2020 to USD 1157.6 billion 2025 (Statista, 2022). The meat alternative market is on the rise, and so the competition. Meat alternatives represent about one-third of 15,061 new products launched in the market from 1996 to 2020 (Boukid & Gagaoua, 2022a). The important players are Amy's Kitchen, Beyond Meat, Impossible Foods Inc., The Kellogg Company, Unilever, Quorn Foods, Meatless B.V., Tyson Foods, Inc., and Sunfed (Research and Markets, 2022). The major protein sources are legumes (soy and pulse), cereal (wheat gluten), potato protein, algae proteins. Plant proteins are also used to make dairy alternatives such as plant-based milk, cheese, yogurt and ice cream (Sethi, Tyagi, & Anurag, 2016), egg alternatives that are to mimic the functionalities of eggs for various food application (stand-alone replacement does not exist) (Boukid & Gagaoua, 2022b; Grizio & Specht, 2018), fish alternatives such as surimi, processed fish products and surimi-based products, and less frequently fish flesh (Kazir & Livney, 2021). Chapter 7 (Boukid & Gagaoua, 2022a) summarizes the main alternative proteins to muscle foods and discusses their advantages and disadvantages, including current challenges and limitations in research and development of such new food products and meat alternatives.

- Food proteins are some of the most complex and unique polymers in nature. The multifaceted functionalities of food proteins hold promise for uses as emerging protein-based biomaterials (Martins, Bourbon, Pinheiro, Fasolin, & Vicente, 2018; Varanko, Saha, & Chilkoti, 2020). Chapter 8 (Huang & Chen, 2022) introduces the major techniques for the fabrication of nanoparticles and nanoemulsions from various food proteins, and the approaches to modulate particle size, bioactive molecules encapsulation and release. The advantages and disadvantages of different techniques were also discussed. Because of their high nutritional value, good functional properties, and health-benefiting effects, nanoparticles and nanoemulsions developed from food proteins show great potentials as novel delivery systems for poorly absorbed and/or labile bioactive molecules to improve their bioavailability.
- In addition to imparting nutritional values and functional properties determining their applications, food proteins are also excellent sources of bioactive peptides. Bioactive peptides are small fragments of a protein, often in the range of 2–20 amino acid residues, and can exert various biological activities, once released from their parent proteins. Bioactive peptides are also present naturally in many foods including the first human food breast milk, or formed during digestion or during food preparation especially in fermented foods. They are extensively explored for their ability to improve human health or prevent chronic diseases. Efficacy therefore is a critical element determining their applications. Chapter 9 (Cao, Liao, & Wang, 2022) provides an overview on the production, bioactivities, bioavailability, safety, and, more importantly, clinical evidence of bioactive peptides in management of nutrition related chronic diseases.

4. Conclusions

Food insecurity and malnutrition remain a daunting challenge, and this has been exacerbated by the COVID-19 pandemic (FAO, IFAD, UNICEF, WFP, & WHO, 2021; Laborde et al., 2020; Rzymski et al., 2021). The global per capita protein availability is estimated to reduce by ~20% by climate change and elevated levels of CO_2 in the atmosphere, which further complicates protein demand (Beach et al., 2019). Alternative proteins are expected to provide the solution to meet the growing protein demand within the environmental limits. Indeed, there is a surging interest in using plant proteins as animal alternatives, growing animal proteins using cellular agriculture and precision fermentation, and identifying emerging protein sources. Human consumption of alternative proteins remains limited, due largely to inferior organoleptic properties. There are other concerns for plant-based analogs: (1) they are ultra-processed and are high in nutrients of concern such as carbohydrate, sodium, and saturated fat (Clark & Bogdan, 2019; Hu et al., 2019); (2) although plant protein production is more sustainable than animal protein at farm gate, greenhouse gas emissions for processing plant-based analogs fall between those of pork and chicken, and energy use is much higher than for pork or chicken (Rubio, Xiang, & Kaplan, 2020); and (3) current post-harvest processing of plant-based analogs accounts for as much as 94.3% of retail costs, which are higher than their animal-based counterparts in a retail setting, restricting affordability (Saget et al., 2021); most of these costs are from the processing of meat analogs after the raw protein sources are harvested. Despite these challenges, the value of the global alternative protein market is expected to reach USD 140 billion by 2030 (Frank, 2019). Alternative proteins are new to some cultures but not to the humankind. Instead of focusing largely on mimicking the structure and taste of animal proteins, we need to learn from the wisdom of the traditional methods of processing and utilization to develop culturally diversified alternative proteins to mitigate global protein malnutrition. Consumer acceptance of alternative proteins also changes as we develop better understanding of culturally diversified alternative proteins.

References

Akharume, F. U., Aluko, R. E., & Adedeji, A. A. (2021). Modification of plant proteins for improved functionality: A review. *Comprehensive Reviews in Food Science and Food Safety*, *20*(1), 198–224.

Akhtar, Y., & Isman, M. B. (2018). Insects as an alternative protein source. In R. Y. Yada (Ed.), *Proteins in food processing* (pp. 263–288). Woodhead Publishing.

Alexandratos, N., & Bruinsma, J. (2012). *World agriculture towards 2030/2050: The 2012 revision. ESA Working Paper No. 12-03*. Rome: Food and Agriculture Organization of the United Nations.

Beach, R. H., Sulser, T. B., Crimmins, A., Cenacchi, N., Cole, J., Fukagawa, N. K., et al. (2019). Combining the effects of increased atmospheric carbon dioxide on protein, iron, and zinc availability and projected climate change on global diets: A modelling study. *The Lancet. Planetary Health*, *3*(7), e307. https://doi.org/10.1016/S2542-5196(19)30094-4.

Bekhit, A. E. D. A., Riley, W. W., & Hussain, M. A. (Eds.). (2022). *Alternative proteins: Safety and food security considerations*, CRC Press.

Bodie, A. R., Micciche, A. C., Atungulu, G. G., Rothrock, M. J., Jr., & Ricke, S. C. (2019). Current trends of rice milling byproducts for agricultural applications and alternative food production systems. *Frontiers in Sustainable Food Systems*, *3*, 47.

Boukid, F., & Gagaoua, M. (2022a). Meat alternatives: A proofed commodity? In F. Toldrá (Series Ed.) & J. Wu (Vol. Ed.), *Advances in Food and Nutrition Research: Vol. 101. Emerging sources and applications of alternative proteins* (pp. 213–236). Amsterdam: Elsevier.

Boukid, F., & Gagaoua, M. (2022b). Vegan egg: A future-proof food ingredient? *Foods*, *11*(2), 161.

Bouvard, V., Loomis, D., Guyton, K. Z., Grosse, Y., El Ghissassi, F., Benbrahim-Tallaa, L., et al. (2015). Carcinogenicity of consumption of red and processed meat. *The Lancet. Oncology*, *16*(16), 1599–1600.

Boye, J., Zare, F., & Pletch, A. (2010). Pulse proteins: Processing, characterization, functional properties and applications in food and feed. *Food Research International*, *43*(2), 414–431.

Bresciani, A., & Marti, A. (2019). Using pulses in baked products: Lights, shadows, and potential solutions. *Foods*, *8*(10), 451.

Canada's Food Guide. (2022). *Eat protein foods*. Retrieved from https://food-guide.canada.ca/en/healthy-eating-recommendations/make-it-a-habit-to-eat-vegetables-fruit-whole-grains-and-protein-foods/eat-protein-foods/. [Online Resource].

Cellular Agriculture Society (2021). Retrieved from https://www.cellag.org/. [Online Resource].

Cao, X., Liao, W., & Wang, S. (2022). Food protein-derived bioactive peptides for the management of nutrition related chronic diseases. In F. Toldrá (Series Ed.) & J. Wu (Vol. Ed.), *Advances in Food and Nutrition Research: Vol. 101. Emerging sources and applications of alternative proteins* (pp. 277–307). Amsterdam: Elsevier.

Chaudhry, N. A., Jacobs, C., Green, P. H., & Rampertab, S. D. (2021). All things gluten: A review. *Gastroenterology Clinics*, *50*(1), 29–40.

Chibbar, R. N., Ambigaipalan, P., & Hoover, R. (2010). Molecular diversity in pulse seed starch and complex carbohydrates and its role in human nutrition and health. *Cereal Chemistry*, *87*(4), 342–352.

Circus, V. E., & Robison, R. (2018). Exploring perceptions of sustainable proteins and meat attachment. *British Food Journal*, *121*(2), 533–545.

Clark, L. F., & Bogdan, A. M. (2019). The role of plant-based foods in Canadian diets: A survey examining food choices, motivations and dietary identity. *Journal of Food Products Marketing*, *25*(4), 355–377. Retrieved from https://doi.org/10.1080/10454446.2019.1566806. [Online Resource].

Crimarco, A., Springfield, S., Petlura, C., Streaty, T., Cunanan, K., Lee, J., et al. (2020). A randomized crossover trial on the effect of plant-based compared with animal-based meat on trimethylamine-N-oxide and cardiovascular disease risk factors in generally healthy adults: Study with appetizing plantfood—meat eating alternative trial (SWAP-MEAT). *The American Journal of Clinical Nutrition*, *112*(5), 1188–1199.

da Rosa Machado, C., & Thys, R. C. S. (2019). Cricket powder (Gryllus assimilis) as a new alternative protein source for gluten-free breads. *Innovative Food Science & Emerging Technologies, 56*, 102180.

De la Cruz-Ares, S., Gutiérrez-Mariscal, F. M., Alcalá-Díaz, J. F., Quintana-Navarro, G. M., Podadera-Herreros, A., Cardelo, M. P., et al. (2021). Quality and quantity of protein intake influence incidence of type 2 diabetes mellitus in coronary heart disease patients: From the CORDIOPREV study. *Nutrients, 13*(4), 1217.

Derbyshire, E., & Delange, J. (2021). The nutritional value of whole pulses and pulse fractions. In B. K. Tiwari, A. Gowen, & B. McKenna (Eds.), *Pulse foods* (pp. 9–29). Academic Press.

Dunkel, F. V., & Payne, C. (2016). Introduction to edible insects. In A. T. Dossey, J. A. Morales-Ramos, & M. G. Rojas (Eds.), *Insects as sustainable food ingredients* (pp. 1–27). Academic Press.

Ellies-Oury, M.-P., Chriki, S., & Hocquette, J.-F. (2022). Should and will "cultured meat" become a reality in our plates? In F. Toldrá (Series Ed.) & J. Wu (Vol. Ed.), *Advances in Food and Nutrition Research: Vol. 101. Emerging sources and applications of alternative proteins* (pp. 181–212). Amsterdam: Elsevier.

FAO, IFAD, UNICEF, WFP and WHO. 2021. The State of Food Security and Nutrition in the World 2021. Transforming food systems for food security, improved nutrition and affordable healthy diets for all. Retrieved from https://doi.org/10.4060/cb4474en. [Online Resource].

Food and Agriculture Organization (FAO) of the United Nations. (2020). *Food Supply*. Retrieved from http://www.fao.org/faostat/en/#data/FBS. [Online Resource].

Frank, T. (2019). *Alternative meat to become $140 billion industry in a decade, Barclays predicts*. CNBC. Retrieved from https://www.cnbc.com/2019/05/23/alternative-meat-to-become-140-billion-industry-barclays-says.html. [Online Resource].

Ghosh, S., Suri, D., & Uauy, R. (2012). Assessment of protein adequacy in developing countries: Quality matters. *British Journal of Nutrition, 108*(S2), S77–S87.

Global Pulse Confederation. (2022). *A brief history of pulses*. Retrieved from https://pulses.org/. [Online Resource].

GRAS. (2010). *GRAS notice No. 327. GRAS notification for cruciferin-rich and napin-rich protein isolates derived from canola/rapeseed (PurateinVR and SuperteinTM)*. Retrieved from https://www.accessdata.fda.gov/scripts/fdcc/?set=GRASNotices&id=327&sort=GRN_No&order=DESC&startrow=1&type=basic&search=327/. [Online Resource].

GRAS. (2011). *GRAS notice No. 386. GRAS notification for BioExx canola proteins: Canola protein isolate (IsolexxTM) and hydrolyzed canola protein isolate (VitalexxTM)*. Retrieved from https://www.accessdata.fda.gov/scripts/fdcc/?set=GRASNotices&id=386&sort=GRN_No&order=DESC&startrow=1&type=basic&search=386/. [Online Resource].

GRAS. (2016). *GRAS notice No. 683. The safety and the generally recognized as safe (GRAS) status of the proposed uses of canola/rapeseed protein isolate in human food*. Retrieved from https://www.fda.gov/media/103212/download/. [Online Resource].

Gravely, E., & Fraser, E. (2018). Transitions on the shopping floor: Investigating the role of Canadian supermarkets in alternative protein consumption. *Appetite, 130*, 146–156.

Grizio, M., & Specht, L. (2018). *Plant-based egg alternatives: Optimizing for functional properties and applications*. Retrieved from https://gfi.org/wp-content/uploads/2021/02/Plantbasedeggalternatives.pdf. [Online Resource].

Gupta, J., Wilson, B. W., & Vadlani, P. V. (2016). Evaluation of green solvents for a sustainable zein extraction from ethanol industry DDGS. *Biomass & Bioenergy, 85*, 313–319.

Hartmann, C., & Siegrist, M. (2017). Consumer perception and behaviour regarding sustainable protein consumption: A systematic review. *Trends in Food Science & Technology, 61*, 11–25.

Hosseininezhad, M., & Hussain, M. A. (2022). Single-cell protein–A group of alternative proteins. In A. E. D. A. Bekhit, W. W. Riley, & M. A. Hussain (Eds.), *Alternative proteins* (pp. 49–63). CRC Press.

Hu, F. B., Otis, B. O., & McCarthy, G. (2019). Can plant-based meat alternatives be part of a healthy and sustainable diet? *The Journal of the American Medical Association, 322*(16), 1547–1548.

Huang, J., Liao, L. M., Weinstein, S. J., Sinha, R., Graubard, B. I., & Albanes, D. (2020). Association between plant and animal protein intake and overall and cause-specific mortality. *JAMA Internal Medicine, 180*(9), 1173–1184.

Huang, W., & Chen, L. (2022). Fabrication of protein nanomaterials as delivery systems. In F. Toldrá (Series Ed.) & J. Wu (Vol. Ed.), *Advances in Food and Nutrition Research: Vol. 101. Emerging sources and applications of alternative proteins* (pp. 237–275). Amsterdam: Elsevier.

Iannuzzi, E., Sisto, R., & Nigro, C. (2019). The willingness to consume insect-based food: An empirical research on Italian consumers. *Agricultural Economics, 65*(10), 454–462.

IndexBox. (2022). *World—Wheat gluten—Market analysis, forecast, size, trends and insights. Global Wheat Gluten Market Report 2022—Prices, Size, Forecast, and Companies (indexbox.io)/. [Online Resource].*

Ivarsson, E., & Neil, M. (2018). Variations in nutritional and antinutritional contents among faba bean cultivars and effects on growth performance of weaner pigs. *Livestock Science, 212*, 14–21.

Jones, K. (2022). *Cultured meat 101: the next generation of food.* Retrieved from https://www.visualcapitalist.com/cultured-meat-101-the-next-generation-of-food/. [Online Resource].

Jongema, Y. (2017). *World list of edible insects* (pp. 1–100). Wageningen University. Retrieved from https://www.wur.nl/upload_mm/8/a/6/0fdfc700-3929-4a74-8b69-f02fd35a1696_Worldwide%20list%20of%20edible%20insects%202017.pdf. [Online Resource].

Kazir, M., & Livney, Y. D. (2021). Plant-based seafood analogs. *Molecules, 26*(6), 1559.

Laborde, D., Martin, W., Swinnen, J., & Vos, R. (2020). COVID-19 risks to global food security. *Science, 369*(6503), 500–502.

Lasztity, R. (2017). *The chemistry of cereal proteins* (2nd ed.). CRC Press.

Liceaga, A. M. (2022). Edible insects, a valuable protein source from ancient to modern times. In F. Toldrá (Series Ed.) & J. Wu (Vol. Ed.), *Advances in Food and Nutrition Research: Vol. 101. Emerging sources and applications of alternative proteins* (pp. 129–152). Amsterdam: Elsevier.

Liu, Y., Liu, S., Huang, C., Ge, X., Xi, B., & Mao, J. (2022). Chinese Baijiu distiller's grains resourcing: Current progress and future prospects. *Resources, Conservation and Recycling, 176*, 105900.

Markandya, A., Salcone, J., Hussain, S., Mueller, A., & Thambi, S. (2021). Covid, the environment and food systems: Contain, cope and rebuild better. *Frontiers in Environmental Science, 9*, 245. https://doi.org/10.3389/fenvs.2021.674432.

Martins, J. T., Bourbon, A. I., Pinheiro, A. C., Fasolin, L. H., & Vicente, A. A. (2018). Protein-based structures for food applications: From macro to nanoscale. *Frontiers in Sustainable Food Systems, 2*, 77. https://doi.org/10.3389/fsufs.2018.00077.

McKinsey & Company. (2021). *Cultured meat could be a $25-billion global industry by 2030.* Retrieved from https://www.consulting.us/news/6313/cultivated-meat-could-be-a-25-billion-global-industry-by-2030/. [Online Resource].

Mussatto, S. I. (2014). Brewer's spent grain: A valuable feedstock for industrial applications. *Journal of the Science of Food and Agriculture, 94*(7), 1264–1275.

Nasrabadi, M. N., Doost, A. S., & Mezzenga, R. (2021). Modification approaches of plant-based proteins to improve their techno-functionality and use in food products. *Food Hydrocolloids, 118*, 106789.

Onwezen, M. C., Bouwman, E. P., Reinders, M. J., & Dagevos, H. (2021). A systematic review on consumer acceptance of alternative proteins: Pulses, algae, insects, plant-based meat alternatives, and cultured meat. *Appetite*, *159*, 105058.

Poutanen, K. S., Kårlund, A. O., Gómez-Gallego, C., Johansson, D. P., Scheers, N. M., Marklinder, I. M., et al. (2022). Grains–A major source of sustainable protein for health. *Nutrition Reviews*, *80*(6), 1648–1663.

Research and Markets. (2022). *Meat substitute market size, share & trends analysis report by source (plant-based protein, mycoprotein, soy-based), by distribution channel (foodservice, retail), by region, and segment forecasts, 2022–2030.* Retrieved from https://www.researchandmarkets.com/reports/4613435/meat-substitute-market-size-share-and-trends?utm_source=GNOM&utm_medium=PressRelease&utm_code=8dp6cg&utm_campaign=1685739+-+Global+%24234.7+Billion+Meat+Substitute+Markets+to+2030&utm_exec=chdo54prd. [Online Resource].

Ritala, A., Häkkinen, S. T., Toivari, M., & Wiebe, M. G. (2017). Single cell protein—State-of-the-art, industrial landscape and patents 2001–2016. *Frontiers in Microbiology*, *8*, 2009.

Ritchie, H., & Roser, M. (2020). Agricultural production. Published online at OurWorldInData.org. Retrieved from: 'https://ourworldindata.org/agricultural-production' [Online Resource].

Ritchie, H., & Roser, M. (2021a). *Environmental impacts of food production.* Published online at OurWorldInData.org. Retrieved from: 'https://ourworldindata.org/environmental-impacts-of-food' [Online Resource].

Ritchie, H., & Roser, M. (2021b). Crop yields. Published online at OurWorldInData.org. Retrieved from: 'https://ourworldindata.org/crop-yields' [Online Resource].

Roser, M., & Ritchie, H. (2013). *Food Supply.* Published online at OurWorldInData.org. Retrieved from: 'https://ourworldindata.org/food-supply' [Online Resource].

Royal Entomological Society. (2022). Retrieved from https://www.royensoc.co.uk/understanding-insects/facts-and-figures/. [Online Resource].

Rubio, N. R., Xiang, N., & Kaplan, D. L. (2020). Plant-based and cell-based approaches to meat production. *Nature Communications*, *11*(1), 1–11.

Rzymski, P., Kulus, M., Jankowski, M., Dompe, C., Bryl, R., Petitte, J. N., et al. (2021). COVID-19 pandemic is a call to search for alternative protein sources as food and feed: A review of possibilities. *Nutrients*, *13*(1), 150.

Saget, S., Costa, M. P., Santos, C. S., Vasconcelos, M., Styles, D., & Williams, M. (2021). Comparative life cycle assessment of plant and beef-based patties, including carbon opportunity costs. *Sustainable Production and Consumption*, *28*, 936–952.

Saldanha do Carmo, C., Silventoinen, P., Nordgård, C. T., Poudroux, C., Dessev, T., Zobel, H., et al. (2020). Is dehulling of peas and faba beans necessary prior to dry fractionation for the production of protein- and starch-rich fractions? Impact on physical properties, chemical composition and techno-functional properties. *Journal of Food Engineering*, *278*, 109937. https://doi.org/10.1016/j.jfoodeng.2020.109937.

Sethi, S., Tyagi, S. K., & Anurag, R. K. (2016). Plant-based milk alternatives an emerging segment of functional beverages: A review. *Journal of Food Science and Technology*, *53*(9), 3408–3423.

Sha, L., & Xiong, Y. L. (2020). Plant protein-based alternatives of reconstructed meat: Science, technology, and challenges. *Trends in Food Science & Technology*, *102*, 51–61.

Shen, Y., Hong, S., & Li, Y. (2022). Pea protein composition, functionality, modification, and food applications: A review. In F. Toldrá (Series Ed.) & J. Wu (Vol. Ed.), *Advances in Food and Nutrition Research: Vol. 101. Emerging sources and applications of alternative proteins* (pp. 71–127). Amsterdam: Elsevier.

Sluik, D., Brouwer-Brolsma, E. M., Berendsen, A. A., Mikkilä, V., Poppitt, S. D., Silvestre, M. P., et al. (2019). Protein intake and the incidence of pre-diabetes and diabetes in 4 population-based studies: The PREVIEW project. *The American Journal of Clinical Nutrition*, *109*(5), 1310–1318.

Song, M., Fung, T. T., Hu, F. B., Willett, W. C., Longo, V. D., Chan, A. T., et al. (2016). Association of animal and plant protein intake with all-cause and cause-specific mortality. *JAMA Internal Medicine, 176*(10), 1453–1463.

Souza Filho, P. F. (2022). Fungal protein. In F. Toldrá (Series Ed.) & J. Wu (Vol. Ed.), *Advances in Food and Nutrition Research: Vol. 101. Emerging sources and applications of alternative proteins* (pp. 153–179). Amsterdam: Elsevier.

Statista. (2022). *Meat industry value worldwide in 2020 and forecast for 2021 and 2025*. Retrieved from https://www.statista.com/statistics/502286/global-meat-and-seafood-market-value/. [Online Resource].

USDA. (2022). *PSD online*. Retrieved from https://apps.fas.usda.gov/psdonline/app/index.html#/app/advQuery/. [Online Resource].

USDA & HHS. (2020). *Dietary Guidelines for Americans, 2020–2025*. Retrieved from https://www.dietaryguidelines.gov/sites/default/files/2020-12/Dietary_Guidelines_for_Americans_2020-2025.pdf. [Online Resource].

Vahedifar, A., & Wu, J. (2022). Extraction, nutrition, functionality and commercial applications of canola proteins as an underutilized plant protein source for human nutrition. In F. Toldrá (Series Ed.) & J. Wu (Vol. Ed.), *Advances in Food and Nutrition Research: Vol. 101. Emerging sources and applications of alternative proteins* (pp. 17–69). Amsterdam: Elsevier.

van der Weele, C., Feindt, P., van der Goot, A. J., van Mierlo, B., & van Boekel, M. (2019). Meat alternatives: An integrative comparison. *Trends in Food Science & Technology, 88*, 505–512.

van Huis, A., van Itterbeeck, J., Klunder, H., et al. (2013). *Edible insects: Future prospects for food and feed security, FAO, Rome, 2013*. Retrieved from https://www.fao.org/3/i3253e/i3253e.pdf. [Online Resource].

Van Vliet, S., Kronberg, S. L., & Provenza, F. D. (2020). Plant-based meats, human health, and climate change. *Frontiers in Sustainable Food Systems, 4*, 128. https://doi.org/10.3389/fsufs.2020.00128.

Varanko, A., Saha, S., & Chilkoti, A. (2020). Recent trends in protein and peptide-based biomaterials for advanced drug delivery. *Advanced Drug Delivery Reviews, 156*, 133–187. https://doi.org/10.1016/j.addr.2020.08.008.

Vogelsang-O'Dwyer, M., Zannini, E., & Arendt, E. K. (2021). Production of pulse protein ingredients and their application in plant-based milk alternatives. *Trends in Food Science & Technology, 110*, 364–374.

WHO. (2007). Protein and amino acid requirements in human nutrition. Report of a joint FAO/WHO/UNU expert consultation. *World Health Organization Technical Report Series, 935*, 1.

Willett, W., Rockström, J., Loken, B., Springmann, M., Lang, T., Vermeulen, S., et al. (2019). Food in the Anthropocene: The EAT–lancet commission on healthy diets from sustainable food systems. *The Lancet, 393*(10170), 447–492.

Wu, G., Fanzo, J., Miller, D. D., Pingali, P., Post, M., Steiner, J. L., et al. (2014). Production and supply of high-quality food protein for human consumption: sustainability, challenges, and innovations. *Annals of the New York Academy of Sciences, 1321*(1), 1–19.

CHAPTER TWO

Extraction, nutrition, functionality and commercial applications of canola proteins as an underutilized plant protein source for human nutrition

Amir Vahedifar and Jianping Wu*
Department of Agricultural, Food and Nutritional Science, University of Alberta, Edmonton, AB, Canada
*Corresponding author: e-mail address: jwu3@ualberta.ca

Contents

Abstract

Concerns about sustainability and nutrition security have encouraged the food sector to replace animal proteins in food formulations with underutilized plant protein sources and their co-products. In this scenario, canola protein-rich materials produced after oil extraction, including canola cold-pressed cakes and meals, offer an excellent opportunity, considering their nutritional advantages such as a well-balanced amino acid

Advances in Food and Nutrition Research, Volume 101
ISSN 1043-4526
https://doi.org/10.1016/bs.afnr.2022.04.001

composition and their potential bioactivity. However, radical differences among major proteins (i.e., cruciferin and napin) in terms of the physicochemical properties, and the presence of a wide array of antinutritional factors in canola, impede the production of a highly pure protein extract with a reasonable extraction yield. In this manuscript, principles regarding the extraction methods applicable for the production of canola protein concentrates and isolates are explored in detail. Alkaline and salt extraction methods are presented as the primary isolation methods, which result in cruciferin-rich and napin-rich isolates with different nutritional and functional properties. Since a harsh alkaline condition would result in an inferior functionality in protein isolates, strategies are recommended to reduce the required solvent alkalinity, including using a combination of salt and alkaline and employing membrane technologies, application of proteases and carbohydrases to facilitate the protein solubilization from biomass, and novel green physical methods, such as ultrasound and microwave treatments. In terms of the commercialization progress, several canola protein products have received a GRAS notification so far, which facilitates their incorporation in food formulations, such as bakery, beverages, salad dressings, meat products and meat analogues, and dairies.

1. Introduction

Meeting the growing global protein demand while being sustainable is a challenge for the agri-food sector. Animal proteins, as a solution, are not sustainable and would place enormous pressure on the environment, owing to their impacts on the carbon and nitrogen cycles and biodiversity (Grasso et al., 2021). Besides, a high animal protein diet may cause several health issues, including an increased risk of obesity, type 2 diabetes, and cardiovascular diseases (Hu, Otis, & McCarthy, 2019). Hence, a transition to alternative proteins, such as plant, insect, fungal and microbial proteins, would be a necessity to feed a projected population of 10 billion people in 2050 (Willett et al., 2019). Among these, plant proteins are vital, considering their established supply chain and satisfactory consumer acceptance (Fasolin et al., 2019), along with their demonstrated nutritional value and their sustainability.

Oilseeds are the most popular plant protein sources, owing to their widespread cultivation for oil production. Canola/Rapeseed is the second most cultivated oilseed worldwide, after soybean, with an annual production of 19.4 million metric tons in Canada, the world's largest producer, in 2020/2021 (Canola Council of Canada, 2021). The name "canola" is specifically denoted to seeds of the genus *Brassica*, including *Brassica napus*, *Brassica rapa*, or *Brassica juncea*, with <2% erucic acid and <30 μmol/g meal of total glucosinolates. After oil extraction, the meal contains ~40% protein, making it as a viable material for protein extraction (Daun, Eskin, & Hickling, 2015).

Although canola protein has traditionally been exploited for animal feeding, its nutritional advantages, including the well-balanced amino acid composition, a high biological value, and a high protein efficiency ratio in isolated protein fractions, encourage their utilization for human consumption (Bos et al., 2007; Chmielewska et al., 2021; Delisle et al., 1984). Additionally, canola protein quality has been established in the last 50 years. The incorporation of canola proteins in human food is, however, restricted due to their inferior techno-functional properties and a significant content of antinutritional factors (ANFs), such as phenolics, glucosinolates, phytic acid, and fibers. Furthermore, an efficient method of canola protein extraction is not available due to a robust cell wall structure and the presence of diverse proteins with different physicochemical properties.

Although there are a number of recent review articles/book chapters discussing canola proteins and their uses in foods (Chmielewska et al., 2021; Rivera, Rommi, Fernandes, Lantto, & Tzanov, 2016; Tan, Mailer, Blanchard, & Agboola, 2011a, 2011b; Mupondwa, Li, & Wanasundara, 2018), here the authors aim to overview the most important characteristics of canola proteins, including their structure and function, methods of extraction, and their nutritional, biological and functional properties, all of which are imperative for their utilization in human nutrition.

2. Canola protein characterization

Proteins play a variety of roles in plants, including structural, catalytic, and storage functions. Storage proteins are the recoverable proteins that are accumulated during the seed-filling stage in order to restrain reduced nitrogen for embryo germination (Rommi et al., 2014; Xu & Diosady, 2012). These, which account for around 90% of canola seed proteins, are deposited primarily in the matrix and crystalloid structures found in the storage vacuoles of cotyledon cells, which are surrounded by a cellulose, hemicellulose, and pectin network (Pustjens, Schols, Kabel, & Gruppen, 2013; Wanasundara, Tan, Alashi, Pudel, & Blanchard, 2017).

Canola proteins can be categorized into distinct fractions based on the Svedberg unit, which is a measure of macromolecules' sedimentation speed, related to the molecular weight, in a centrifugal field. These fractions include 12S (or 11S), 7S, and a split 2S (or 1.7S and 1.8S) (Alashi, Blanchard, Mailer, & Agboola, 2013; Hoglund, Rodin, Larsson, & Rask, 1992). Among these, 12S and 2S fractions are the principals, which, according to the general naming practice of botanical species or families, are named as "Cruciferin" (or "Brassin") and "Napin" (or "Conbrassin"), respectively (Prakash & Narasinga Rao, 1986).

Additionally, canola proteins can be divided into four fractions, based on their solvent solubility: water-soluble albumins, salt-soluble globulins, ethanol-soluble prolamins, and alkaline-soluble glutelins. Although the relative proportion of these fractions is highly variable and depends on numerous factors, such as the canola variety, cultivation condition, and plant metabolism, the majority of canola proteins are reported as globulins (~60–70%), albumins (10–15%) and prolamins (20%) (Salunkhe, Adsule, Chavan, & Kadam, 1992). While canola protein extracts mainly include globulin and albumin fractions, prolamins are the structural proteins associated with oil bodies and oil extracts.

2.1 Napin and cruciferin

2.1.1 Structural characteristics

Napin has several isoforms with a molecular weight ranging from 12.5 to 14.5 kDa, including Napin-1, Napin-2, Napin-3, Nap1, Napin-1A, and Napin-B. (Perera, McIntosh, & Wanasundara, 2016). Mature napin, with a hydrodynamic diameter of 5–6 nm, consists of a short (5 kDa) and a long (9 kDa) polypeptide chain, linked together by two interchain disulfide bonds. Long-chain polypeptides are also stabilized by two intrachain disulfide bonds (Shewry, Napier, & Tatham, 1995). In terms of the secondary structure, napin has been reported to have a high content of α-helix (37–59%) and a relatively low content of β-sheets (7–20%) (Rico, Bruix, González, Monsalve, & Rodríguez, 1996; Schmidt et al., 2004).

Despite most 12S plant globulins consisting of disulfide-stabilized subunits, cruciferin has an unusual polypeptide pattern based on both disulfide bonds and non-covalent interactions (Aluko & McIntosh, 2001; Schwenke, 1994). The hexameric (consisting of six subunits) quaternary structure of cruciferin is built on two trimers primarily stabilized by non-covalent interactions. Each subunit or protomers is built by two polypeptides that are linked together by one disulfide bond: acidic α-chain, which is the larger polypeptide with a molecular weight of ~30–40 kDa, and a basic β-chain, which is the smaller polypeptide with a molecular weight of ~20 kDa. The identified subunits are named as CRUA, CRU1, CRU2, CRU3, CRUA, and CRU5 (Aluko & McIntosh, 2001; Wanasundara, 2011). In terms of secondary structure, like other members of the cupin superfamily, β-barrels are predominant structures. This is supported by cruciferin's high β-conformation content (25–27 β-sheets, accounting for ~30–50% of polypeptide chains), and low α-helix content (7 α-helices, accounting for 11% of polypeptide chains). (Schwenke, Raab, Plietz, & Damaschun, 1983; Tandang-Silvas et al., 2010). Regarding the glycoprotein nature of cruciferin, contradictory results have been reported, some of which account the cruciferin as a glycoprotein, while others reported no glycosylated state (Schwenke, 1994).

Due to the role of non-covalent interactions in the stabilization of two assembling trimers, cruciferin structure would be expected to have a high sensitivity to environmental conditions. Although Schwenke and Linow (1982) reported the 12S fraction stability at high ionic strength ($\geq$0.05), they reported their dissociation into 7S components (as the trimeric half of the hexamers) after dialysis against water, freeze-drying, and reconstitution in weakly alkaline water (pH 8.0). Additionally, Bhatty, McKenzie, and Finlayson (1968) reported a reversible pH-induced structural destabilization from 12S into 7S followed by a shift from 7S to 3S monomers by decreasing the pH value up to pH 3.6, which results in increased hydrophobicity and promotes aggregation at these pH values (Folawiyo & Apenten, 1996; Schwenke, 1994).

2.1.2 Isoelectric point (pI)

Due to the high content of basic amino acids in napin, it exhibits basic properties, with an pI of 11.0. However, becasue of the various napin isoforms present, the ionizable amino acids on the protein surface could differ, leading to an altered pI value (Cheung, Wanasundara, & Nickerson, 2015). Yoshie-Stark, Wada, and Wäsche (2008) found that the protein extraction condition may affect the isoform composition of napin-rich extracts, resulting in varied reported pI values, as Cheung et al. (2015) found an electrically neutral surface charge at pH 6.6 for a napin protein isolate. In addition, the method of determination also affects the reported pI value.

Cruciferin extracted from *B. napus*, isolated by salt extraction and purified by chromatography, gave a single fraction in the isoelectric focusing with the pI of 7.2 (Schwenke, 1990; Schwenke et al., 1983). However, cruciferin-rich isolates, either prepared by alkaline extraction or salt extraction methods, have been shown to have pIs ranging from pH 4.0–10.0, which is attributed to the complexity of the canola protein composition and the presence of other interacting components (Paulson & Tung, 1987; Wanasundara, Abeysekara, McIntosh, & Falk, 2012). Dong et al. (2011) investigated the distribution of the pI in 12S fraction using two-dimensional electrophoresis, and they concluded that the pI was distributed in two regions: a pI range of

8–10 for weakly alkaline components with molecular weights around 20 kDa and a pI range of 6–7 for slightly acidic components with molecular weights around 30 kDa.

2.1.3 Molecular weight

Regarding the molecular weight, napin is a 14–15 kDa polypeptide, which shows one corresponding band in non-reducing SDS-PAGE analysis. Since napin consists of small (5 kDa) and large (9 kDa) polypeptides linked by disulfide bonds, two corresponding major bands would be appeared in reducing SDS-PAGE analysis. Although napin is commonly observed in monomer form, the dimer could result in another band around 27–28 kDa in non-reducing conditions (Tan et al., 2011a, 2011b).

Due to the complexity of cruciferin polypeptides, molecular weight range of 300 to 350 kDa has been recorded, depending on the isolation and the employed analytical method. However, lower molecular weights, in the range of 230–240 kDa, were also reported, which may be attributed to the dissociated 7S half-molecules induced by pH and ionic strength alterations (Dalgalarrondo, Robin, & Azanza, 1986; Schwenke, 1994; Wu & Muir, 2008). In terms of the molecular weight of subunits, non-reducing SDS-PAGE analysis showed major bands between 40 and 60 kDa for cruciferin, corresponding to the subunits built by acidic α-chain (~30 kDa) and basic β-chain (~20 kDa) polypeptide chains linked by disulfide bonds (Dalgalarrondo et al., 1986; Ramlan et al., 2002). Under reducing conditions, these bands disappeared and new bands appeared between 18 and 23 kDa and 30–37 kDa, showing the role of disulfide bonds in stabilization of cruciferin α- and β-polypeptide chains (Akbari & Wu, 2015).

2.1.4 Structural stability against heating, pH alteration, and reducing agents

Heating induces structural changes in proteins by disrupting the bonds that stabilize them. The heating condition required for both protein denaturation and protein unfolding illustrates the protein heat stability.

The presence of several interchain and intrachain disulfide bridges in napin endowed it with a compact rigid structure capable of remaining intact under harsh environmental conditions. Thermal stability of napin is characterized by a lower unfolding temperature and a higher irreversible thermal transition corresponding to the denaturation temperature. Although reports on napin thermal stability at different pH values vary in the literature, Wu and Muir (2008) observed a relatively high denaturation temperature of

110 °C at pH 7.4. Krzyzaniak, Burova, Haertlé, and Barciszewski (1998) also reported a thermal denaturation peak of 100.3 °C at pH 6 and 80 °C at pH 3.0. In addition to the denaturation temperature, the irreversible/reversible conformational changes characterize further the thermal stability of proteins. In this sense, an unfolding temperature of 63 °C has been reported for napin at neutral pHs, which decreased to 39 °C and 45 °C by pH alteration to pH 1.0 and pH 10.0, respectively (Folawiyo & Apenten, 1996). Regarding the interactions involved, Wu and Muir (2008) showed that reducing the disulfide bonds in napin results in a significantly lower denaturation temperature and enthalpy compared to the untreated napin, demonstrating the importance of disulfide bonds in their thermal stability.

Cruciferin has a lower denaturation temperature than napin, which is aligned with reported denaturation temperatures for most of globulins (88 to 97 °C) (Rommi et al., 2014; Wu & Muir, 2008). In this regard, Ismond and Welsh (1992) reported a denaturation temperature of 88–89 °C for a cruciferin-rich isolate dissolved in sodium phosphate buffer (0.1 M, pH 6.5–5.5), and Folawiyo and Apenten (1996) reported a denaturation temperature of 86–96 °C at pH ranging from 7.0 to 10.0. Lower denaturation temperatures have also been reported, which increases by pH elevation from 75.4 °C (pH 5.0) to 81.2 °C (pH 10.0) (Léger & Arntfield, 1993). Folawiyo and Apenten (1996) also observed a conformational alteration at 20 °C at pH range of 1.0–3.0, which may be attributed to the unfolding of protein and exposure of initially buried non-polar domains. Regarding the effect of interactions involved, Wu and Muir (2008) concluded that although reducing the disulfide bonds in cruciferin decreased the thermal stability substantially, the enthalpy of denaturation was not affected, which demonstrates the pronounced role of non-covalent interactions in protein stabilization.

2.2 Other proteins

Oil body proteins are another major protein in canola, effective in stabilizing oil bodies. These structural proteins, including oleosins, caleosins, and steroleosins, with a low molecular weight of 18–25, 27, and 39–41 kDa, respectively, represent about 2–8% of the total canola seed proteins. Resembling the amphiphilic characteristic of surfactants, the long hydrophobic regions in oleosins, the dominant oil body protein, allow them to anchor into the oil droplet while their hydrophilic regions reside on the oil droplet surface (Huang, 1992). In addition, canola contains several minor proteins, including thionins, trypsin inhibitors, and lipid transfer proteins, which are excluded from the current manuscript (Bérot, Compoint, Larré, Malabat, & Guéguen, 2005).

3. Canola protein extraction

Table 1 shows the proximate composition of canola seeds, and the protein-rich products of the oil extraction process (press cakes and meal) in comparison to those of soybean. The protein content in the canola seeds and canola meal is lower than soybean counterparts, still canola represents a great material for protein extraction. However, producing a highly pure canola protein product with a high yield of extraction, under a reasonable production cost is challenging.

Canola seeds contain numerous secondary plant metabolites, mainly including phenolic compounds, glucosinolates, and phytic acid, which due to their nutritional concerns are considered as antinutritional factors (ANFs). Although some of these compounds are reported as nutritionally

Table 1 Proximate composition of seeds, press cakes, and meals in canola and soybean.

	Seed[a]		Press cakes[b]		Meals[c]	
Component	**Canola (*B. napus*)**	**Soybean**	**Canola (*B. napus*)**	**Soybean**	**Canola (*B. napus*)**	**Soybean**
Moisture	6–9	5–14	6–11	8–9	8–11	9–11
Crude protein	17–26	31–43	32–40	43–49	34–43	45–55
Oil	38–50	15–24	15–18	9–15	2–4	<1
Crude fibers	8–16	6–11	11–14	4–11	12–21	21
Total carbohydrates	10–36	6–33	30–35	14–21	35–40	30–40
Ash	3–8	3–7	4–6	5–6	6–8	6
Glucosinolates (μmol/g)	23	–	67	–	4–13	–
Phytic acid	2.6	2.3	1.4	–	0.8–2.5	1–2

[a]Compiled from Xu and Diosady (2012), Uppstrom (1995), Mustafa, Christensen, McKinnon, and Newkirk (2000), Ebrahimi-Mahmoudabad and Taghinejad-Roudbaneh (2011), and Anwar, Kamal, Nadeem, and Shabir (2016).
[b]Compiled from Teh, Bekhit, Carne, and Birch (2014), Rommi et al. (2014), Mailer (2004), Arrutia, Binner, Williams, and Waldron (2020), Ancuța and Sonia (2020), and Anwar et al. (2016).
[c]Compiled from Gerzhova, Mondor, Benali, and Aider (2016b), Tan et al. (2011a, 2011b), Gerzhova, Mondor, Benali, and Aider (2015), and Akbari and Wu (2015).

beneficial, such as phenolic compounds with antioxidant activity, they induce undesirable sensory and nutritional issues in canola protein extracts, including a dark color and a negative flavor in the case of phenolics and glucosinolates, and the adverse effect of phytic acid on protein bioavailability (Bell, 1993). Table 2 collected the reported methods applied for the removal of ANFs in canola/rapeseed products.

Due to the diversity of canola protein fractions, and their subunits in terms of the physicochemical properties, only one specific extraction condition would bring a limited yield of extraction. Regarding the conventional alkaline extraction-pI precipitation method, differing pI values of cruciferin (pI of ~7) and napin (pI of ~11) would result in a low extraction yield. By contrast, due to the dominant globulin fraction in soy proteins and the narrow pH range of pI, soy protein extraction yield is high using an alkaline extraction-pI precipitation method (Sari, Bruins, & Sanders, 2013). Comparing the yield of protein extraction in canola to other plants using the conventional alkaline extraction method, a yield of 71–84%, 40–70%, and 29–41% have been reported for soybean, rapeseed, and sunflower proteins, respectively (Ghodsvali, Khodaparast, Vosoughi, & Diosady, 2005; Jung, 2009; Sari et al., 2013).

Additionally, protein extractability and the purity of extract depend significantly on the starting raw material, which could vary based on the canola cultivar, genotype, growing condition (Si, Mailer, Galwey, & Turner, 2003), and the oil extraction method applied prior to protein extraction process. Regarding the effect of seeds, black and yellow cultivars of *B. napus* have been reported to differ significantly in the content of oil, crude protein, and fiber. Comparing the protein content in different canola varieties, meals from yellow-seeded *B. juncea* and *B. napus* were observed to contain more proteins than black-seeded *B. napus* (Mejicanos, 2015; Radfar, Rogiewicz, Hickling, & Slominski, 2015).

Processing parameters during the oil extraction also play a major role in the chemical, nutritional and functional quality of the final protein product. Based on the oil extraction method, residual seed biomass are named as (1) press cakes, produced solely by a physical pressing process, which could be accompanied with an intense heating step up to 160 °C (hot-pressed) or a gentle heating process around 50–60 °C (cold-pressed), and (2) meals, obtained after extraction of the residual oil in press cakes using solvents. In the latter case, desolventizing the meal requires a heating process of ~95–115 °C, which increases the risk of protein denaturation and decreases the protein solubility, resulting in lower protein extractability

Table 2 Reported methods for removal of antinutritional factors (ANFs) in canola meal/press cakes.

Method	Targeted compound	Description	References
Electro-activation	Minerals	Removal of base-causing ions by using electro-activation method for pH adjustment	Gerzhova et al. (2015)
Heating	Glucosinolates, phytic acid, phenolics, and trypsin inhibitor	Using boiling, toasting, steaming, pressurized heating, or autoclaving on canola meals for deactivation of myrosinase that prevents from glucosinolate decomposition, deactivation of trypsin inhibitor and enhances the removal of phenolic and phytic acid	Yoshie-Stark, Wada, Schott, and Wäsche (2006); Khattab and Arntfield (2009); Mansour et al. (1993)
Addition of salts, such as NaCl and KCl, $CaCl_2$	Phytic acid, phenolics	Using $CaCl_2$ to break down the protein-phytic acid complexes at acidic pH, followed by ultrafiltration/diafiltration to remove phytic acid	Tzeng, Diosady, & Rubin, 1988a, 1988b; Tzeng, Diosady, and Rubin (1990); Serraino and Thompson (1984)
		Breaking the ionically bonded phenolics by using NaCl prior to ultrafiltration/diafiltration process	Xu and Diosady (2002)
Organic solvent washing	Residual oil, glucosinolates, and phenolics	Using absolute organic solvents, such as methanol, ethanol, acetone, diethyl ether, petroleum ether and hexane, or an aqueous organic solvent to remove phenolics and glucosinolates before protein extraction from canola meal	Jia, Curubeto, Rodríguez-Alonso, Keppler, and van der Goot (2021); Chabanon, Chevalot, Framboisier, Chenu, and Marc (2007); Das Purkayastha et al. (2014)
		Using different solvent mixtures (acetone or methanol) combined with water or an acid (hydrochloric, acetic, perchloric, trichloroacetic, phosphoric)	Das Purkayastha et al. (2013)

Enzyme treatment	Glucosinolates, phytic acid	Using an Aspergillus niger enzyme mixture for cell wall degradation and removal of glucosinolates	Jensen, Olsen, and Sørensen (1990)
		Using phytase for removal of phytic acid	Serraino and Thompson (1984)
Fermentation	Phenolics, phytic acid, fiber, and glucosinolates	using microorganisms such as Aspergillus sojae, Aspergillus ficuum, *Lactobacillus salivarius*, and Rhizopus oligosporus for removing ANFs in canola meal	Olukomaiya, Fernando, Mereddy, Li, and Sultanbawa (2020), Aljuobori et al. (2014)
Germination	Glucosinolates, phenolics, and phytic acid	Removal of ANFs by germination	Mahajan and Dua (1997)
Microwave	Phenolics, glucosinolates, and phytic acid	Using microwave to reduce the phenolics in extract	Yang et al. (2013)
Using oxidants	Phenolics	Using sodium sulfite, as an oxidant, to prevent oxidation of phenolics and formation of phenolic-protein bonds at alkaline condition	Blaicher, Elstner, Stein, and Mukherjee (1983); Tzeng et al. (1990)
		Using hydrogen peroxide to remove phenolic compounds in canola meal	El-Kadiri, Khelifi, and Aider (2013)
Using complexing compounds	Phenolics, phytic acid	Removal of phenolic compounds by complaxation with insoluble poly(vinylpyrrolidone)	Blaicher et al. (1983)
		Removal of phytic acid by complexation with EDTA or SDS in canola protein extracts	Serraino and Thompson (1984); Akbari and Wu (2015)

Continued

Table 2 Reported methods for removal of antinutritional factors (ANFs) in canola meal/press cakes.—cont'd

Method	Targeted compound	Description	References
Dehulling	Fibers	Physical removal of hull before oil extraction	Rommi et al. (2014)
Membrane technologies (ultrafiltration and nanofiltration)	Glucosinolates, phenolics, phytic acid, and minerals	Applying nanofiltration on acid-soluble napin-rich fractions to remove phenolics and minerals	Bérot et al. (2005)
		Removal of 80–90% of phenolic compounds by applying a ultrafiltration/diafiltration process on an alkaline-extracted protein extract prior to isoelectric precipitation	Xu and Diosady (2002)
Acidic washing	Phenolics, and phytic acid	Dispersing meal at acidified aqueous solution and subsequent centrifugation results in partially removal of phenolics and phytic acid at the supernatant	Akbari and Wu (2015)
Resins	Phenolics, and phytic acid	Using base anionic and anorganic adsorber resins for removal of phytic acid and phenolics, respectively, after acid precipitation of alkaline protein extracts	Von Der Haar, Müller, Bader-Mittermaier, and Eisner (2014)

(Leming & Lember, 2005; Newkirk, 2009). In this regard, Dietz, King, and Harris (1991) found that boiling rapeseed seeds for 5 min significantly decreased the nitrogen solubility by 60–70%. Comparing the protein extractability of hot- and cold-pressed cakes, and solvent-extracted meals, the highest yield is reported for cold-pressed cakes, followed by hot-pressed cakes and solvent-extracted meals (Fetzer, Herfellner, & Eisner, 2019; Östbring, Malmqvist, Nilsson, Rosenlind, & Rayner, 2020). Additionally, intense heating conditions may result in the uncovering of histidine groups, which makes the protein more basic and more potent to interact with negatively charged phytic acid, leading to more stable phytate-protein complexes with lower solubility and extractability (Gillberg & Tornell, 1976a; Serraino & Thompson, 1984).

Next step after selection of an appropriate starting material is to choose an efficient protein extraction method, depending on the desirable compositional, nutritional, and functional properties. Based on the protein content, canola protein extracts could be categorized into protein concentrates (containing up to 55–80% of proteins) and isolates (containing more than 80% of proteins) (Oreopoulou & Tzia, 2007), which will be described more in the following sections.

3.1 Canola protein concentrate

Dry fractionation technologies are applied when the goal is to extract a protein product with a low concentration. In this regard, the starting materials are subjected to milling process, resulting in the detachment of different components from the cellular matrix, followed by a physical separation process without the interference of any solvents. Partanen et al. (2016) prepared a protein-rich fraction (46% protein content) with improved color and higher dispersion stability compared to a rapeseed protein isolate (90% protein content) using an air classification technique. In addition, novel dry fractionation methods, such as electrostatic separation and turbo separation, have been successfully employed for producing protein-rich fractions from canola/rapeseed meals (Laguna et al., 2018).

Furthermore, wet extraction methods could be applied for the production of canola protein concentrates, through which undesirable compounds are intended to be removed. Table 2 summarized the reported methods for removal of various ANFs in canola meal/press cakes. Aqueous alcoholic washing process is a common method that yields a protein concentrate with a protein content of ~50–70%, while simultaneously removes remaining oil, soluble sugars, phenolics (>75%), and other anti-nutrients such as the trace amounts of glucosinolates (Chabanon et al., 2007; Kozlowska, Zadernowski, & Sosulski, 1983; Von Der Haar et al., 2014). The protein extract from alcohol-washed meal/press cakes could also be used for further protein isolation, which results in highly pure protein isolates with acceptable sensory properties (Das Purkayastha et al., 2014). Aside from the undesirable effects of an intense alcohol treatment combined with high temperature on the functionality of protein extract, especially protein solubility, the process needs a large volume of solvents, making it uneconomical and infeasible in an industrial scale (Xu & Diosady, 2012).

3.2 Canola protein isolate

Although extraction methods for canola protein isolation vary significantly, they could be mainly divided into alkaline extraction and salt extraction procedures, based on the application of suitable extraction solvents. Membrane technologies have applications in both of these methods, resulting in the removal of ANFs and producing different protein products with varied yield and purity.

3.2.1 Alkaline extraction

The most common extraction technique for the preparation of plant protein isolates is the alkaline extraction method (Fig. 1). In this procedure, first, the proteins are solubilized at high pH conditions far from the pI, where the strong alkalinity disrupts the cell wall structure and increases the protein solubility due to the strong electrostatic repulsions and breakage of disulfide bonds (Mittal, Katahira, Himmel, & Johnson, 2011), followed by a solid/liquid separation stage to separate the insoluble fraction from the protein extract. Next, the extracted proteins are precipitated at pH values at which the desired proteins have a low solubility, commonly including the pI value. Presumably, high nitrogen yield and high purity are the main motivation for the wide applications of this method (Pedroche et al., 2004). The extract is expected to contain predominantly cruciferin and a little quantity of napin due to napin's basic character and its high solubility at the acid precipitation condition (Akbari & Wu, 2015; Aluko & McIntosh, 2001). Although theoretically proteins extracted at alkaline conditions are referred to as glutelins, it would be safe to conclude the existence of a combination of Osborne fractions in these protein isolates due to the overlapping solubility of canola proteins in different solvents (Tan et al., 2011a, 2011b).

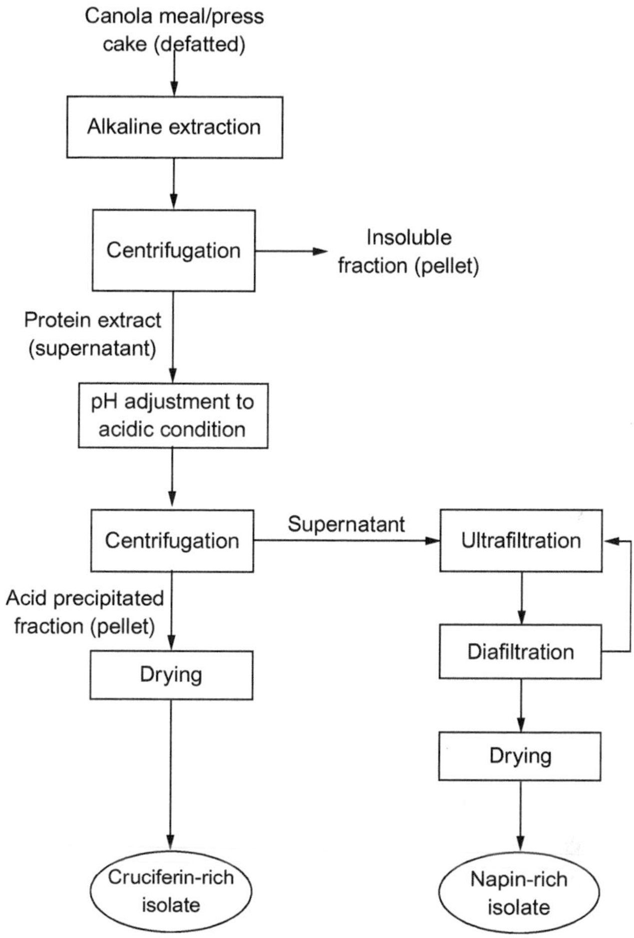

Fig. 1 Conventional alkaline extraction-acid precipitation method.

3.2.1.1 Solubilization stage

Among the parameters influencing the yield and purity of protein extracts, major factors during the solubilization stage are the base concentration/pH value, solvent to biomass ratio, temperature, and extraction time.

Although varied values of protein extractability have been reported at alkaline conditions, ranging from 36% to >85% at an extraction pH of 12.0, it is generally accepted that increasing the extraction pH from 9.0 to 12.0 leads in a significant increase in the protein solubilization (Ghodsvali et al., 2005; Gillberg & Tornell, 1976a; Tzeng et al., 1988a, 1988b). Besides the industrial and environmental limitations of an extraction pH higher than pH 9.0, a highly alkaline condition is associated with adverse effects on nutritional and functional properties of the final extract, including the damaged amino acids (e.g., lysine and cysteine), reduced digestibility, and reduced protein solubility (Fetzer et al., 2019; Pedroche et al., 2004). In addition, such highly alkaline conditions decrease the purity of the extract, considering the oxidation of phenols and their subsequence binding to proteins at alkaline conditions (Xu & Diosady, 2000). Karaca, Low, and Nickerson (2011) reported a high content of glucosinolates, phytic acid, and sinapine at increasing extraction pH from 8.0 to 12.0. They attributed this to the hydrophobic interactions of glucosinolates to proteins, the formation of ternary phytate-cation-protein complexes, and electrostatic interactions between sinapin and proteins under alkaline conditions.

Although a minimum amount of an aqueous solvent is required for ensuring the fast protein diffusion during the extraction, an excess amount of that encourages a higher protein extraction, to the point that the maximal protein solubility in the solvent is not reached (Sari, Mulder, Sanders, & Bruins, 2015). Gerzhova et al. (2016b) reported that the optimum solvent to meal ratio is dependent on the extraction pH; at pH 10.0, a lower solvent to meal ratio results in a higher protein extractability, while at pH 11.0 this effect reduced and at pH 12.0, total extractability slightly decreased. Additionally, a higher solvent to meal ratio has been reported to cause a lower phytic acid content and improved color in the final extract (Das Purkayastha & Mahanta, 2015).

In terms of the influence of time and extraction temperature, it has been shown that an increase in duration or extraction temperature contributes to a higher protein yield. While a longer extraction period allows for a facilitated protein diffusion, an extremely long extraction time raises the risk of microbial growth (Sari et al., 2015). On the other hand, Radwan and Lu (1976) reported an improved extraction yield by increasing the temperature at a wide pH range (1.0–13.0), which was attributed to the limited hydrolysis of proteins or release of proteins upon the matrix destabilization.

3.2.1.2 Precipitation stage

After the dissolution of proteins and separation of insoluble fractions, downstream processes are employed to isolate solubilized proteins (Fig. 1). Acid precipitation is a common approach, including precipitation of proteins at the pH condition with a low solubility, which is generally associated with the pI of proteins. Regarding the canola proteins, pH range of 3.5 to 5.5

has been reported as the optimum condition for protein precipitation, which would result in the recovery of >50% of solubilized proteins (Aluko & McIntosh, 2001; Ghodsvali et al., 2005; Pedroche et al., 2004; Tan et al., 2011a, 2011b). Considering the diverse pI values for cruciferin subunits (pH of 4.0–10.0), and the sensitivity of cruciferin to acidic conditions, this low-solubility pH condition could be attributed to the pI of disrupted cruciferin (cruciferin's subunits) with higher hydrophobicity. Also, this could be explained by considering the formation of phytate-protein electrostatic complexes at this pH condition (Gillberg & Tornell, 1976a). Besides this pH range, another pH condition around the pH of 7.0–8.0 has been reported with a low solubility, which is attributed to the pI of cruciferin (~7.2) (Gillberg & Tornell, 1976a; Quinn & Jones, 1976).

Despite the high portion of proteins solubilized from canola meal (80–99%) using various techniques such as the countercurrent extraction procedure (Diosday, Tzeng, & Rubin, 1984; El Nockrashy, Mukherjee, & Mangold, 1977; Ismond & Welsh, 1992; Tzeng et al., 1988a, 1988b), only a small fraction of dissolved proteins is reported to be recovered, ranging from a protein yield as low as 5% to 87.5%. This attributes to the diverse pI among extracted proteins and the presence of other components that could interact with protein and change their solubility pattern.

Therefore, approaches have been examined to increase protein recovery. To investigate the role of phytic acid in the formation of complexes with proteins and promoting protein precipitation at acidic conditions, Gillberg and Tornell (1976a) added sodium phytate at the extraction pH of 11.1 before acid precipitation and reported a significant increase in nitrogen recovery from about 35% to 75%. They showed that an extraction pH of 11.7 yields a moderate yield due to the least solubility of phytic acid at this extraction pH. However, due to the adverse nutritional effect of phytic acid (Bell, 1993), other precipitation aids have been examined. In this regard, Gillberg and Tornell (1976b) examined precipitation of alkaline-extracted proteins by different anionic compounds, including sodium hexametaphosphate (SHMP), carboxymethyl cellulose (CMC), polygalactouronic acid, alginate, and kappa-, iota- and lambda-carrageenan. They observed an increase from ~40% to >80% using all the precipitation aids at optimum concentrations, pH values, and ionic strengths, which were attributed to the formation of neutral complexes with proteins. In another study, Chen and Rohani (1992) investigated the yield of alkaline extraction in the presence of SHMP, CMC, ammonium sulfate, and acidic condition, and reported that ammonium sulfate caused the highest precipitation of dissolved proteins

(~94%), which were attributed to the dehydration of proteins, while only 53% of proteins recovered at the acidic condition (pH of 3.7).

3.2.1.3 Simultaneous production of napin-rich and cruciferin-rich isolates

To avoid the loss of acid-soluble albumin fraction (containing napin) at the supernatant after acid precipitation, an ultrafiltration (UF) process could be applied on the supernatant, resulting in acid-soluble napin-rich isolates (Fig. 1). Moreover, the incorporation of diafiltration (DF) would promote the removal of soluble low molecular weight antinutritional factors, leading to isolates of high purity (>85%) (Diosday et al., 1984; Ghodsvali et al., 2005; Tzeng et al., 1988a, 1988b).

Wanasundara and McIntosh (2013) developed a patent for acidic extraction of napin at pH 2.5–5.0 from Brassicaceae seed meals, followed by alkaline extraction of the cruciferin-rich fraction at pH values of 7.0–13.0. This acidic washing process could be used exclusively for napin extraction, as Albe-Slabi, Defaix, Beaubier, Galet, and Kapel (2022) optimized the condition for napin extraction with the highest recovery, extract purity, and superior functionality. Furthermore, Akbari and Wu (2015) used the acidic washing stage before alkaline extraction to early separation of acid-soluble fraction and phytic acid and phenolics, which was followed by UF/DF, resulting in highly pure protein fractions with reasonable extraction yield.

3.2.2 Salt extraction

To take advantage of the large portion of salt-soluble globulin fractions among canola proteins, a dilute salt solution could be used for canola protein extraction. Protein solubilization, in this case, is attributed to the salting-in effect (loosening effect of ions at low concentrations on protein-protein interactions), while at higher ion concentrations, the protein solubility would decrease due to the salting-out effect (because of formation of ion-water hydration complexes) (Gerzhova et al., 2016b). After the separation of insoluble fractions from solubilized proteins, the supernatant would be dialyzed, aiming to remove the salts, which makes the salt-soluble fraction insoluble and possible to separate. Regarding the protein fractions present in the salt extract, an electrophoretic analysis of the salt-extracted isolates showed the presence of a majority (68.8%) of globulin fraction (cruciferin) and a minority of albumin fraction (napin) (25.8%) (Tang & Ghosh, 2021). Noteworthy that although the protein content of salt-extracted isolates is commonly higher than alkaline-extracted protein isolates (78.5–98%), the yield of extraction is less and it needs a dialysis step to remove the salt,

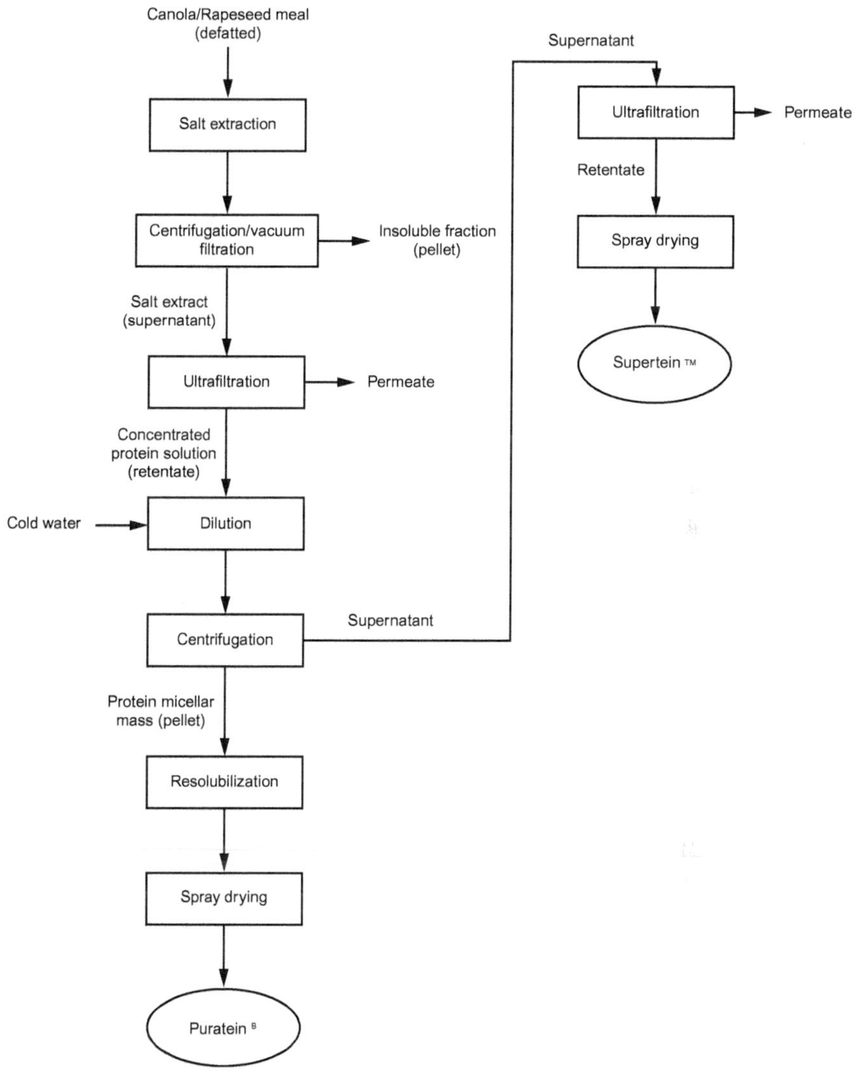

Fig. 2 Burcon's production method for Puratein and Supertein.

which complicates the extraction process (Ismond & Welsh, 1992; Kim, Varankovich and Nickerson, 2016).

A method that used a combination of salt extraction and UF to produce a highly pure protein isolate has been introduced by a series of patents (Mupondwa et al., 2018) (Fig. 2). The developed method with 40–70% yield includes four stages of salt extraction, UF, dilution, and precipitation, which results in an amorphous, sticky, gelatinous gluten-like protein micellar mass (PMM) with >80% protein content (Cheung, Wanasundara, & Nickerson, 2014; Ismond & Welsh, 1992). Additionally, the supernatant of the extract was further concentrated by UF, and dried, resulting in another protein isolate (>90% protein). Although the results are promising, the low flux, originated from the high required protein concentration in the retentate, would contribute to high costs of extraction whether due to the need for a high membrane surface area or pumping expenses (Xu & Diosady, 2012).

Salt extraction could be combined with alkaline extraction process, which would lead to a comparable extraction yield at lower alkalinity of the extraction condition. Gerzhova et al. (2016b) studied the protein extractability in salt extraction method at pH range of 10.0–12.0, concluding that the addition of NaCl significantly increases the extractability only at pH 10. Jia, Rodriguez-Alonso, et al. (2021), in another study, used a combination of alkaline (pH 8.0) and salt (0.2 M NaCl) extraction methods with UF (cutoff of 3 kDa) and dialysis processes, in order to produce a highly pure protein isolate (94% protein content), compromised by a low yield of extraction (36%).

In addition, recovered proteins by salt and alkaline extraction methods could be collected separately. This method, called as Osborne method, includes the application of aqueous, alcoholic, saline, and alkaline solutions to prepare albumin, globulin, prolamin, and glutelin protein fractions, respectively. Tan et al. (2011a, 2011b) compared the Osborne method to alkaline extraction method, and reported a higher cumulative protein yield with lower protein content (lower purity) in different fractions, compared to alkaline-extracted isolates.

4. Extraction-assisting technologies

To overcome the challenges associated with the extraction of canola protein, various assisting techniques have been recommended, including biochemical (e.g., enzyme-assisted extraction), physical (e.g., microwave-assisted, ultrasound-assisted, pressurized liquid extraction, pulsed electric field, extrusion, and supercritical fluid extraction), physical-chemical (e.g., extrusion), and electro-chemical methods, to be incorporated into two principal protein extraction methods. Although the application of these technologies is not fully investigated with regard to the canola, their promising applicability is reported in the extraction of various plant proteins. However, due to the irreversible alterations in proteins caused by these methods, such as hydrolysis, denaturation, and formation of undesirable compounds, and also challenges regarding their large-scale feasibilities, further studies are required for integrating them into the canola protein extraction process. Only the most practiced assisting approach (i.e., enzyme-assisted techniques) is discussed in this section, and interested readers are referred to the review articles in this area (del Mar Contreras et al., 2019; Pojić, Mišan, & Tiwari, 2018).

With regard to the advances in biotechnology, enzyme-assisted procedures have been increasingly applied for enhancing the extraction process outcomes. In this scenario, different carbohydrases and proteases can be employed to break down the lignocellulosic cell wall structures or the proteinous materials, in order to increase protein solubilization from the starting biomass. This could result in a lowered extraction pH, and avoid harsh protein denaturing conditions. However, the high delicacy of the applied condition, long extraction time, and high operational costs, as well as irreversible disruption of carbohydrates and proteins, limits their applicability (Sari et al., 2015).

Regarding the proteases, endopeptidases are favored over exopeptidases since they cleave the peptide bonds located within the polypeptide chains (Rivera et al., 2016). Sari et al. (2013) studied the utilization of Protex enzymes for increasing the yield of extraction in rapeseed, soybean, and microalgae meals. They observed a superior performance among alkaline proteases comparing to acidic ones, with increasing the protein extractability from 15–30% to 50–80% for rapeseed and microalgae meals, and from 80% to 90% for soybean meal. Pedroche et al. (2004) examined Alcalase enzyme in *B. carinata*, and reported an increase in the protein extractability from 33.5% to 57.5% at pH 7.0, and from 41.6% to 62% at pH 9.0. Due to the compactness and globular conformation of napin, the released peptides were more of cruciferin hydrolysates (Chabanon et al., 2007). As a drawback, the protein content in these extracts is low, as Sari et al. (2013) reported only a 40–58% protein content in the extract, which is related to the difficulties in down-stream purification of hydrolyzed proteins due to the irreversible protein structure disruption.

Due to the limiting effect of cell wall on protein extraction, cell wall-degrading carbohydrases may be useful for boosting the extraction yield. In this regard, Rodrigues, Carvalho, and Rocha (2014) prepared a rapeseed protein concentrate from meals by removing carbohydrates using commercially available carbohydrase enzyme mixtures, applied at pH 3.5–4.0 to minimize the protein solubilization. Viscozyme® L was reported as the most effective enzyme, resulting in an increase in protein content from 41% to 68%. Additionally, Rommi et al. (2014) examined the effectiveness of pectinolytic, xylanolytic, and cellulolytic enzyme preparations in increasing the protein extractability in cold-pressed cakes. They observed the disintegration of embryonic cell walls via pectinase activity and an increase, up to 1.7-fold, in protein solubilization. Rommi et al. (2015) showed that although pectinase-aided aqueous extraction could be considered as an alternative for conventional alkaline extraction method, application of pectinase at alkaline conditions did not improve the protein solubilization, and even it hindered the protein precipitation at acidic conditions due to the interactions between low-molecular-weight hydrolysates of pectin and cruciferin at pH 4.0.

Carbohydrases could also be used in combination with proteases, as Niu et al. (2012) employed a mixture of carbohydrase (Viscozyme L) and protease (Alcalase 2.4 L) enzymes in a rapeseed cold-pressed cake, and achieved an extraction yield of 82.1% and 71.89% for protein and oil, respectively. Likewise, Zhang, Wang, and Xu (2007) developed an enzymatic method for the simultaneous extraction of rapeseed oil and proteins. In this method, emulsions prepared after enzymatic hydrolysis (a mixture of pectinase, cellulase, and betaglucanase, and Alcalase 2.4 L) and washing of the wet precipitates were demulsified by storage-centrifugation and freezing-thawing procedures. This approach resulted in the yield of 88–90% and 94–97% for total free oil and protein hydrolysates, respectively, with a 66.7% protein recovery and protein content of 73.51% in the protein-rich fraction.

5. Protein quality: Nutritional value and digestibility, and bioactivity

5.1 Nutritional aspects and digestibility

Amino acid profile is of fundamental importance for assessing protein quality. Table 3 presents the content of essential amino acids present in canola protein meals, concentrates, and isolates and compared it to soy protein and casein isolates. As could be noticed, the content of essential amino acids in most of the canola protein products exceeds 400 mg/g protein, which demonstrates their well-balanced amino acid composition. Comparing the content of individual essential amino acids in canola protein products to the WHO/FAO/UNU (2007) pattern, it could be concluded that canola proteins meet the requirements for almost all essential amino acids in different age groups, especially the adults. Interestingly, the content of

Table 3 Essential amino acid composition of canola meal and canola protein products, compared to the international recommendations, and soy protein isolate and casein.

Amino acid	Meal		Protein concentrate	Napin-rich isolate	Cruciferin-rich isolate			International recommendation[a]					
					Salt -extracted	Alkaline-extracted							
	Tzeng et al. (1988a, 1988b)	Mansour et al. (1993)	Mansour et al. (1993)	Tzeng et al. (1988a, 1988b)	Kim, Varankovich and Nickerson, 2016, Kim, Varankovich, Stone, & Nickerson, 2016)	Tzeng et al. (1988a, 1988b)	Mansour et al. (1993)	Infant	Child (3–10 year)	Child (11–14 year)	Adults	Soy protein isolate[b]	Casein[c]
His	3.1	4.1	4.0	3.4	3.1	3.1	5.0	2.0	1.6	1.6	1.5	2.4	2.7
Ile	4.4	3.7	4.1	4.0	3.4	5.1	2.7	3.2	3.1	3.0	3.0	4.0	4.9
Leu	7.4	6.6	6.8	7.7	6.6	9.2	8.2	6.6	6.1	6.0	5.9	7.6	8.4
Lys	6.6	8.8	10.8	5.9	4.9	5.6	7.3	5.7	4.8	4.8	4.5	6.4	7.1
Met+Cys	2.2+2.0	1.8+1.9	1.5+1.8	2.3+1.0	1.8+2.2	2.6+0.3	1.5+3.0	2.8	2.4	2.3	2.2	1.1+1.1	2.6+0.04
Phe+Tyr	4.6+3.1	3.8+2.6	3.3+1.8	4.0+2.4	3.8+1.9	5.1+3.9	3.7+2.7	5.2	4.1	4.1	3.8	4.9+3.7	4.5+5.5
Thr	4.8	5.5	4.3	4.1	2.9	5.3	3.4	3.1	2.5	2.5	2.3	3.9	3.7
Trp	ND	1.2	1.6	ND	1.3	ND	2.7	8.5	6.6	6.5	0.6	2.6	ND
Val	5.6	4.4	3.9	4.8	3.8	5.8	3.6	4.3	4.0	4.0	3.9	4.1	6.0
Total	43.8	43.4	43.9	39.6	35.7	46	43.8	41.4	35.2	34.8	27.7	41.8	45.4

[a]WHO/FAO/UNU (2007).
[b]Fleddermann et al. (2013).
[c]Wang, Hettiarachchy, Qi, Burks, and Siebenmorgen (1999).

sulfur-containing amino acids (i.e., methionine and cysteine) in canola protein products exceeds the requirements established by WHO/FAO/UNU (2007), which is advantageous considering the limited content of these in many other plant proteins (Young & Pellett, 1994).

In terms of the effect of the extraction procedure on amino acid composition, Mansour et al. (1993) found that canola protein isolates have a lower content of lysine, along with isoleucine and valine, when compared to protein concentrates and canola meal. This could be related to the sensitivity of lysine to oxidation and possibility of their engagement in some deteriorating reactions such as Maillard reaction. However, it is reported that the content of lysinoalanine, a potentially toxic compound produced at highly alkaline conditions, in alkaline-extracted isolates is <100 ppm (Deng, Barefoot, Diosady, Rubin, & Tzeng, 1990; Fleddermann et al., 2013; Tan et al., 2011a, 2011b). Furthermore, the isolation method reduces cysteine content, indicating the negative effect of long processing times at high pH values (Tzeng et al., 1988a, 1988b). Comparing the canola protein isolates rich in napin and cruciferin, a higher content of lysine- and sulfur-containing amino acids is evident in napin-rich isolates (Albe-Slabi et al., 2022: Monsalve, Lopez-Otin, Villalba, & Rodríguez, 1991). However, total content of essential amino acids in alkaline-extracted cruciferin-rich isolates is much higher than napin-rich isolates.

A more precise method for evaluation of protein quality is to assess Protein Digestibility-Corrected Amino Acid Score (PDCAAS) values for individual amino acids. PDCASS is based on the comparison between the content of essential amino acids in a given protein to a reference essential amino acid pattern, corrected by the true digestibility determined by a rat assay (Schaafsma, 2005). Calculated PDCASS values for essential amino acids in a cruciferin-rich canola protein isolate showed that this protein source is not adequate to meet the dietary requirement for infants and preschool children (2–5 years) in several amino acids, while efficient for 10–12 years olds and adults, suggesting mixing them with other plant protein sources as a complementary strategy (Klockeman, Toledo, & Sims, 1997).

Protein digestibility depends on their availability to digestive enzymes, which is associated with the protein composition and the presence of ANFs. In canola meal, high content of fiber, presence of glucosinolates and their breakdown products, as well as phytic acid and phenolics, contribute to the reduced activity of proteolytic enzymes. Regarding the *in vitro* digestibility of canola meals, it is shown that a toasting process applied on meals could improve the *in vitro* digestibility, which was attributed to the effect of heating on the destruction of trypsin inhibitors and tannic acid. Mansour et al. (1993) and Ivanova, Chalova, Uzunova, Koleva, and Manolov (2016) investigated the *in vitro* digestibility of rapeseed meals compared to protein concentrates and isolates, and reported higher values for the latter. Concerning *in vivo* protein digestibility of canola protein products in animal models, although the older studies reported poor protein digestibility in animals (Grala, Verstegen, Jansman, Huisman, & Van Leeusen, 1998), Fleddermann et al. (2013) reported a high true digestibility in rat models for canola protein isolate (93.3%) and canola protein hydrolysates (97.35%), which was comparable to egg (98%), casein (95%), and collagen (95%), and higher than other plant protein sources (wheat 91%; beans 78%).

Studies on canola protein digestibility in human are rare. Bos et al. (2007) reported a low bioavailability of rapeseed proteins by measuring real ileal digestibility, which was inferior to milk, soy, wheat, pea, and lupin. They did, however, report comparable postprandial retention of released amino acids to milk proteins, indicating a significant postprandial biological value due to the large amount of essential amino acids. Likewise, Fleddermann et al. (2013) examined the amino acid bioavailability of canola protein isolate and hydrolysates in human subjects and reported a high dietary value and a postprandial amino acid response comparable to soy proteins.

5.2 Canola proteins bioactivity

The bioactivity of canola proteins and their hydrolysates have been widely investigated. In this section, the most reported bioactivity of canola protein hydrolysates, including Angiotensin-Converting Enzyme (ACE)-inhibitory and antioxidant activity, are discussed in detail, while other activities such as bile acid-binding, anti-inflammatory, anti-thrombotic, anti-microbial, and effect on cell growth have also been reported in the literature and discussed in depth by Fetzer et al. (2019), Wanasundara, McIntosh, Perera, Withana-Gamage, and Mitra (2016), and Chmielewska et al. (2021).

5.2.1 ACE-inhibitory activity

Renin-angiotensin system is a regulatory hormonal mechanism that controls blood pressure. Due to the involvement of ACE and renin in this regulatory system, inhibition of their activity could be considered as an alternative strategy for lowering the blood pressure (Hartmann & Meisel, 2007).

ACE-inhibitory effect of canola meal and canola protein concentrates were investigated by Yoshie-Stark et al. (2006). Although they did not observe an ACE-inhibitory effect in non-hydrolyzed canola meal, they reported an ACE-inhibition effect in defatted meal, canola protein concentrates and steamed concentrates hydrolyzed by pepsin or pepsin-pancreatin. Wu, Aluko, and Muir (2008) also observed an ACE-inhibitory activity for Alcalase hydrolyzed heat-treated canola meals, which increased in parallel with the hydrolysis time and degree of hydrolysis up to a certain level. They purified two ACE-inhibitory peptides, including VSV and FL, located in the napin and cruciferin sequences. Comparing the ACE-inhibitory effect of cruciferin-rich and napin-rich isolates, Yoshie-Stark et al. (2008) reported a greater activity in cruciferin-rich isolates hydrolyzed by a pepsin–pancreatin mixture.

Various studies reported a relation between the ACE-inhibitory activity and molecular size of the released peptides. Alashi et al. (2014a, 2014b) compared the *in vitro* and *in vivo* antihypertensive activities of canola protein hydrolysates produced by different enzymes. They reported the highest *in vitro* inhibition of ACE activity and an appreciable antihypertensive effect in spontaneously hypertensive rats for Alcalase hydrolysates, which was attributed to the small size of the peptides released by the enzyme activity. He et al. (2013) also investigated the ACE-inhibitory effect of rapeseed protein hydrolysates using different enzymes followed by size fractionation, and reported the highest *in vitro* ACE- and renin-inhibitory activity in <5 kDa fractions, compared to 5–10 kDa, almost in all hydrolysates produced by different enzymes. It is reported that di- and tripeptides have lower osmolarities compared to free amino acids, and therefore they can be absorbed better than free amino acids or intact proteins (Matthews, 1977).

In addition, amino acid composition has also been demonstrated to have a significant effect, as it is reported that a lower content of basic amino acids and a high content of bulky amino acids, including aromatic and hydrophobic amino acids, are responsible for ACE-inhibitory activity (Wu, Aluko, & Nakai, 2006). This is demonstrated by He et al. (2013), working on rapeseed protein hydrolysates prepared by Alcalase, who reported three ACE-inhibitory peptides (LY, TF, and RALP), among which the higher number of hydrophobic amino acids in LY and RALP was concluded to result in higher *in vitro* and *in vivo* activities. The importance of basic and bulky amino acids could also be confirmed considering the amino acid composition of four ACE-inhibitory peptides (IY, RIY, VW, and VWIS), examined *in vitro* and *in vivo* and isolated from a subtilisin hydrolysates of rapeseed proteins (Marczak et al., 2003).

5.2.2 Antioxidant activity

Antioxidants are effective substances against oxidation of easily oxidized compounds such as free radicals, as well as unsaturated fats. Free radicals (e.g., superoxide, hydrogen peroxide, and hydroxyl) are generated commonly throughout the body as physiological metabolites that, in excessive amounts, could damage biological macromolecules, leading to various ailments, including cardiovascular diseases, hypertension, and cancer (Hartmann & Meisel, 2007).

The effectiveness of enzymatic hydrolysis on increasing the antioxidant activity has been reported by Yoshie-Stark et al. (2006), who observed 2.3–3.0 times higher DPPH radical scavenging activity in hydrolyzed canola protein concentrates compared to non-hydrolyzed samples. On the other hand, canola protein isolates, either napin-rich or cruciferin-rich isolates, could be the origin of antioxidant peptides, as Pan, Jiang, and Pan (2011) documented the dose-dependent *in vitro* antioxidant activity of Alcalase hydrolyzed cruciferin-rich rapeseed isolate, and Mäkinen, Johannson, Gerd, Pihlava, and Pihlanto (2012) reported the antioxidant activity of a napin-rich isolate hydrolyzed sequentially using Alcalase and Flavorzyme. Comparing the antioxidant activity between cruciferin-rich and napin-rich isolates, a stronger DPPH radical-scavenging activity has been reported for an alkaline-extracted cruciferin-rich isolate and its hydrolysates, where both protein isolates and their hydrolysates have a remarkable stronger antioxidant activity compared to the de-oiled soybean and its hydrolysates (Yoshie-Stark et al., 2008).

The importance of enzyme specificity and the peptide size on antioxidant activity of hydrolysates have been shown by various studies. Fractions containing small-sized peptides (e.g., <1 kDa and <3 kDa) have been observed to have a higher capacity to inhibit oxidation (He, Girgih, Malomo, Ju, & Aluko, 2013; He, Alashi, et al., 2013a; Mäkinen et al., 2012). Also, enzymes with numerous cleavage sites, such as Alcalase and Flavourzyme, have been documented to release stronger antioxidant peptides (Mäkinen et al., 2012; He, Girgih, et al., 2013c; Cumby, Zhong, Naczk, & Shahidi, 2008). Additionally, it has shown that the amino acid composition of protein hydrolysates has a significant impact on antioxidant activity, suggesting the general effect of tryptophan or tyrosine at C-terminus, and the importance of hydrophobic amino acid in exhibiting antioxidant activity (Saito et al., 2003; Wang et al., 2016; Zhang, Wang, Xu, & Gao, 2009). However, Alashi et al. (2014a, 2014b) observed that the amino acid composition do not influence the antioxidant activity of canola protein hydrolysates, while specific sequences of amino acids are supposed to play the major role in determining the antioxidant activity of peptides.

6. Functional properties

Functional properties of canola proteins dictate their potential applications as an ingredient in food formulations. In addition to variations in the starting material, a variety of parameters applied during oil extraction and protein recovery procedures alter the canola protein functional properties. Since these parameters, such as the applied heat and pressure treatments, pH adjustment, and the addition of salts, differ substantially among various studies, contradictory functional properties have been documented for canola proteins. Once a harsh extraction condition was applied, deterioration of functional properties would be inevitable due to the irreversible structural changes (Tan et al., 2011a, 2011b). In this regard, recent studies attempted to avoid extensive purification of plant proteins by studying the incorporation of mildly purified protein products, such as protein concentrates, in different food formulations. Also, techniques are being introduced for improving the functional properties in canola protein products, including chemical (e.g., succinylation and acetylation), biochemical (e.g., enzymatic hydrolysis and cross-linking), and physical (e.g., high-pressure treatment) methods (Wanasundara et al., 2016). As an example, a limited hydrolysis process has been shown to improve solubility and emulsifying/foaming properties, while an extensive hydrolysis condition has been reported to have a negative impact on the latter (Alashi, Blanchard, Mailer, & Agboola, 2011; Pedroche et al., 2004).

6.1 Solubility

Protein solubility is an instrumental feature that determines other functional properties. Because protein solubility is determined by the balance between protein-protein and protein–solvent interactions, parameters that promote or weaken these interactions, such as the pH and ionic strength affecting the protein charge, or heating process resulting in increased hydrophobicity, would affect the protein solubility. Furthermore, the presence of protein-interacting components may substantially affect the solubility as well (Kroll, 1991).

6.1.1 Solubility of canola protein concentrates

Jia, Curubeto, et al. (2021) compared the solubility of an alcohol-washed protein concentrate to a salt-extracted cruciferin-rich isolate at pH ~6, and observed a solubility of <20% in rapeseed protein concentrates, which was attributed to the high protein denaturation due to alcoholic treatment, while they reported a high nitrogen solubility (76%) and protein nativity in rapeseed protein isolates. However, alcohol-washed rapeseed protein concentrates exhibited much higher solubility in another study (Yoshie-Stark et al., 2006).

6.1.2 Solubility of canola protein isolates

The solubility of cruciferin-rich isolates differs significantly along with the pH range, as an acid-precipitated cruciferin-rich isolate has been reported to have a minimum around pH 4.0, but high at extreme acidic and alkaline conditions. They also observed a low solubility (<40%) throughout the pH range, except for pH 12.0 (73.5%), which was increased to 50–70% by addition of NaCl, showing the salting-in effect of the added ions (Klockeman et al., 1997). Similarly, Paulson and Tung (1987) found that adding NaCl to acid-precipitated cruciferin-rich isolates had a solubilizing impact at pH values of 3.5–8.0, but a weakening effect at higher pHs, which were attributed to the salting-out effect.

Regarding the salt-extracted cruciferin-rich isolates, Karaca et al. (2011) reported a significantly higher solubility (~80%) at pH 7.0 compared to acid-precipitated isolates (<5%), which is related to the milder extraction condition in the salt extraction procedure. Cheung et al. (2014) investigated the influence of pH and ionic strength on the solubility of salt-extracted cruciferin-rich isolates, and they observed a significant increase in solubility at pH values of 5.0 and 7.0 by increasing the NaCl concentration to 100 mM, while the solubility at pH 3.0 was high (~90%) regardless of the NaCl concentration. The low solubility of salt-extracted isolates at pH 7.0 in this study contradicts the high solubility (~80%) of salt-extracted isolates prepared by Karaca et al. (2011) at the same pH value. This could be related to the starting material, since the latter desolventized the meals by air-drying (~20 °C), while the former used a drying temperature of ~75 °C, showing the effect of heating on denaturation of proteins and reducing their solubility.

Acid-soluble napin-rich isolates are reported to have high solubility at a wide range of pH (2.0–10.0), due to their basic pI value, their hydrophilic nature, low hydrophobicity, high surface charge, and their soluble form maintained during the extraction process (Cheung et al., 2015; Wanasundara et al., 2012; Xu & Diosady, 1994). Comparing the napin-rich and cruciferin-rich isolates, Yoshie-Stark et al. (2008) reported a significantly higher solubility for acid-soluble napin-rich isolates (~90%) compared to acid-precipitated cruciferin-rich isolates (~20%) at pH range of 5.0–9.0. Tan, Mailer, Blanchard, Agboola, and Day (2014) compared the solubility between napin-rich albumin, cruciferin-rich globulin, and acid-precipitated cruciferin-rich isolates, and observed a high solubility at pH values of 4.0, 7.0, and 9.0 for albumin and globulin fractions (>70%), but a low solubility (<40%) for acid-precipitated isolates, confirming the effect of alkaline extraction and acid-precipitation on reducing the solubility in cruciferin-rich isolates.

6.1.3 Solubility comparison to other protein sources

Comparing the solubility of canola protein products to that of other plant proteins, Xu and Diosady (1994) reported that at pH values of 6.0–12.0, an alkaline-extracted acid-precipitated rapeseed protein isolate had a lower solubility than soy protein isolates, but a higher solubility at pH values of 2.0–5.0. Regarding the salt-extracted cruciferin-rich isolates, Karaca et al. (2011) observed a lower solubility of canola protein isolates (~80%) at pH 7.0 compared to flaxseed protein isolates (~95%), both of which were lower than WPI (~100%). Chang, Tu, Ghosh, and Nickerson (2015) compared the solubility of salt-extracted canola protein isolates to commercial pea, soy, and lentil protein isolates at pH range of 3.0–7.0, and reported the highest solubility for canola proteins at pH 3.0 (~90%) and 5.0 (~40%), and the lowest one at pH 7.0 (~20%), in contrast to other protein isolates having the minimum around pH 5.0.

Napin-rich isolates are superior in terms of their solubility over a wide range of pH compared to other plant proteins. It is advantageous since most proteins, including plant and animal proteins, are low in solubility around the pH values of 4.0–5.0 due to their pI (Xu & Diosady, 1994; Yoshie-Stark et al., 2008), which makes napin-rich isolates as a great option for plant-based acidic beverages.

6.2 Emulsifying

Canola proteins could be utilized as emulsion-stabilizing ingredients in several food formulations, such as emulsion-type meat products, salad, and mayonnaise. To stabilize oil droplets, either through electrostatic or steric mechanisms, protein-based emulsifiers act by diffusing to the interface, re-orienting on the interface, and forming the interfacial viscoelastic film (Morris & Gunning, 2008). Therefore, to generate strong viscoelastic interfacial films, a potent plant protein should be strongly charged to provide repulsions between oil droplets, and have a good solubility, good conformational flexibility, and reasonably high hydrophobicity to allow for fast diffusion, and proper rearrangement on the interface (Chang et al., 2015).

6.2.1 Emulsifying of canola protein concentrates

In comparison to protein isolates, canola protein concentrates contain less protein and include some other major components such as fibers, which theoretically may deteriorate their emulsifying ability. In this sense, Yoshie-Stark et al. (2006) investigated the emulsifying capacity of isopropanol-washed rapeseed protein concentrates, and reported an emulsion capacity of 325–373 mL/g at pH 5.0, which was lower than lupin (800 mL/g) and egg white (450 mL/g).

6.2.2 Emulsifying of canola protein isolates

Regarding the canola protein isolates, the effect of thermal treatment applied during the oil extraction process on emulsifying activity is imperative, as Östbring et al. (2020) showed that an alkaline-extracted isolate produced from cold-pressed cakes has the highest emulsifying properties, compared to hot-pressed and solvent-extracted meals. Others also confirmed the deteriorating effect of heating applied on canola seeds or meals on their interfacial properties (Khattab & Arntfield, 2009; Teh et al., 2014). The applied condition during the isolation process may also affect the emulsifying ability, as it has demonstrated that an extreme alkali condition during the alkaline extraction induces denaturation and reduces protein solubility, resulting in a decreased emulsifying activity (Pedroche et al., 2004; Teh et al., 2014). Hence, the emulsifying properties of alkaline-extracted isolates might be greatly improved by using a mild alkali extraction pH along with an adjusted precipitation pH.

Tan, Mailer, Blanchard, and Agboola (2014) investigated the effect of pH on the emulsifying properties of salt-extracted and acid-precipitated cruciferin-rich isolates, and found that salt-extracted isolates had a significant higher emulsion activity at pH values of 4.0, 7.0, and 9.0. Chang et al. (2015) reported that salt-extracted isolates formed smaller droplet sizes at pH 3.0 compared to the pH values of 5.0 and 7.0. The acid-precipitated isolates were reported to have the largest oil droplets at pH 4.0, which decreased at higher pHs (Alashi et al., 2018; Tan, Mailer, Blanchard, & Agboola, 2014). These larger droplets at pH 4.0–5.0 relate to the charge neutralization and promotion of flocculation due to the intensified attractive forces among the adsorbed proteins on the droplets. Regarding the interrelation between the pH and ionic strength on emulsifying properties of a cruciferin-rich isolate, Cheung et al. (2014) reported no effect of salt addition at pH 3.0 on emulsifying activity, but a reducing effect at pH values of 5.0 and 7.0. In the case of emulsion stability, salt addition had a reducing effect, which was attributed to the reduced surface charge and induced flocculation and coalescence.

Comparing the emulsifying activity of napin and cruciferin, a recent study showed that only napins could adsorb onto the interface in a mixture of napin and cruciferin extracted by alkaline extraction from canola seeds (Ntone et al., 2021). Also, some other studies reported a higher emulsification activity for the napin-rich isolates (Fetzer et al., 2019; Karaca et al., 2011; Krause & Schwenke, 2001) compared to the cruciferin-rich isolates. The superiority of napin in emulsifying could be associated with its higher solubility, its Janus-like structure with 45% hydrophobic amino acids located on one side of the protein, and its low molecular weight, which enable it to diffuse fast toward the oil-water interface and diminish the interfacial tension effectively (Cheung et al., 2015; Ntone et al., 2021). However, the presence of cruciferin is also essential in emulsifying activity, as Ntone et al. (2021) concluded that formation of cruciferin aggregates in the intervening continuous phase between the droplets increased the local viscosity and contributed to emulsion stability. Tan, Mailer, Blanchard, and Agboola (2014) compared the emulsifying activity of napin-rich isolates to salt-extracted and alkaline-extracted cruciferin-rich isolates, and reported the best emulsifying properties for salt-extracted isolates, which emphasizes on the extraction method affecting the emulsifying properties, and the synergistic effect of napin and cruciferin on emulsifying activity.

6.2.3 Comparison to other protein sources

Chang et al. (2015) compared salt-extracted canola protein isolates to commercial pea, lentil, and soy protein isolates. They observed similar droplet sizes induced by canola proteins and soy proteins at pH 3.0, and significantly lower sizes compared to lentil and pea protein isolates. At pH 5.0, the droplet size was lower in salt-extracted isolates compared to other plant proteins, while it was higher at pH 7.0 than others due to the pI of cruciferin (~7.2). Results of emulsion stability, however, showed a good stability at pH 3.0, while the emulsion was unstable at pH 5.0 and 7.0. Tan, Mailer, Blanchard, and Agboola (2014) compared the emulsifying activity of salt-extracted cruciferin-rich isolates to soy protein isolates, and concluded that the activity is higher for salt-extracted canola isolates at pH values <7.0, while soy protein isolates are superior at pH values >7.0.

6.3 Foaming

Foaming is the other functional property, besides emulsifying, that relates to the interfacial activity of proteins. Foams are two-phase systems consisting of air bubbles stabilized by surfactants/proteins in a continuous liquid lamellar phase. To minimize the interfacial free energy, proteins denature once they adsorped to the interface, to adopt the correct conformation (Tan et al., 2011a, 2011b). This depends on various factors, such as the protein structure, the extent of protein-protein and protein-solvent interactions, and the presence of other non-protein components in the protein isolates.

6.3.1 Foaming of canola protein isolates

Comparing the foaming properties of napin-rich and cruciferin-rich isolates, it is shown that the napin-rich fraction is more suitable for foaming properties. Xu and Diosady (1994) compared an alkaline-extracted acid-precipitated cruciferin-rich isolate to an acid-soluble napin-rich isolate, and reported a similar foaming capacity, but higher foam stability for the napin-rich isolate. Akbari and Wu (2015) also reported a higher foaming capacity and stability for the napin-rich isolate, compared to an alkaline-extracted cruciferin-rich isolate. The superiority of napin-rich isolates in foaming properties relates to their good solubility as a prerequisite, high degree of α-helical structures aiding at accumulation at the air-water interface, and their low molecular weight leading to the faster diffusion and adsorption to the interface (Fetzer et al., 2019; Krause & Schwenke, 2001).

In terms of the relation between the napin's foaming properties and pH value/ionic strength, Ivanova et al. (2018) reported a high foaming capacity for an acid-soluble napin-rich isolate, irrespective of pH values (2.0 to 10.0) and NaCl concentration (0.03 and 0.25 M). However, the foam stability decreased by >70% after the addition of 0.25 M NaCl at all pH values, attributed to the weakened viscoelastic protein film in effect of charge neutralization.

6.3.2 Comparison to other plant proteins

Dev and Mukherjee (1986) reported a higher foaming capacity and stability for alkaline-extracted cruciferin-rich isolates compared to commercial soy protein isolates. Using an ultrasound-assisted method, Dong et al. (2011) also reported a better foaming capacity for napin-rich acid-soluble and acid-precipitated cruciferin-rich isolates than soy protein isolates. Additionally, the foaming capabilities of a SHMP-precipitated isolate were compared to soy protein isolate and egg white by Thompson, Liu, and Jones (1982), who reported a higher foaming capacity and stability in the SHMP-precipitated isolate. However, since the protein composition of these isolates is unknown, it is impossible to conclude about the proteins involved in their foaming activity.

6.4 Gelation

Gelling ability refers to the capability of proteins to form a water-holding three-dimensional matrix, which happens based on a balance between repulsive and attractive interactions, including disulfide bonds, hydrogen bonds, and electrostatic interactions (Léger & Arntfield, 1993).

Generally, globular proteins could yield cold- and heat-set gels. The former involves gelation induction at room temperature, using ions, acid, etc., after denaturation and limited aggregation of proteins far from the pI. However, the latter include the denaturation and subsequent intensive aggregation of proteins by applying a thermal treatment around the pI of proteins at higher protein concentrations (Abaee, Mohammadian, & Jafari, 2017). Considering the wide range of pI among canola proteins, a cold-set gelation strategy is not applicable, and therefore heat-set gelation have been reported in most of studies on canola protein gelation.

6.4.1 Gelation of canola protein concentrates

Gelation of canola protein concentrates has not been studied extensively, although they are advantageous over canola protein isolates in terms of their lower processing level. Jia, Curubeto, et al. (2021) compared the apparent complex modulus of a texturized ethanol-washed protein concentrate to a salt-extracted cruciferin-rich isolate, soy protein concentrates/isolates, and wheat gluten. They witnessed a high apparent complex modulus in texturized rapeseed protein concentrates compared to other examined proteins, which was attributed to the strong mechanical strength in the structure, suitable for incorporation in meat analogue formulations.

6.4.2 Gelation of canola protein isolates

Thermal gelation of cruciferin was first reported by Gill and Tung (1978), using a chromatographically purified alkaline-extracted cruciferin from *B. campestris*. They observed that at pH $\geq$4.0 and 4.5% protein concentration, self-association during the heating results in gelation, with a progression in network ordering by increasing the pH from 6.0 to 10.0. Likewise, Léger and Arntfield (1993) reported an elastic modulus of 2.5 kPa at pH 8, among the analyzed pH values

(4.0–11.0), in heat-set gels prepared from a salt-extracted rapeseed protein isolate at a concentration of 6%. Interactions involved in the cruciferin gelation were analyzed by several research groups, and they reported that hydrophobic interactions are in charge for the establishment of the gel structure, while disulfide and hydrogen bonds contribute to the gel stabilization (Kim, Varankovich, Stone, & Nickerson, 2016; Yang, Wang, Vasanthan, & Chen, 2014).

Regarding the optimal gelation condition, Ramlan et al. (2002) examined different pH values (6.0, 7.6, and 9.0), protein concentration (5% and 7.5%), ionic strength (0.06 and 0.5), and heating temperatures (80, 90, and 100 °C), for gelation of a salt-extracted cruciferin-rich isolate, and reported that opaque gels with no elasticity are formed with higher hardness at higher protein concentration, higher heating temperature, higher pH, and lower ionic strength. This is aligned with a number of other studies confirming the effect of alkaline condition in obtaining stronger gelling properties in cruciferin-rich isolates (Kim, Varankovich and Nickerson, 2016; Léger & Arntfield, 1993; Yang et al., 2014). Superior heat-gelation properties at alkaline conditions are attributed to the pI of canola proteins located at high pHs, resulting in the promotion of aggregation due to the weakened electrostatic repulsions, and facilitated disruption of disulfide bonds at basic pHs (Yang et al., 2014).

Considering the influence of the extraction method on the physicochemical properties of cruciferin, differences in gelation properties of cruciferin-rich isolates prepared by either a salt extraction or an alkaline extraction method are expected. In this regard, frequency and strain sweep analysis showed that a salt-extracted cruciferin-rich isolate formed a stronger gel, compared to alkaline-extracted cruciferin-rich isolates, at pH values of 7.0 and 9.0 and 0 to 0.4 NaCl concentrations (Tan, Mailer, Blanchard, & Agboola, 2014; Tan, Mailer, Blanchard, Agboola, & Day, 2014). Despite the promising gelling properties of cruciferin, which relates to their high molecular weight and their potential for chain entanglement, the gelling ability of napin-rich isolates is under question (Schwenke, Dahme, & Wolter, 1998; Yang et al., 2014), as Tan, Mailer, Blanchard, Agboola, and Day (2014) reported a weak gel for napin-rich isolate, while a strong one for salt-extracted cruciferin-rich isolates at different pHs (4.0, 7.0, and 9.0) and NaCl concentrations (0–0.4 M).

6.4.3 Comparison to other plant proteins

Comparing the gelling properties of canola proteins to other plant proteins, Ramlan et al. (2002) studied the gelation of cruciferin with regard to glycinin (11S soy globulin), and reported the formation of opaque and not elastic gels by heat-gelation of cruciferin at pH 7.6 and ionic strength of 0.5, while glycinin produced a transparent and elastic gel. Kim, Varankovich and Nickerson, 2016 compared the gelation of a salt-extracted cruciferin-rich isolate to the commercial soy protein isolate. Analyzing the rheological properties, they reported a more rigid gel at optimal pH values for cruciferin-rich isolates (pH 9.0) compared to the soy protein isolate (pH 5.0).

7. Commercial canola protein products

Commercialization of canola/rapeseed protein products is encouraged by environmental concerns and the availability of cheap canola protein raw materials (pressed cakes or meals) produced by the oil extraction industry, as well as their high nutritional value and good functional properties. This is also supported by the reports related to the recent growing consumer interest and acceptance toward rapeseed protein-incorporated food products (Banovic & Sveinsdóttir, 2021).

However, commercialization of canola proteins in food products is suffered from the high investment and operational costs of wet extraction procedures due to the high water and energy use of the current methods. Additionally, the availability of starting material is another issue. In the case of soy protein extraction, because soybean oil accounts for only 40% of total revenue in soybean plants, an emphasis is being placed on the diversification of defatted materials suitable for protein extraction aimed for human consumption, which contain less damaged proteins due to the application of technologies like flash desolventizers. By contrast, canola oil accounts for about 70–80% of total revenue in canola processing plants, and therefore production of high-quality starting materials for canola protein extraction is not a priority and the co-streams are readily absorbed by the animal feeding market (Mupondwa et al., 2018). In addition, the strong tendency of antinutritional factors, including phytic acid and phenolics, to interact with proteins makes it difficult to separate them completely, resulting in an undesirable color and some off-notes in the protein extracts and the final food products.

Although the diversity of physicochemical and structural properties among canola proteins results in complicated protein recovery methods, these difference endowed them with diversified functional properties suitable for a range of applications. In this regard, cruciferin are proper proteins for use in meat substitutes, dressings and sauces, and baked products, due to the good emulsifying, gelling, and binding properties, while the napin is applicable in confectionery, beverages, and aerated desserts due to the superior foaming properties. Table 4 described the studied applications of canola/rapeseed protein concentrates, napin-rich and cruciferin-rich isolates, and protein hydrolysates in different food products.

Regarding the commercialized canola protein extracts, Table 5 summarized the protein composition, functional properties, and recommended

Table 4 Reported applications of canola protein products in food formulations.

Protein-rich product	Food Formulation	Description of the study	References
Canola protein concentrate	Sausage	• Limited substitution (2%) of sodium caseinate and soy protein concentrate with an alcohol-washed concentrate • Comparable emulsifying and water binding capacity but slightly lower firmness • Strawy smell, oily taste, and darker color in formulated products	Von Der Haar et al. (2014)
		• Substitution of casein with an alcohol-washed concentrate • Improved taste, good texture, and characteristic aroma in sausages	Yoshie-Stark et al. (2006)
	Meat analogue	• Induction of fibrous structure in an alcohol-washed concentrate and their mixture with wheat gluten at 140–150 °C with 40% dry matter using shear cell technology as the texturizer • Improved fibrous structure and the whiteness in texturized protein concentrates in the presence of wheat gluten	Jia, Curubeto, et al. (2021)
	Pound cake	• Substitution of egg with an alcohol-washed concentrate • Comparable volume expansion but slightly darker color in formulated products	Von Der Haar et al. (2014)
	Gluten-free bread	• Incorporation of protein concentrates (alkaline extract of canola meal without acidic precipitation) in bread made of rice flour or wheat flour in 3, 6, and 9% levels • Lower mass volume and honeycomb structure in supplemented wheat bread compared to control wheat bread • Higher mass volume in supplemented rice bread compared to control rice bread	Salah, Olkhovatov, and Aïder (2019)

Continued

Table 4 Reported applications of canola protein products in food formulations.—cont'd

Protein-rich product	Food Formulation	Description of the study	References
	Bread	• Substitution of 5% wheat flour with protein isolates and concentrates • Decreased loaf volume, which improved in the presence of 0.5% emulsifier	Kodagoda, Nakai, and Powrie (1973)
	Meringue	• Substitution of 3% egg white protein with protein isolates and concentrates • Decreased specific volume, with improved foam stability	Kodagoda et al. (1973)
Canola protein isolate	Sausage	• Lower cooking loss in formulations with incorporated cruciferin-rich and napin-rich isolates compared to the reference product, with similar sensory properties compared to conventional and sodium caseinate-incorporated formulations	Von Der Haar et al. (2014)
	Wiener	• Incorporation of 2% SHMP extracted concentrates, or soy protein isolates, in the conventional formulation • Increased emulsion stability, firmness and shrinkage • Similar results compared to soy protein isolate-incorporated formulations, but poorer color and flavor	Thompson et al. (1982)
	Patties	• Incorporation of 3% SHMP extracted concentrates, or soy protein isolates, in the conventional formulation • Increased cooking yield and firmness, reduced shrinkage and • Similar results compared to soy protein isolate-incorporated formulations, but poorer color and flavor	
	Mayonnaise	• Similar firmness after preparation using napin-rich isolate, compared to egg protein-incorporated product, but firmer during the storage time • Softer texture in products contained cruciferin-rich isolate during storage compared to egg protein-incorporated product • Similar sensory properties, except for the color (more brownish)	Von Der Haar et al. (2014)

	Biscuit	• Electroactivation-extracted isolates blended with rice and buckwheat flours in biscuits • Higher thickness and diameter and more aerated texture, resulting in a lower spread ratio in the formulated product • Less hardness and moisture content but higher springiness in protein-supplemented biscuits	Gerzhova, Mondor, Benali, and Aider (2016a)
	Gluten-free bread	• Substitution of corn and potato starches with 6–15% cruciferin-rich isolates • Formation of weak gel with reduced G' and G" in dough samples, and reduced viscosity	Witczak, Chmielewska, Ziobro, Korus, and Juszczak (2021)
Canola protein hydrolysates	Meat model	• Incorporation of 0.5% and 1% protein hydrolysates to a meat model containing 85% ground pork • Improved cooking yield in hydrolysates-incorporated formulations • The highest decrease observed in Flavourzyme hydrolysates-incorporated formulations	Cumby et al. (2008)
	Mayonnaise	• Substitution of 20–50% egg yolk with protein hydrolysates (DH of 7–14%) without emulsion breakdown • Increased levels of hydrolysates reduced the whiteness, increased the droplet size and lower the viscosity	Aluko and McIntosh (2005)

Table 5 Commercialized canola protein products.

Company	Product name	Protein composition	Functional feature	Recommended applications[a]	Patent number[b]	GRAS notification
Burcon	Supertein®	> 80% of napin fraction	High solubility (>90%) at a diverse pH range, high foaming capacity, and high heat stability[a]	Powdered egg/egg substitute (60%), dairy products (5%), processed meats (2%), grain products (2%), fruit and vegetable beverages (10%), salad dressings (2%), protein supplement powders (95%), and meal replacement (50%)	US Patent 5,844,086 and 6,005,076	GRAS (2010)
	Puratein®	> 80% of cruciferin fraction	Good emulsification, thickening, and heat-induced gelation[a]			
BioExx	Isolexx™	60–65% of globulin, 30–35% of albumins	Excellent water solubility and emulsification and foaming properties	Bakery products (3%), snack foods (20%), beverages and soups (5%), dairy products (4%), dry instant protein drinks (9%), processed meat products (2%), meat analogues (20%), and meal replacement (30%)	US Patent 8,623,445	GRAS (2011)
	Vitalexx™	a highly hydrolyzed protein isolates	Small peptides intended for applications with high required adsorption, ease of digestion and high solubility	Beverages and soups (5%), instant powdered beverages (15%), and meal replacement (25%)	US Patent 8,529,981	
DSM	CanolaPro	40–65% cruciferins and 35–60% napins[a]	High water solubility at neutral pH, good foaming and emulsifying	Prepared food (10%), meat analogues (30%), bakery products (5%), protein enriched bakery products (30%), sport nutrition (10%), weight management (30%), beverages (5%), dairy products (5%), medical nutrition (10%), and elderly nutrition (10%)	WO2017102535 A1	GRAS (2016)

[a]Mupondwa et al. (2018).
[b]The recommended maximum substitution level is mentioned in parentheses, based on the corresponding GRAS notifications.

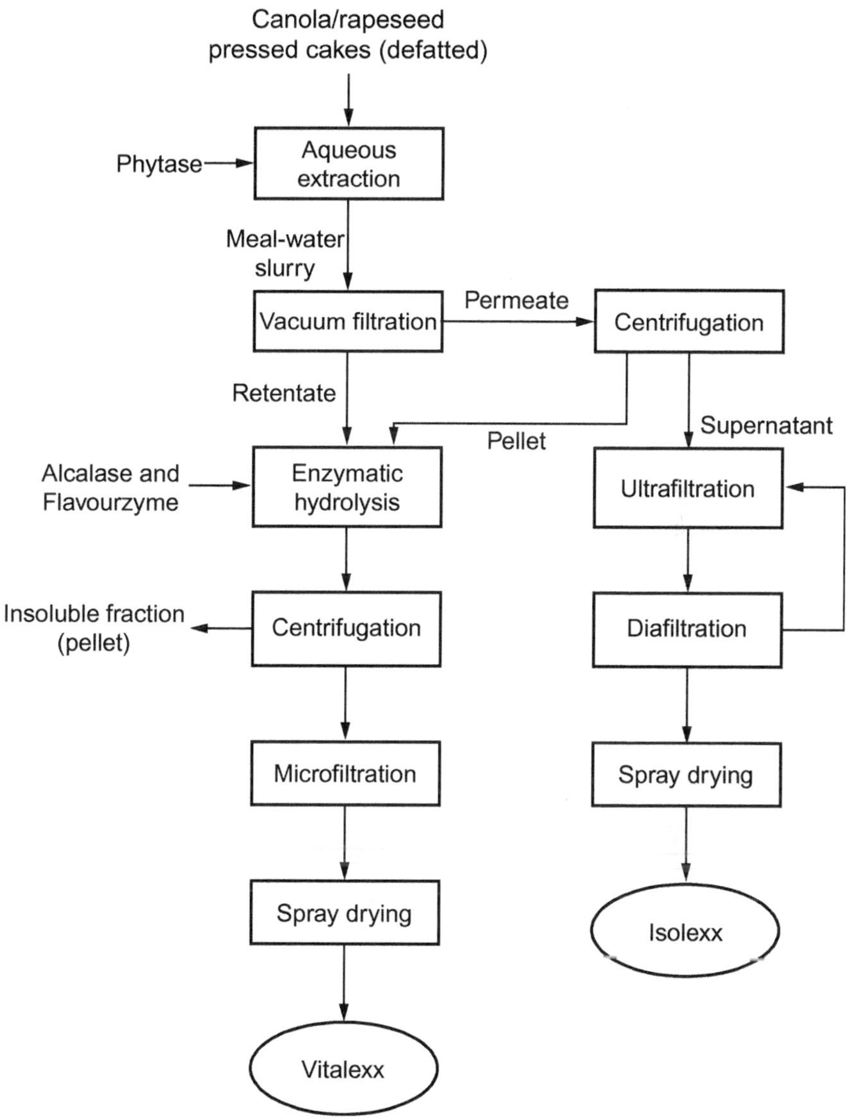

Fig. 3 BioExx's production method for Vitalexx and Isolexx.

applications of protein extracts that obtained the Generally Approved As Safe (GRAS) status from US Food and Drug Administration (FDA), and Figs. 2–4 depicted the flowcharts for their production. The reasonable content of ANFs in these protein extract products, their balanced amino acid

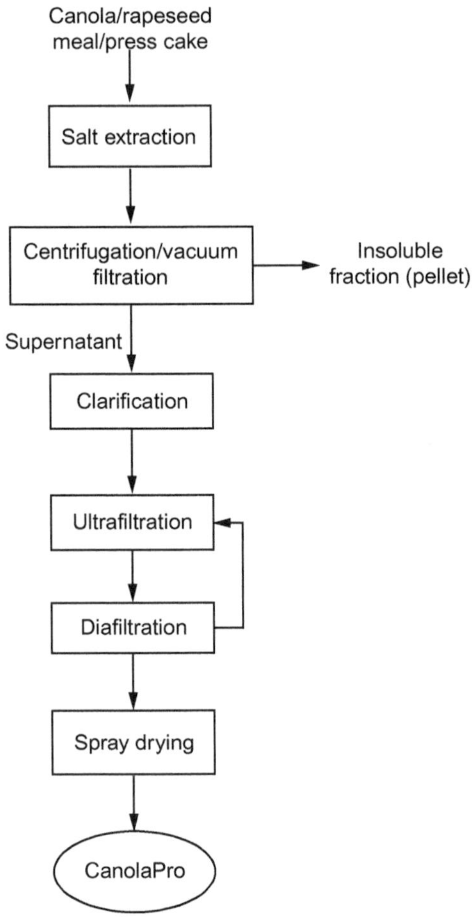

Fig. 4 DSM's production method for CanolaPro.

composition, as well as the promising digestibility and toxicity assessment results, suggest them as sources of plant proteins in food formulations (GRAS, 2010, 2011, 2016).

8. Conclusion

Environmental and nutritional concerns associated with animal protein consumption have motivated the food industry to look for new protein choices. Plant proteins are spectacular options, considering their developed supply chain compared to other alternative protein sources, and their long-time presence in human diet. Canola is a great potential source of protein with a well-balanced amino acid composition and promising functional properties suitable for uses in food formulations.

However, the presence of different canola proteins with distinct structures and physicochemical properties as well as numerous ANFs are considered as the major inherent challenges for canola protein extraction. Canola protein extraction is further complicated and negatively affected by harsh extraction conditions applied during oil extraction. Developing an efficient method of protein extraction remains a challenge. Although canola protein extraction methods, mainly including salt and alkaline extraction techniques coupled with membrane technologies, are widely applied, the major differences in the starting materials have resulted in canola proteins with varied purity and yield, and functionality. This urges the researchers to be alerted on the starting material and the ratio of napin to cruciferin in their extract, as the major parameters contributing to the properties of canola protein extracts. Regarding the functional properties, the large physicochemical differences between main canola proteins (i.e., napin and cruciferin) provide a range of functional features that are suitable for different food applications. Overall, napin is advantageous in terms of the solubility at a wide range of pHs, and interfacial properties, while cruciferin is better for gelling applications. However, canola proteins produced by various extraction procedures contain both napin and cruciferin, showing different functional properties depending on the ratio of napin and cruciferin. This highlights a necessity for further research on the effect of interactions between napin and cruciferin on the functional properties of canola proteins. The major application of plant proteins is to replace animal proteins, in particular meat alternatives, further studies are urgently warranted to understand the texturization behavior of canola proteins.

References

Abaee, A., Mohammadian, M., & Jafari, S. M. (2017). Whey and soy protein-based hydrogels and nano-hydrogels as bioactive delivery systems. *Trends in Food Science & Technology*, *70*, 69–81.

Akbari, A., & Wu, J. (2015). An integrated method of isolating napin and cruciferin from defatted canola meal. *LWT- Food Science and Technology*, *64*(1), 308–315.

Alashi, A., Blanchard, C., Mailer, R., & Agboola, S. (2011). Improving the emulsifying properties of canola meal protein isolate by enzymatic modification. In *17th Australian research assembly on brassicas (ARAB)*.

Alashi, A. M., Blanchard, C. L., Mailer, R. J., & Agboola, S. O. (2013). Technological and bioactive functionalities of canola meal proteins and hydrolysates. *Food Reviews International*, *29*(3), 231–260.

Alashi, A. M., Blanchard, C. L., Mailer, R. J., Agboola, S. O., Mawson, J. A., & Aluko, R. E. (2018). Influence of enzymatic hydrolysis, pH and storage temperature on the emulsifying properties of canola protein isolate and hydrolysates. *International Journal of Food Science & Technology, 53*(10), 2316–2324.

Alashi, A. M., Blanchard, C. L., Mailer, R. J., Agboola, S. O., Mawson, A. J., He, R., et al. (2014a). Blood pressure lowering effects of Australian canola protein hydrolysates in spontaneously hypertensive rats. *Food Research International, 55*, 281–287.

Alashi, A. M., Blanchard, C. L., Mailer, R. J., Agboola, S. O., Mawson, A. J., He, R., et al. (2014b). Antioxidant properties of Australian canola meal protein hydrolysates. *Food Chemistry, 146*, 500–506.

Albe-Slabi, S., Defaix, C., Beaubier, S., Galet, O., & Kapel, R. (2022). Selective extraction of napins: Process optimization and impact on structural and functional properties. *Food Hydrocolloids, 122*, 107105.

Aljuobori, A., Abdullah, N., Zulkifli, I., Soleimani, A. F., Liang, J. B., & Oskoueian, E. (2014). Lactobacillus salivarius fermentation reduced glucosinolate and fibre in canola meal. *Journal of Food Research, 3*(5), 95.

Aluko, R. E., & McIntosh, T. (2001). Polypeptide profile and functional properties of defatted meals and protein isolates of canola seeds. *Journal of the Science of Food and Agriculture, 81*(4), 391–396.

Aluko, R. E., & McIntosh, T. (2005). Limited enzymatic proteolysis increases the level of incorporation of canola proteins into mayonnaise. *Innovative Food Science & Emerging Technologies, 6*(2), 195–202.

Ancuţa, P., & Sonia, A. (2020). Oil press-cakes and meals valorization through circular economy approaches: A review. *Applied Sciences, 10*(21), 7432.

Anwar, F., Kamal, G. M., Nadeem, F., & Shabir, G. (2016). Variations of quality characteristics among oils of different soybean varieties. *Journal of King Saud University-Science, 28*(4), 332–338.

Arrutia, F., Binner, E., Williams, P., & Waldron, K. W. (2020). Oilseeds beyond oil: Press cakes and meals supplying global protein requirements. *Trends in Food Science & Technology, 100*, 88–102.

Banovic, M., & Sveinsdóttir, K. (2021). Importance of being analogue: Female attitudes towards meat analogue containing rapeseed protein. *Food Control, 123*, 107833.

Bell, J. M. (1993). Factors affecting the nutritional value of canola meal: A review. *Canadian Journal of Animal Science, 73*(4), 689–697.

Bérot, S., Compoint, J. P., Larré, C., Malabat, C., & Guéguen, J. (2005). Large scale purification of rapeseed proteins (Brassica napus L.). *Journal of Chromatography B, 818*(1), 35–42.

Bhatty, R. S., McKenzie, S. L., & Finlayson, A. J. (1968). The proteins of rapeseed (Brassica napus L.) soluble in salt solutions. *Canadian Journal of Biochemistry, 46*(10), 1191–1197.

Blaicher, F. M., Elstner, F., Stein, W., & Mukherjee, K. D. (1983). Rapeseed protein isolates: Effect of processing on yield and composition of protein. *Journal of Agricultural and Food Chemistry, 31*(2), 358–362.

Bos, C., Airinei, G., Mariotti, F., Benamouzig, R., Bérot, S., Evrard, J., et al. (2007). The poor digestibility of rapeseed protein is balanced by its very high metabolic utilization in humans. *The Journal of Nutrition, 137*(3), 594–600.

Canola Council of Canada. (2021). *Canadian canola production statistics*. https://www.canolacouncil.org/markets-stats/production/.

Chabanon, G., Chevalot, I., Framboisier, X., Chenu, S., & Marc, I. (2007). Hydrolysis of rapeseed protein isolates: Kinetics, characterization and functional properties of hydrolysates. *Process Biochemistry, 42*(10), 1419–1428.

Chang, C., Tu, S., Ghosh, S., & Nickerson, M. T. (2015). Effect of pH on the inter-relationships between the physicochemical, interfacial and emulsifying properties for pea, soy, lentil and canola protein isolates. *Food Research International, 77*, 360–367.

Chen, M., & Rohani, S. (1992). Recovery of canola meal proteins by precipitation. *Biotechnology and Bioengineering*, *40*(1), 63–68.
Cheung, L., Wanasundara, J., & Nickerson, M. T. (2014). The effect of pH and NaCl levels on the physicochemical and emulsifying properties of a cruciferin protein isolate. *Food Biophysics*, *9*(2), 105–113.
Cheung, L., Wanasundara, J., & Nickerson, M. T. (2015). Effect of pH and NaCl on the emulsifying properties of a napin protein isolate. *Food Biophysics*, *10*(1), 30–38.
Chmielewska, A., Kozłowska, M., Rachwał, D., Wnukowski, P., Amarowicz, R., Nebesny, E., et al. (2021). Canola/rapeseed protein–nutritional value, functionality and food application: A review. *Critical Reviews in Food Science and Nutrition*, *61*(22), 3836–3856.
Cumby, N., Zhong, Y., Naczk, M., & Shahidi, F. (2008). Antioxidant activity and water-holding capacity of canola protein hydrolysates. *Food Chemistry*, *109*(1), 144–148.
Dalgalarrondo, M., Robin, J. M., & Azanza, J. L. (1986). Subunit composition of the globulin fraction of rapeseed (Brassica napus L.). *Plant Science*, *43*(2), 115–124.
Das Purkayastha, M., Das, S., Manhar, A. K., Deka, D., Mandal, M., & Mahanta, C. L. (2013). Removing antinutrients from rapeseed press-cake and their benevolent role in waste cooking oil-derived biodiesel: Conjoining the valorization of two disparate industrial wastes. *Journal of Agricultural and Food Chemistry*, *61*(45), 10746–10756.
Das Purkayastha, M., Gogoi, J., Kalita, D., Chattopadhyay, P., Nakhuru, K. S., Goyary, D., et al. (2014). Physicochemical and functional properties of rapeseed protein isolate: Influence of antinutrient removal with acidified organic solvents from rapeseed meal. *Journal of Agricultural and Food Chemistry*, *62*(31), 7903–7914.
Das Purkayastha, M., & Mahanta, C. L. (2015). Statistically designed optimal process conditions for recuperation of protein from rapeseed meal. *Journal of Food Science and Technology*, *52*(6), 3203–3218.
Daun, J. K., Eskin, M. N., & Hickling, D. (Eds.). (2015). *Canola: Chemistry, production, processing, and utilization* Elsevier.
del Mar Contreras, M., Lama-Muñoz, A., Gutiérrez-Pérez, J. M., Espínola, F., Moya, M., & Castro, E. (2019). Protein extraction from Agri-food residues for integration in biorefinery: Potential techniques and current status. *Bioresource Technology*, *280*, 459–477.
Delisle, J., Amiot, J., Goulet, G., Simard, C., Brisson, G. J., & Jones, J. D. (1984). Nutritive value of protein fractions extracted from soybean, rapeseed and wheat flours in the rat. *Plant Foods for Human Nutrition*, *34*(4), 243–251.
Deng, Q. Y., Barefoot, R. R., Diosady, L. L., Rubin, L. J., & Tzeng, Y. M. (1990). Lysinoalanine concentrations in rapeseed protein meals and isolates. *Canadian Institute of Food Science and Technology Journal*, *23*(2–3), 140–142.
Dev, D. K., & Mukherjee, K. D. (1986). Functional properties of rapeseed protein products with varying phytic acid contents. *Journal of Agricultural and Food Chemistry*, *34*(5), 775–780.
Dietz, H. M., King, R. D., & Harris, R. V. (1991). The aqueous extraction of glucosinolates from rapeseed. *International Journal of Food Science & Technology*, *26*(1), 53–63.
Diosday, L. L., Tzeng, Y. M., & Rubin, L. J. (1984). Preparation of rapeseed protein concentrates and isolates using ultrafiltration. *Journal of Food Science*, *49*(3), 768–770.
Dong, X. Y., Guo, L. L., Wei, F., Li, J. F., Jiang, M. L., Li, G. M., et al. (2011). Some characteristics and functional properties of rapeseed protein prepared by ultrasonication, ultrafiltration and isoelectric precipitation. *Journal of the Science of Food and Agriculture*, *91*(8), 1488–1498.
Ebrahimi-Mahmoudabad, S. R., & Taghinejad-Roudbaneh, M. (2011). Investigation of electron beam irradiation effects on anti-nutritional factors, chemical composition and digestion kinetics of whole cottonseed, soybean and canola seeds. *Radiation Physics and Chemistry*, *80*(12), 1441–1447.

El Nockrashy, A. S., Mukherjee, K. D., & Mangold, H. K. (1977). Rapeseed protein isolates by countercurrent extraction and isoelectric precipitation. *Journal of Agricultural and Food Chemistry, 25*(1), 193–197.

El-Kadiri, I., Khelifi, M., & Aider, M. (2013). The effect of hydrogen peroxide bleaching of canola meal on product colour, dry matter and protein extractability and molecular weight profile. *International Journal of Food Science & Technology, 48*(5), 1071–1085.

Fasolin, L. H., Pereira, R. N., Pinheiro, A. C., Martins, J. T., Andrade, C. C. P., Ramos, O. L., et al. (2019). Emergent food proteins–towards sustainability, health and innovation. *Food Research International, 125*, 108586.

Fetzer, A., Herfellner, T., & Eisner, P. (2019). Rapeseed protein concentrates for non-food applications prepared from pre-pressed and cold-pressed press cake via acidic precipitation and ultrafiltration. *Industrial Crops and Products, 132*, 396–406.

Fleddermann, M., Fechner, A., Rößler, A., Bähr, M., Pastor, A., Liebert, F., et al. (2013). Nutritional evaluation of rapeseed protein compared to soy protein for quality, plasma amino acids, and nitrogen balance–a randomized cross-over intervention study in humans. *Clinical Nutrition, 32*(4), 519–526.

Folawiyo, Y. L., & Apenten, R. K. O. (1996). Effect of pH and ionic strength on the heat stability of rapeseed 12S globulin (cruciferin) by the ANS fluorescence method. *Journal of the Science of Food and Agriculture, 70*(2), 241–246.

Gerzhova, A., Mondor, M., Benali, M., & Aider, M. (2015). A comparative study between the electro-activation technique and conventional extraction method on the extractability, composition and physicochemical properties of canola protein concentrates and isolates. *Food Bioscience, 11*, 56–71.

Gerzhova, A., Mondor, M., Benali, M., & Aider, M. (2016a). Incorporation of canola proteins extracted by electroactivated solutions in gluten-free biscuit formulation of rice–buckwheat flour blend: Assessment of quality characteristics and textural properties of the product. *International Journal of Food Science & Technology, 51*(3), 814–827.

Gerzhova, A., Mondor, M., Benali, M., & Aider, M. (2016b). Study of total dry matter and protein extraction from canola meal as affected by the pH, salt addition and use of zeta-potential/turbidimetry analysis to optimize the extraction conditions. *Food Chemistry, 201*, 243–252.

Ghodsvali, A., Khodaparast, M. H., Vosoughi, M., & Diosady, L. L. (2005). Preparation of canola protein materials using membrane technology and evaluation of meals functional properties. *Food Research International, 38*(2), 223–231.

Gill, T. A., & Tung, M. A. (1978). Thermally induced gelation of the 12S rapeseed glycoprotein. *Journal of Food Science, 43*(5), 1481–1485.

Gillberg, L., & Tornell, B. (1976a). Preparation of rapeseed protein isolates. Dissolution and precipitation behavior of rapeseed proteins. *Journal of Food Science, 41*(5), 1063–1069.

Gillberg, L., & Tornell, B. (1976b). Preparation of rapeseed protein isolates. Precipitation of rapeseed proteins in the presence of polyacids. *Journal of Food Science, 41*(5), 1070–1075.

Grala, W., Verstegen, M. W. A., Jansman, A. J. M., Huisman, J., & Van Leeusen, P. (1998). Ileal apparent protein and amino acid digestibilities and endogenous nitrogen losses in pigs fed soybean and rapeseed products. *Journal of Animal Science, 76*(2), 557–568.

GRAS. (2010). *GRAS notice no. 327. GRAS notification for cruciferin-rich and napin-rich protein isolates derived from canola/rapeseed (Puratein VR and Supertein TM)*. https://www.accessdata.fda.gov/scripts/fdcc/?set=GRASNotices&id=327&sort=GRN_No&order=DESC&startrow=1&type=basic&search=327.

GRAS. (2011). *GRAS notice no. 386. GRAS notification for BioExx canola proteins: Canola protein isolate (IsolexxTM) and hydrolyzed canola protein isolate (VitalexxTM)*. https://www.accessdata.fda.gov/scripts/fdcc/?set=GRASNotices&id=386&sort=GRN_No&order=DESC&startrow=1&type=basic&search=386.

GRAS. (2016). *GRAS notice no. 683. The safety and the generally recognized as safe (GRAS) status of the proposed uses of canola/rapeseed protein isolate in human food.* https://www.fda.gov/media/103212/download.

Grasso, A. C., Olthof, M. R., van Dooren, C., Broekema, R., Visser, M., & Brouwer, I. A. (2021). Protein for a healthy future: How to increase protein intake in an environmentally sustainable way in older adults in the Netherlands. *The Journal of Nutrition, 151*(1), 109–119.

Hartmann, R., & Meisel, H. (2007). Food-derived peptides with biological activity: From research to food applications. *Current Opinion in Biotechnology, 18*(2), 163–169.

He, R., Alashi, A., Malomo, S. A., Girgih, A. T., Chao, D., Ju, X., et al. (2013). Antihypertensive and free radical scavenging properties of enzymatic rapeseed protein hydrolysates. *Food Chemistry, 141*(1), 153–159.

He, R., Girgih, A. T., Malomo, S. A., Ju, X., & Aluko, R. E. (2013). Antioxidant activities of enzymatic rapeseed protein hydrolysates and the membrane ultrafiltration fractions. *Journal of Functional Foods, 5*(1), 219–227.

He, R., Malomo, S. A., Alashi, A., Girgih, A. T., Ju, X., & Aluko, R. E. (2013). Purification and hypotensive activity of rapeseed protein-derived renin and angiotensin converting enzyme inhibitory peptides. *Journal of Functional Foods, 5*(2), 781–789.

Hoglund, A. S., Rodin, J., Larsson, E., & Rask, L. (1992). Distribution of napin and cruciferin in developing rape seed embryos. *Plant Physiology, 98*(2), 509–515.

Hu, F. B., Otis, B. O., & McCarthy, G. (2019). Can plant-based meat alternatives be part of a healthy and sustainable diet? *JAMA, 322*(16), 1547–1548.

Huang, A. H. (1992). Oil bodies and oleosins in seeds. *Annual Review of Plant Biology, 43*(1), 177–200.

Ismond, M. A. H., & Welsh, W. D. (1992). Application of new methodology to canola protein isolation. *Food Chemistry, 45*(2), 125–127.

Ivanova, P., Chalova, V., Uzunova, G., Koleva, L., & Manolov, I. (2016). Biochemical characterization of industrially produced rapeseed meal as a protein source in food industry. *Agriculture and Agricultural Science Procedia, 10*, 55–62.

Ivanova, P., Kalaydzhiev, H., Dessev, T. T., Silva, C. L., Rustad, T., & Chalova, V. I. (2018). Foaming properties of acid-soluble protein-rich ingredient obtained from industrial rapeseed meal. *Journal of Food Science and Technology, 55*(9), 3792–3798.

Jensen, S. K., Olsen, H. S., & Sørensen, H. (1990). Aqueous enzymatic processing of rapeseed for production of high quality products. In *Canola and rapeseed* (pp. 331–343). Boston, MA: Springer.

Jia, W., Curubeto, N., Rodríguez-Alonso, E., Keppler, J. K., & van der Goot, A. J. (2021). Rapeseed protein concentrate as a potential ingredient for meat analogues. *Innovative Food Science & Emerging Technologies, 72*, 102758.

Jia, W., Rodriguez-Alonso, E., Bianeis, M., Keppler, J. K., & van der Goot, A. J. (2021). Assessing functional properties of rapeseed protein concentrate versus isolate for food applications. *Innovative Food Science & Emerging Technologies, 68*, 102636.

Jung, S. (2009). Aqueous extraction of oil and protein from soybean and lupin: A comparative study. *Journal of Food Processing and Preservation, 33*(4), 547–559.

Karaca, A. C., Low, N., & Nickerson, M. (2011). Emulsifying properties of canola and flaxseed protein isolates produced by isoelectric precipitation and salt extraction. *Food Research International, 44*(9), 2991–2998.

Khattab, R. Y., & Arntfield, S. D. (2009). Functional properties of raw and processed canola meal. *LWT- Food Science and Technology, 42*(6), 1119–1124.

Kim, J. H., Varankovich, N. V., Stone, A. K., & Nickerson, M. T. (2016). Nature of protein-protein interactions during the gelation of canola protein isolate networks. *Food Research International, 89*, 408–414.

Kim, J. H. J., Varankovich, N. V., & Nickerson, M. T. (2016). The effect of pH on the gelling behaviour of canola and soy protein isolates. *Food Research International, 81*, 31–38.
Klockeman, D. M., Toledo, R., & Sims, K. A. (1997). Isolation and characterization of defatted canola meal protein. *Journal of Agricultural and Food Chemistry, 45*(10), 3867–3870.
Kodagoda, L. P., Nakai, S., & Powrie, W. D. (1973). Some functional properties of rapeseed protein isolates and concentrates. *Canadian Institute of Food Science and Technology Journal, 6*(4), 266–269.
Kozlowska, H., Zadernowski, R., & Sosulski, F. W. (1983). Phenolic acids in oilseed flours. *Food/Nahrung, 27*(5), 449–453.
Krause, J. P., & Schwenke, K. D. (2001). Behaviour of a protein isolate from rapeseed (Brassica napus) and its main protein components—Globulin and albumin—At air/solution and solid interfaces, and in emulsions. *Colloids and Surfaces B: Biointerfaces, 21*(1–3), 29–36.
Kroll, J. (1991). Selected functional properties of detoxified rapeseed protein preparations effected by phytic acid. *Food/Nahrung, 35*(6), 619–624.
Krzyzaniak, A., Burova, T., Haertlé, T., & Barciszewski, J. (1998). The structure and properties of Napin-seed storage protein from rape (Brassica napus L.). *Food/Nahrung, 42*(03–04), 201–204.
Laguna, O., Barakat, A., Alhamada, H., Durand, E., Baréa, B., Fine, F., et al. (2018). Production of proteins and phenolic compounds enriched fractions from rapeseed and sunflower meals by dry fractionation processes. *Industrial Crops and Products, 118*, 160–172.
Léger, L. W., & Arntfield, S. D. (1993). Thermal gelation of the 12S canola globulin. *Journal of the American Oil Chemists' Society, 70*(9), 853–861.
Leming, R., & Lember, A. (2005). Chemical composition of expeller-extracted and cold-pressed rapeseed cake. *Agraarteadus, 16*(2), 96–109.
Mahajan, A., & Dua, S. (1997). Nonchemical approach for reducing antinutritional factors in rapeseed (Brassica campestris var. Toria) and characterization of enzyme phytase. *Journal of Agricultural and Food Chemistry, 45*(7), 2504–2508.
Mailer, R. (2004). Canola meal: Limitations and opportunities. *Australian Oilseed Federation.*
Mäkinen, S., Johannson, T., Gerd, E. V., Pihlava, J. M., & Pihlanto, A. (2012). Angiotensin I-converting enzyme inhibitory and antioxidant properties of rapeseed hydrolysates. *Journal of Functional Foods, 4*(3), 575–583.
Mansour, E. H., Dworschák, E., Lugasi, A., Gaál, Ö., Barna, E., & Gergely, A. (1993). Effect of processing on the antinutritive factors and nutritive value of rapeseed products. *Food Chemistry, 47*(3), 247–252.
Marczak, E. D., Usui, H., Fujita, H., Yang, Y., Yokoo, M., Lipkowski, A. W., et al. (2003). New antihypertensive peptides isolated from rapeseed. *Peptides, 24*(6), 791–798.
Matthews, D. M. (1977). Memorial lecture: Protein absorption-then and now. *Gastroenterology, 73*(6), 1267–1279.
Mejicanos, G. A. (2015). *Tail-end dehulling of canola meal: Chemical composition and nutritive value of dehulled meal for broiler chickens and weaned pigs (Master's thesis).*
Mittal, A., Katahira, R., Himmel, M. E., & Johnson, D. K. (2011). Effects of alkaline or liquid-ammonia treatment on crystalline cellulose: Changes in crystalline structure and effects on enzymatic digestibility. *Biotechnology for Biofuels, 4*(1), 1–16.
Monsalve, R. I., Lopez-Otin, C., Villalba, M., & Rodríguez, R. (1991). A new distinct group of 2 S albumins from rapeseed: Amino acid sequence of two low molecular weight napins. *FEBS Letters, 295*(1–3), 207–210.
Morris, V. J., & Gunning, A. P. (2008). Microscopy, microstructure and displacement of proteins from interfaces: Implications for food quality and digestion. *Soft Matter, 4*(5), 943–951.

Mupondwa, E., Li, X., & Wanasundara, J. P. (2018). Technoeconomic prospects for commercialization of Brassica (cruciferous) plant proteins. *Journal of the American Oil Chemists' Society*, *95*(8), 903–922.

Mustafa, A. F., Christensen, D. A., McKinnon, J. J., & Newkirk, R. (2000). Effects of stage of processing of canola seed on chemical composition and in vitro protein degradability of canola meal and intermediate products. *Canadian Journal of Animal Science*, *80*(1), 211–214.

Newkirk, R. (2009). *Canola meal feed industry guide*. Canadian International Grains Institute.

Niu, Y. X., Li, W., Zhu, J., Huang, Q., Jiang, M., & Huang, F. (2012). Aqueous enzymatic extraction of rapeseed oil and protein from dehulled cold-pressed double-low rapeseed cake. *International Journal of Food Engineering*, *8*(3).

Ntone, E., van Wesel, T., Sagis, L. M., Meinders, M., Bitter, J. H., & Nikiforidis, C. V. (2021). Adsorption of rapeseed proteins at oil/water interfaces. Janus-like napins dominate the interface. *Journal of Colloid and Interface Science*, *583*, 459–469.

Olukomaiya, O. O., Fernando, W. C., Mereddy, R., Li, X., & Sultanbawa, Y. (2020). Solid-state fermentation of canola meal with aspergillus sojae, aspergillus ficuum and their co-cultures: Effects on physicochemical, microbiological and functional properties. *LWT- Food Science and Technology*, *127*, 109362.

Oreopoulou, V., & Tzia, C. (2007). Utilization of plant by-products for the recovery of proteins, dietary fibers, antioxidants, and colorants. In *Utilization of by-products and treatment of waste in the food industry* (pp. 209–232). Boston, MA: Springer.

Östbring, K., Malmqvist, E., Nilsson, K., Rosenlind, I., & Rayner, M. (2020). The effects of oil extraction methods on recovery yield and emulsifying properties of proteins from rapeseed meal and press cake. *Food*, *9*(1), 19.

Pan, M., Jiang, T. S., & Pan, J. L. (2011). Antioxidant activities of rapeseed protein hydrolysates. *Food and Bioprocess Technology*, *4*(7), 1144–1152.

Partanen, R., Sibakov, J., Rommi, K., Hakala, T., Holopainen-Mantila, U., Lahtinen, P., et al. (2016). Dispersion stability of non-refined turnip rapeseed (Brassica rapa) protein concentrate: Impact of thermal, mechanical and enzymatic treatments. *Food and Bioproducts Processing*, *99*, 29–37.

Paulson, A. T., & Tung, M. A. (1987). Solubility, hydrophobicity and net charge of succinylated canola protein isolate. *Journal of Food Science*, *52*(6), 1557–1561.

Pedroche, J., Yust, M. M., Lqari, H., Girón-Calle, J., Alaiz, M., Vioque, J., et al. (2004). Brassica carinata protein isolates: Chemical composition, protein characterization and improvement of functional properties by protein hydrolysis. *Food Chemistry*, *88*(3), 337–346.

Perera, S. P., McIntosh, T. C., & Wanasundara, J. P. (2016). Structural properties of cruciferin and napin of Brassica napus (canola) show distinct responses to changes in pH and temperature. *Plants*, *5*(3), 36.

Pojić, M., Mišan, A., & Tiwari, B. (2018). Eco-innovative technologies for extraction of proteins for human consumption from renewable protein sources of plant origin. *Trends in Food Science & Technology*, *75*, 93–104.

Prakash, V., & Narasinga Rao, M. S. (1986). Physicochemical properties of oilseed protein. *Critical Reviews in Biochemistry*, *20*(3), 265–363.

Pustjens, A. M., Schols, H. A., Kabel, M. A., & Gruppen, H. (2013). Characterisation of cell wall polysaccharides from rapeseed (Brassica napus) meal. *Carbohydrate Polymers*, *98*(2), 1650–1656.

Quinn, J. R., & Jones, J. D. (1976). Rapeseed protein. pH solubility and electrophoretic characteristics. *Canadian Institute of Food Science and Technology Journal*, *9*(1), 47–51.

Radfar, M., Rogiewicz, A., Hickling, D., & Slominski, B. A. (2015). *Chemical composition and nutritive value of meals from yellow-seeded canola*. Saskatoon, SK: Proc. Intl Rapeseed Congr.

Radwan, M. N., & Lu, B. Y. (1976). Solubility of rapeseed protein in aqueous solutions. *Journal of the American Oil Chemists' Society*, *53*(4), 142–144.
Ramlan, M., Maruyama, N., Adachi, M., Hontani, N., Saka, S., Kato, N., et al. (2002). Comparison of protein chemical and physicochemical properties of rapeseed cruciferin with those of soybean glycinin. *Journal of Agricultural and Food Chemistry*, *50*(25), 7380–7385.
Rico, M., Bruix, M., González, C., Monsalve, R. I., & Rodríguez, R. (1996). 1H NMR assignment and global fold of napin BnIb, a representative 2S albumin seed protein. *Biochemistry*, *35*(49), 15672–15682.
Rivera, D., Rommi, K., Fernandes, M. M., Lantto, R., & Tzanov, T. (2016). Enzyme-aided processing of rapeseed oil industry by-products into added value ingredients for food, cosmetic and pharmaceutical applications. In *Rapeseed: Chemical composition, production and health benefits* (pp. 33–48). Nova Science Publishers.
Rodrigues, I. M. M. A., Carvalho, M. G. V. S., & Rocha, J. M. S. (2014). Increasing the protein content of rapeseed meal by enzymatic hydrolysis of carbohydrates. *BioResources*, *9*(2), 2010–2025.
Rommi, K., Ercili-Cura, D., Hakala, T. K., Nordlund, E., Poutanen, K., & Lantto, R. (2015). Impact of total solid content and extraction pH on enzyme-aided recovery of protein from defatted rapeseed (Brassica rapa L.) press cake and physicochemical properties of the protein fractions. *Journal of Agricultural and Food Chemistry*, *63*(11), 2997–3003.
Rommi, K., Hakala, T. K., Holopainen, U., Nordlund, E., Poutanen, K., & Lantto, R. (2014). Effect of enzyme-aided cell wall disintegration on protein extractability from intact and dehulled rapeseed (Brassica rapa L. and Brassica napus L.) press cakes. *Journal of Agricultural and Food Chemistry*, *62*(32), 7989–7997.
Saito, K., Jin, D. H., Ogawa, T., Muramoto, K., Hatakeyama, E., Yasuhara, T., et al. (2003). Antioxidative properties of tripeptide libraries prepared by the combinatorial chemistry. *Journal of Agricultural and Food Chemistry*, *51*(12), 3668–3674.
Salah, K., Olkhovatov, E. A., & Aïder, M. (2019). Effect of canola proteins on rice flour bread and mathematical modelling of the baking process. *Journal of Food Science and Technology*, *56*(8), 3744–3753.
Salunkhe, D. K., Adsule, R. N., Chavan, J. K., & Kadam, S. S. (1992). *World oilseeds*. Springer Science & Business Media.
Sari, Y. W., Bruins, M. E., & Sanders, J. P. (2013). Enzyme assisted protein extraction from rapeseed, soybean, and microalgae meals. *Industrial Crops and Products*, *43*, 78–83.
Sari, Y. W., Mulder, W. J., Sanders, J. P., & Bruins, M. E. (2015). Towards plant protein refinery: Review on protein extraction using alkali and potential enzymatic assistance. *Biotechnology Journal*, *10*(8), 1138–1157.
Schaafsma, G. (2005). The protein digestibility-corrected amino acid score (PDCAAS)—A concept for describing protein quality in foods and food ingredients: A critical review. *Journal of AOAC International*, *88*(3), 988–994.
Schmidt, I., Renard, D., Rondeau, D., Richomme, P., Popineau, Y., & Axelos, M. A. V. (2004). Detailed physicochemical characterization of the 2S storage protein from rape (Brassica napus L.). *Journal of Agricultural and Food Chemistry*, *52*(19), 5995–6001.
Schwenke, K. D. (1990). Structural studies on native and chemically modified storage proteins from rapeseed (Brassica napus L.) and related plant proteins. *Food/Nahrung*, *34*(3), 225–240.
Schwenke, K. D. (1994). Rapeseed proteins. In *New and developing sources of food proteins* (pp. 281–306). Boston, MA: Springer.
Schwenke, K. D., Dahme, A., & Wolter, T. (1998). Heat-induced gelation of rapeseed proteins: Effect of protein interaction and acetylation. *Journal of the American Oil Chemists' Society*, *75*(1), 83–87.

Schwenke, K. D., & Linow, K. J. (1982). A reversible dissociation of the 12 S globulin from rapeseed (Brassica napus L.) depending on ionic strength. *Food/Nahrung*, *26*(1), K5–K6.

Schwenke, K. D., Raab, B., Plietz, P., & Damaschun, G. (1983). The structure of the 12 S globulin from rapeseed (Brassica napus L.). *Food/Nahrung*, *27*(2), 165–175.

Serraino, M. R., & Thompson, L. U. (1984). Removal of phytic acid and protein-phytic acid interactions in rapeseed. *Journal of Agricultural and Food Chemistry*, *32*(1), 38–40.

Shewry, P. R., Napier, J. A., & Tatham, A. S. (1995). Seed storage proteins: Structures and biosynthesis. *The Plant Cell*, 7(7), 945.

Si, P., Mailer, R. J., Galwey, N., & Turner, D. W. (2003). Influence of genotype and environment on oil and protein concentrations of canola (Brassica napus L.) grown across southern Australia. *Australian Journal of Agricultural Research*, *54*(4), 397–407.

Tan, S. H., Mailer, R. J., Blanchard, C. L., & Agboola, S. O. (2011a). Canola proteins for human consumption: Extraction, profile, and functional properties. *Journal of Food Science*, *76*(1), R16–R28.

Tan, S. H., Mailer, R. J., Blanchard, C. L., & Agboola, S. O. (2011b). Extraction and characterization of protein fractions from Australian canola meals. *Food Research International*, *44*(4), 1075–1082.

Tan, S. H., Mailer, R. J., Blanchard, C. L., & Agboola, S. O. (2014). Emulsifying properties of proteins extracted from Australian canola meal. *LWT- Food Science and Technology*, *57*(1), 376–382.

Tan, S. H., Mailer, R. J., Blanchard, C. L., Agboola, S. O., & Day, L. (2014). Gelling properties of protein fractions and protein isolate extracted from Australian canola meal. *Food Research International*, *62*, 819–828.

Tandang-Silvas, M. R. G., Fukuda, T., Fukuda, C., Prak, K., Cabanos, C., Kimura, A., et al. (2010). Conservation and divergence on plant seed 11S globulins based on crystal structures. *Biochimica et Biophysica Acta (BBA)-Proteins and Proteomics*, *1804*(7), 1432–1442.

Tang, Y. R., & Ghosh, S. (2021). Stability and rheology of canola protein isolate-stabilized concentrated oil-in-water emulsions. *Food Hydrocolloids*, *113*, 106399.

Teh, S. S., Bekhit, A. E. D., Carne, A., & Birch, J. (2014). Effect of the defatting process, acid and alkali extraction on the physicochemical and functional properties of hemp, flax and canola seed cake protein isolates. *Journal of Food Measurement and Characterization*, *8*(2), 92–104.

WHO/FAO/UNU. (2007). *Protein and amino acid requirements in human nutrition. Vol. 935* (p. 1). World Health Organization Technical Report Series.

Thompson, L. U., Liu, R. F. K., & Jones, J. D. (1982). Functional properties and food applications of rapeseed protein concentrate. *Journal of Food Science*, *47*(4), 1175–1180.

Tzeng, Y. M., Diosady, L. L., & Rubin, L. J. (1988a). Preparation of rapeseed protein isolates using ultrafiltration, precipitation and diafiltration. *Canadian Institute of Food Science and Technology Journal*, *21*(4), 419–424.

Tzeng, Y. M., Diosady, L. L., & Rubin, L. J. (1988b). Preparation of rapeseed protein isolate by sodium hexametaphosphate extraction, ultrafiltration, diafiltration, and ion-exchange. *Journal of Food Science*, *53*(5), 1537–1541.

Tzeng, Y. M., Diosady, L. L., & Rubin, L. J. (1990). Production of canola protein materials by alkaline extraction, precipitation, and membrane processing. *Journal of Food Science*, *55*(4), 1147–1151.

Uppstrom, B. (1995). Seed chemistry. *Brassica Oilseeds: Production and Utilization*, 217–242.

Von Der Haar, D., Müller, K., Bader-Mittermaier, S., & Eisner, P. (2014). Rapeseed proteins–production methods and possible application ranges. *Oilseeds and Fats, Crops and Lipids*, *21*(1), D104.

Wanasundara, J. P. (2011). Proteins of Brassicaceae oilseeds and their potential as a plant protein source. *Critical Reviews in Food Science and Nutrition*, *51*(7), 635–677.

Wanasundara, J. P., Abeysekara, S. J., McIntosh, T. C., & Falk, K. C. (2012). Solubility differences of major storage proteins of Brassicaceae oilseeds. *Journal of the American Oil Chemists' Society*, *89*(5), 869–881.

Wanasundara, J. P., & McIntosh, T. C. (2013). *U.S. patent no. 8,557,963*. Washington, DC: U.S. Patent and Trademark Office.

Wanasundara, J. P., McIntosh, T. C., Perera, S. P., Withana-Gamage, T. S., & Mitra, P. (2016). Canola/rapeseed protein-functionality and nutrition. *Oilseeds and Fats, Crops and Lipids*, *23*(4), D407.

Wanasundara, J. P. D., Tan, S., Alashi, A. M., Pudel, F., & Blanchard, C. (2017). Proteins from canola/rapeseed: Current status. In *Sustainable protein sources* (pp. 285–304). Academic Press.

Wang, M., Hettiarachchy, N. S., Qi, M., Burks, W., & Siebenmorgen, T. (1999). Preparation and functional properties of rice bran protein isolate. *Journal of Agricultural and Food Chemistry*, *47*(2), 411–416.

Wang, L., Yang, J., Wang, Y., Zhang, J., Gao, Y., Yuan, J., et al. (2016). Study on antioxidant activity and amino acid analysis of rapeseed protein hydrolysates. *International Journal of Food Properties*, *19*(9), 1899–1911.

Willett, W., Rockström, J., Loken, B., Springmann, M., Lang, T., Vermeulen, S., et al. (2019). Food in the Anthropocene: The EAT–lancet commission on healthy diets from sustainable food systems. *The Lancet*, *393*(10170), 447–492.

Witczak, M., Chmielewska, A., Ziobro, R., Korus, J., & Juszczak, L. (2021). Rapeseed protein as a novel ingredient of gluten-free dough: Rheological and thermal properties. *Food Hydrocolloids*, *118*, 106813.

Wu, J., Aluko, R. E., & Muir, A. D. (2008). Purification of angiotensin I-converting enzyme-inhibitory peptides from the enzymatic hydrolysate of defatted canola meal. *Food Chemistry*, *111*(4), 942–950.

Wu, J., Aluko, R. E., & Nakai, S. (2006). Structural requirements of angiotensin I-converting enzyme inhibitory peptides: Quantitative structure – activity relationship study of di-and tripeptides. *Journal of Agricultural and Food Chemistry*, *54*(3), 732–738.

Wu, J., & Muir, A. D. (2008). Comparative structural, emulsifying, and biological properties of 2 major canola proteins, cruciferin and napin. *Journal of Food Science*, *73*(3), C210–C216.

Xu, L., & Diosady, L. L. (1994). Functional properties of Chinese rapeseed protein isolates. *Journal of Food Science*, *59*(5), 1127–1130.

Xu, L., & Diosady, L. L. (2000). Interactions between canola proteins and phenolic compounds in aqueous media. *Food Research International*, *33*(9), 725–731.

Xu, L., & Diosady, L. L. (2002). Removal of phenolic compounds in the production of high-quality canola protein isolates. *Food Research International*, *35*(1), 23–30.

Xu, L., & Diosady, L. L. (2012). Processing of canola proteins. *Canola and Rapeseed: Production, Processing, Food Quality, and Nutrition*, *59*.

Yang, M., Huang, F., Liu, C., Zheng, C., Zhou, Q., & Wang, H. (2013). Influence of microwave treatment of rapeseed on minor components content and oxidative stability of oil. *Food and Bioprocess Technology*, *6*(11), 3206–3216.

Yang, C., Wang, Y., Vasanthan, T., & Chen, L. (2014). Impacts of pH and heating temperature on formation mechanisms and properties of thermally induced canola protein gels. *Food Hydrocolloids*, *40*, 225–236.

Yoshie-Stark, Y., Wada, Y., Schott, M., & Wäsche, A. (2006). Functional and bioactive properties of rapeseed protein concentrates and sensory analysis of food application with rapeseed protein concentrates. *LWT- Food Science and Technology*, *39*(5), 503–512.

Yoshie-Stark, Y., Wada, Y., & Wäsche, A. (2008). Chemical composition, functional properties, and bioactivities of rapeseed protein isolates. *Food Chemistry*, *107*(1), 32–39.

Young, V. R., & Pellett, P. L. (1994). Plant proteins in relation to human protein and amino acid nutrition. *The American Journal of Clinical Nutrition*, *59*(5), 1203S–1212S.

Zhang, S. B., Wang, Z., & Xu, S. Y. (2007). Downstream processes for aqueous enzymatic extraction of rapeseed oil and protein hydrolysates. *Journal of the American Oil Chemists' Society*, *84*(7), 693–700.

Zhang, S. B., Wang, Z., Xu, S. Y., & Gao, X. F. (2009). Purification and characterization of a radical scavenging peptide from rapeseed protein hydrolysates. *Journal of the American Oil Chemists' Society*, *86*(10), 959–966.

CHAPTER THREE

Pea protein composition, functionality, modification, and food applications: A review

Yanting Shen, Shan Hong, and Yonghui Li*
Department of Grain Science and Industry, Kansas State University, Manhattan, KS, United States
*Corresponding author: e-mail address: yonghui@ksu.edu

Contents

Advances in Food and Nutrition Research, Volume 101
ISSN 1043-4526
https://doi.org/10.1016/bs.afnr.2022.02.002

Abstract

The demand for proteins continues to increase due to their nutritional benefits, the growing world population, and rising protein deficiency. Plant-based proteins represent a sustainable source to supplement costly animal proteins. Pea (*Pisum sativum* L.) is one of the most produced plant legume crops in the world and contributes to 26% of the total pulse production. The average protein content of pea is about 20%–25%. The commercial utilization of pea proteins is limited, partially due to its less desirable functionalities and beany off-flavor. Protein modification may change these properties and broaden the application of pea proteins in the food industry. Functional properties such as protein solubility, water and oil holding capacity, emulsifying/foaming capacity and stability, and gelation can be altered and improved by enzymatic, chemical, and physical modifications. These modifications work by affecting protein chemical structures, hydrophobicity/hydrophilicity balance, and interactions with other food constituents. Modifiers, reaction conditions, and degree of modifications are critical variables for protein modifications and can be controlled to achieve desirable functional attributes that may meet applications in meat analogs, baking products, dressings, beverages, dairy mimics, encapsulation, and emulsions. Understanding pea protein characteristics will allow us to design better functional ingredients for food applications.

Abbreviations

DH degree of hydrolysis
EC emulsion capacity
ES emulsion stability
FC foaming capacity
FS foaming stability
HHP high hydrostatic pressure
MW molecular weight
TSP texturized soy protein
WHC water holding capacity

1. Introduction

According to the Food and Agriculture Organization of the United Nations (FAOSTAT, 2015), the United States ranks the fifth-highest country in per capita meat consumption. The average amount of meat consumption in the United States has exceeded the healthy level set by the 2015–2020 Dietary Guidelines by 20%–60% (Fehrenbach, Righter, & Santo, 2016). Excessive meat consumption can increase risks of cardiovascular disease, stroke, diabetes, and gastrointestinal cancers (Fehrenbach et al., 2016; Micha, Wallace, & Mozaffarian, 2010; Pan et al., 2012; Rohrmann et al., 2013; Sadler, 2004). On the other hand, protein deficiency is a major nutritional problem in developing countries (Batista, Portugal, Sousa, Crespo, & Raymundo, 2005; Ghaly & Alkoaik, 2010). Proteins are essential building blocks of human body (Ghaly & Alkoaik, 2010; López, Galante, Robson, Boeris, & Spelzini, 2018; Wild et al., 2014). For a healthy adult, the Recommended Dietary Allowance (RDA) of protein is 0.8 g protein per kg body weight per day with minimal physical activity (Wu, 2016). Protein malnutrition may result in stunting, physical weakness, vascular dysfunction, and impaired immunity (Wu, 2016). Each year, 60% of the 10.9 million deaths of children under the age of five are reported for suffering malnutrition in the developing countries (WHO, 2002).

Due to the health concerns and high cost of animal proteins, there exists a strong interest in sustainable plant-based proteins. Soy and wheat are the most common plant proteins and have been widely applied to partially replace animal proteins in food applications because of their favorable functional properties. However, due to allergies, some consumers are unable to consume food products containing soy and gluten (Föste, Elgeti, Brunner, Jekle, & Becker, 2015; López-Castejón, Bengoechea, Díaz-Franco, & Carrera, 2020). Thus, developing novel plant proteins is a strategy to solve protein shortage and increase plant protein usage as functional ingredients in food products (Steffolani et al., 2016; Timilsena, Adhikari, Barrow, & Adhikari, 2016).

Plant proteins are used as food ingredients in many common applications, such as meat, bakery, and alternative dairy products (Owusu-Ansah & McCurdy, 1991; Sandberg, 2011). In meat products such as patties, hamburgers, and sausages, plant proteins are usually used as extenders for their good oil and water holding capacity, emulsification, and gelation properties (Asgar, Fazilah, Huda, Bhat, & Karim, 2010; Egbert & Payne, 2009; Jones, 2016; Owusu-Ansah & McCurdy, 1991). In bakery products, plant proteins are used for nutritional improvement, with possible negatively impaired textural properties. For instance, usage of plant proteins may lower bread volume, induce poor crumb texture and sensory properties (Fleming & Sosulski, 1977; Owusu-Ansah & McCurdy, 1991). On the other hand, the addition of plant proteins in pasta may reduce dough stickiness and improve processability (McWatters, 1980; Nielsen, Sumner, & Whalley, 1980). In alternative dairy products, plant proteins can be good natural emulsifiers to replace dairy proteins (Nylander, Arnebrant, Cárdenas, Bos, & Wilde, 2019).

Important functional properties of plant proteins, including solubility, water and oil holding capacity, emulsifying/foaming capacity and stability, viscosity, and gelation, are the intrinsic physicochemical characteristics highly associated with protein behaviors in food items (Adebiyi & Aluko, 2011). The abilities of proteins to form gels or stabilize emulsions, foams, or networks are attributed to their functional properties (Mirmoghtadaie, Shojaee Aliabadi, & Hosseini, 2016). Protein hydration influences proteins' absorption of water and oil, solubility, and thickening characteristics. Protein surface properties such as hydrophilicity, hydrophobicity, or net charge affect protein emulsifying and foaming properties. Protein rheological properties such as gelation, viscosity, and aggregation may be linked to the protein structural changes of size, shape, and amino acid sequences (Egbert & Payne, 2009; Speroni et al., 2009). To date, the utilization of commercial pea proteins as food ingredients is limited by its less desirable solubility, emulsification, and beany off-flavor (Lam, Can Karaca, Tyler, & Nickerson, 2018; Tamnak, Mirhosseini, Tan, Ghazali, & Muhammad, 2016; Zha, Dong, Rao, & Chen, 2019a). To overcome these limitations, protein could be modified to improve the functional properties according to different application purposes.

Chemical, enzymatic, and physical modifications are the common methods to modify proteins. Chemical modification is achieved by modifying the side groups of amino acids, including the ε-amino group of lysine; the thiol group of cysteine; the carboxyl group of glutamic acid and aspartic acid; the hydroxyl group of tyrosine, threonine, and serine; the imidazole group of histidine; the indole group of tryptophan; or the thioether group of methionine (Ustunol, 2015). The involved reactions include glycosylation, acylation, alkylation, and deamination, which would alter protein structures and functionalities depending on the reaction conditions and degree of modifications. Enzymatic modification mostly involves hydrolyzing the protein by breaking down the peptide bonds, which consequently reduces the protein molecular size and improves solubility and interfacial activity (Martínez, Sánchez, Ruíz-Henestrosa, Rodríguez Patino, & Pilosof, 2007; Tamm, Herbst, Brodkorb, & Drusch, 2016; Ustunol, 2015). Another enzymatic approach is to covalently cross-link proteins by enzymes such as transglutaminase and improve protein gelation (Klompong, Benjakul, Kantachote, & Shahidi, 2007; Mirmoghtadaie et al., 2016). Physical modifications include thermal, extrusion, and high-pressure treatments, to name a few. Thermal treatment alters proteins' hierarchy structure and causes protein denaturation, which results in protein aggregation through disulfide bonds and hydrophobic and electrostatic interactions (Barac, Stanojevic,

Jovanovic, & Pesic, 2004). The extrusion process may induce protein crosslinking and aggregation and reduce protein solubility due to the exposure of hydrophobic groups on the protein surface (Mirmoghtadaie et al., 2016). High-pressure treatment has the advantage of avoiding thermal degradation and preserving food due to the absence of heating processes. This approach could change protein functional properties by disrupting the hydrophobic and electrostatic interactions (Chapleau & de Lamballerie-Anton, 2003; Mirmoghtadaie et al., 2016).

Though there have been a few reviews on pea protein functional properties, extractions, and applications (Burger & Zhang, 2019; Ge et al., 2020; Lam et al., 2018; Lu, He, Zhang, & Bing, 2020), this review is more comprehensive by discussing pea protein composition and structures, functional properties, modification methods, and specific food applications. It provides the fundamental knowledge of pea protein structures and functional improvements for potential food applications. Since pea or modified pea proteins have not been widely studied, we also included some references and food applications of other plant proteins to inspire further study of pea protein.

2. Pea protein

Pea (*Pisum sativum* L.) is one of the most popular legume crops in the world, contributing to 26% of the total pulse production in 2014 (Rawal & Navarro, 2019). Peas originated from Southwest Asia and the Mediterranean region. Canada is the largest yellow pea producer, producing 30% of the world's annual dry pea stock (Pownall, Udenigwe, & Aluko, 2010; Sijtsma, Tezera, Hustinx, & Vereijken, 1998; Zhan, Shi, Wang, Li, & Chen, 2019). Pea seeds contain high protein content, carbohydrate, dietary fiber, fat, minerals, and vitamins (Gharsallaoui, Saurel, Chambin, & Voilley, 2012; Liang & Tang, 2014; Sandberg, 2011). The average protein content of pea is about 20%–25%, depending on cultivars, genetics, and environmental factors (Burger & Zhang, 2019; Sandberg, 2011; Zhan et al., 2019). The protein content in pea is strongly related to the amount of nitrogen, phosphorus, and *S*-triazine applications and also soil conditions (Owusu-Ansah & McCurdy, 1991). Pea protein contains high levels of lysine, threonine, and tryptophan (Adebiyi & Aluko, 2011; Boye, Aksay, et al., 2010; Boye, Zare, & Pletch, 2010; Burger & Zhang, 2019; Xiong et al., 2018). It has 18%–25% of water soluble albumin and 55%–65% of salt extractable globulin (Table 1) (Burger & Zhang, 2019; Owusu-Ansah & McCurdy, 1991; Sandberg, 2011). The water soluble albumin consists mainly of enzymic

Table 1 Classification of pea proteins and their molecular characteristics (Lu et al., 2020; Croy, Hoque, Gatehouse, & Boulter, 1984).

Class	Content	Solubility	Protein	Svedberg unit	MW (kDa)	No. of subunit	MW of subunit (kDa)
Globulin	55%–65%	Salt solution	Legumin	11S	320–410	6	60–65
			Vicilin	7S	150	3	48–50
			Convicilin	8S	180–210	3	70
Albumin	18%–25%	Water solution	Albumin	2S	53	2	25
					48	2	24
Prolamin	4%–5%	Alcohol solution	Prolamin	n/a	n/a	n/a	n/a
Glutelin	3%–4%	Insoluble	Glutelin	n/a	n/a	n/a	n/a

n/a, not available.

and metabolic proteins (Burger & Zhang, 2019). The globulin protein consists of two major components, which are legumin (11S) and vicilin (7S), as well as a small quantity of convicilin with a 180–210 kDa molecular weight (Burger & Zhang, 2019; Owusu-Ansah & McCurdy, 1991; Zhan et al., 2019). During seed development, legumin synthesizes faster than vicilin; therefore, legumin is presented in larger quantities than vicilin in the mature seed (Owusu-Ansah & McCurdy, 1991).

Legumin is a hexameric protein with a molecular weight of 320–410 kDa and is comprised of six heterogeneous pairs of subunits that are held by non-covalent interactions (Lu et al., 2020) (Table 1). Each of the pair consists of an acidic (40 kDa) and a basic (20 kDa) chain linked by disulfide bonds. Most of the legumin acidic subunits are located on the surface of the protein, while the hydrophobic core consists of basic subunits (Owusu-Ansah & McCurdy, 1991). By contrast, vicilin is a glycosylated trimeric cluster containing three subunits (48–50 kDa each) with a lower molecular weight of 150 kDa (Lu et al., 2020).

While legumin has high content of sulfur-containing amino acid such as methionine, vicilin is rich in isoleucine, leucine, phenylalanine, and lysine (Owusu-Ansah & McCurdy, 1991) (Table 2). Legumin is less soluble in salt solutions, contains higher amounts of nitrogen and sulfur, and is difficult to coagulate at higher temperatures (Gueguen, Chevalier, And, & Schaeffer, 1988). Generally, legumin stays in its native form at pH 7–9, but it is prone to dissociate at extremely high pH conditions. By contrast, vicilin is soluble at pH 4.8 (Casey, 1982). The ratio of vicilin to legumin in pea protein varies from 0.5 to 1.7 depending on the species and extraction methods (Burger & Zhang, 2019; Mession, Sok, Assifaoui, & Saurel, 2013; Owusu-Ansah & McCurdy, 1991). Different ratios of vicilin and legumin could result in structural and functional differences, and higher content of overall globulins tend to bring about better emulsifying properties (Makri, Papalamprou, & Doxastakis, 2005). In addition, the nutritional quality of pea protein may be improved by increasing the proportion of legumin to vicilin (Owusu-Ansah & McCurdy, 1991).

Pea protein isolate is typically obtained via wet processing. The process starts with solubilization, followed by isoelectric precipitation (Owusu-Ansah & McCurdy, 1991). Specifically, after pea flour is dispersed in water, the pH is adjusted to 9–10 to solubilize the protein, followed by centrifugation of the slurry to separate the protein and carbohydrates. The pH of the supernatant is then adjusted to the isoelectric point (4.3–4.5) to precipitate the protein. Finally, the precipitated protein extract is neutralized and spray dried or

Table 2 Amino acid composition (g/100 g protein) of pea seed meal and its protein fractions (Rubio et al., 2014).

Amino acid	Seed meal	Legumin[a]	Vicilin[b]	Albumin[c]	Prolamin+glutelin[d]
Essential					
Methionine	2.6	1.2	0.8	2.2	1.3
Cystine[e]	?	?	?	?	?
Lysine	8.5	6.4	12.4	12.5	8.2
Histidine	3.8	3.0	3.6	4.4	3.7
Arginine	6.0	3.9	4.6	10.4	5.1
Tyrosine	3.9	3.4	3.9	4.8	5.6
Leucine	6.0	6.3	9.7	4.3	7.0
Isoleucine	3.5	3.3	4.6	3.5	4.2
Threonine	6.3	6.5	7.1	5.4	8.9
Valine	3.9	3.8	4.4	4.5	5.6
Phenylalanine	4.3	3.9	6.1	3.8	4.7
Non-essential					
Asparate	11.5	11.2	16.8	16.0	10.1
Glutamate	15.2	16.2	22.7	18.9	12.0
Serine	4.8	4.5	6.2	5.4	5.1
Glycine	4.4	3.5	3.7	7.0	5.0
Alanine	3.8	3.3	3.5	6.3	4.2
Proline	6.0	5.1	5.5	5.5	10.3

[a]Borate buffer (pH 8) extract, adjust pH to 4.5, collect the sediment, dialysis, and freeze-dry.
[b]Borate buffer (pH 8) extract, adjust pH to 4.5, collect the supernatant, dialysis, centrifuge, collect the sediment, freeze-dry.
[c]Borate buffer (pH 8) extract, adjust pH to 4.5, collect the supernatant, dialysis, centrifuge, collect the supernatant, percipient in 608 g/L $(NH_4)_2SO_4$, centrifuge, collect sediment, dialysis, freeze-dry.
[d]Collect the residue after borate buffer (pH 8) extraction, freeze-dry.
[e]Cysteine was determined as cysteic acid in this reference, and "0 g/100 g protein" was reported.

lyophilized to protein powders (Swanson, 1990). The final yield of protein is determined by the processes, solubilizing agent, and pH of solubilization and precipitation (Owusu-Ansah & McCurdy, 1991). Gueguen and Cerletti (1994) reported that solubilizing agents potassium hydroxide and sodium hydroxide resulted in similar protein yields, while calcium hydroxide may solubilize 10% less pea protein due to the salting-out effect of calcium ions. They also pointed out that protein isolates precipitated below pH 5.3 had lower protein content but higher lipid content, compared with the sample precipitated at pH 5.3 (Gueguen, 1983). Some bond lipids were also co-extracted from this extraction method, as they were not extractable in hexane because of their tight binding to proteins (Owusu-Ansah & McCurdy, 1991).

Pea protein has a beany off-flavor due to the presence of lipoxygenase. Ma et al. (2011) reported that beany flavor volatiles such as alcohols, aldehydes, and ketones were formed partially by lipoxygenase-catalyzed oxidation of unsaturated fatty acids. Gao et al. (2020) found that pea protein extracted at pH 9 had the lowest beany flavor, as it had the lowest lipoxygenase activity. Zha, Dong, Rao, and Chen (2019b) indicated that conjugating pea protein with gum Arabic through Maillard reaction could mitigate the beany flavor. Other drawbacks of pea protein include its gritty texture and a feeling of lumps stuck in the throat during swallowing, which may be caused by the aggregation of pea protein through hydrophobic interactions, hydrogen bonding, and van der Waals forces, electrostatic, and steric interactions (Fang, Xiang, Sun-Waterhouse, Cui, & Lin, 2020). Further research is necessary to eliminate or reduce the off-flavor of pea protein.

3. Functional properties

Some functional properties, including water holding capacity, oil holding capacity, emulsifying properties, and foaming properties of pea protein are summarized in Table 3.

3.1 Water holding capacity

Water holding capacity (WHC), also known as hydration capacity, water binding capacity, or water absorption capacity, refers to the ability of proteins to retain water (Shevkani, Singh, Kaur, & Rana, 2015), or the capacity of per gram protein to absorb water (Boye, Aksay, et al., 2010; Boye, Zare, & Pletch, 2010). It is a critical parameter to determine protein function in terms of water retention, swelling, and gelation. It is also important in food formulations, affecting hydration with dry ingredients (Foegeding & Davis, 2011; Xiong, 2014). Poor water holding capacity may cause liquid loss during processing and lead to textural changes of final products (Lam et al., 2018). Water binding is usually resulted from a combination of ion-dipole, dipole-dipole, and hydrophilic interactions (Lam et al., 2018).

Table 3 Functional properties of pea proteins.

Sources	Treatments	Protein content (%)	Water holding capacity (g/g)	Oil holding capacity (g/g)	Emulsifying capacity (EC)/stability (ES)	Foaming capacity (FC)/stability (FS) (%)	References
Commercial air-classified pea protein-enriched flour	Untreated	55.5	1.18	0.96	ES-40.00%	ND	Wang, Guldiken, Tulbek, House, and Nickerson (2020)
	20% Ethanol wash	62.4	2.53	0.80	ES-40.31%	ND	
	50% Ethanol wash	61.4	2.60	0.77	ES-29.06%	ND	
	80% Ethanol wash	58.2	2.00	0.76	ES-28.13%	ND	
	20% Isopropanol wash	64.2	2.39	0.72	ES-46.88%	ND	
	50% Isopropanol wash	64.3	2.54	0.67	ES-45.31%	ND	
	80% Isopropanol wash	59.3	2.07	0.73	ES-37.19%	ND	
Pea Agassiz	Dehulled, milled, defatted, solubilized at pH 9, and precipitated at pH 4.5	90.9	ND	3.3	ES-95.80%	FS-74.90	Lam, Warkentin, Tyler, and Nickerson (2017)
Pea CDC Dakota		91.0	ND	3.2	ES-95.70%	FS-74.10	
Pea CDC Golden		91.1	ND	3.2	ES-95.10%	FS-75.00	
Pea CDC Striker		92.5	ND	3.1	ES-95.70%	FS-75.20	
Pea -Tetris		91.7	ND	3.3	ES-96.00%	FS-75.30	
Pea Cooper		89.7	ND	3.1	ES-96.10%	FS-73.50	
Commercial pea protein isolate	None	83.5	ND	1.5	ES-79.3%	FS-56.6	Lam et al. (2017)

Grass pea seeds	Solubilize at pH 9.7 and precipitate at pH 4.5	80.0	2.24	0.67	EC-36.73% ES-18.03 min	FC-109.78 FS-121.35	Milad, Reihaneh, Mehdi, and Akram (2019)
	Salt extraction containing 6.4% KCl	77.9	2.39	2.16	EC-36.21% ES-12.90 min	FC-106.23 FS-118.34	
	Ultrafiltration-diafiltration precipitation	90.5	2.83	2.21	EC-38.63% ES-18.86 min	FC-105.06 FS-119.79	
Grass pea seeds	Optimized extraction based on protein content	92.5	2.70	1.37	EC-35.80% ES-29.75%	FC-41.00 FS-100.00	Feyzi, Milani, and Golimovahhed (2018)
	Optimized extraction based on extraction yield	87.5	2.15	1.19	EC-87.50% ES-28.65%	FC-87.00 FS-78.00	

ND, not determined.

A lot of factors are associated with water holding capacity, such as protein extraction methods, protein concentration, pH, temperature, as well as the balance of hydrophilic and hydrophobic amino acid residues.

Wang, Guldiken, et al. (2020) reported that the water holding capacity of pea protein ranged from 1.18 to 2.60 g water/g protein. They found that alcohol-washed and air-classified pea protein had significantly higher water holding capacity than untreated protein. However, water holding capacity was dramatically decreased when increasing the concentration of ethanol or isopropanol up to 80% during protein pretreatment, which may be attributed to charge reduction on the protein surface. Feyzi et al. (2018) reported water holding capacity values of pea protein from 2.15 to 2.70 g water/g protein, with no significant differences among different extraction methods. Milad et al. (2019) found that pea protein extracted by ultrafiltration-diafiltration precipitation exhibited higher water holding capacity of 2.83 g water/g protein than those by isoelectric precipitation and salt extraction methods, which were 2.24 and 2.39 g water/g protein, respectively. It was believed that isoelectric precipitation resulted in higher hydrophobic/hydrophilic ratio than the other methods and greater surface hydrophobicity of proteins than salt extraction. This may be responsible for protein denaturation during acid precipitation and cause interactions between protein and non-protein compounds, thus decreasing protein's solubility and water holding capacity (Krause, Schultz, & Dudek, 2002). In addition to pea protein isolates, Ribéreau, Aryee, Tanvier, Han, and Boye (2018) studied pre-germinated and micronized pea flour and found a 7%–16% increase in water holding capacity compared with untreated pea flour. This may be because both pretreatments altered protein structures and enhanced imbibition, swelling, and retention of water. Furthermore, the micronization process can induce water entrapment in protein due to protein denaturation and starch damage during the process.

3.2 Oil holding capacity

Oil holding capacity is another important functional property that may affect food texture and quality. It is defined as the ability of protein to absorb and retain fat through interactions with lipids in food (Foegeding & Davis, 2011; Xiong, 2014). The interaction of lipid and protein occurs via the binding of aliphatic chains of lipids and the nonpolar side chains of amino acid residues, so proteins with higher hydrophobicity have greater propensities to interact with oil. Oil holding capacity is especially important when protein is used as an ingredient in meat, beverage, and salad dressing applications and is related to other functionalities such as emulsifying properties. It is affected by protein sources and structures, processing conditions, droplet size, and the distribution and stability of lipids.

Several studies reported variable oil holding capacities of pea protein isolates, ranging from 0.67 to 3.30g oil/g protein among pea varieties and different treatments (Feyzi et al., 2018; Lam et al., 2017; Milad et al., 2019; Wang, Guldiken, et al., 2020). Lam et al. (2017) reported that the oil holding capacities of six pea cultivars ranged from 3.0 to 3.3g oil/g protein without significant differences. Feyzi et al. (2018) also found no significant difference in oil holding capacity among different pea protein extraction methods. However, Milad et al. (2019) demonstrated that pea protein extracted from ultrafiltration-diafiltration precipitation exhibited similar oil holding capacity (2.21 g oil/ g protein) as that from salt extraction (2.16g oil/ g protein), while isoelectric precipitation resulted in a lower oil holding capacity (0.67g oil/g protein). The nonpolar side chains of amino acids play an important role in oil holding capacity, thus better oil holding capacity is attributed to higher content and/or availability of hydrophobic amino acid residues. Furthermore, compared with that of untreated protein, the oil holding capacity of alcohol-washed, air-classified pea protein was reported to be lowered from 0.96 to 0.67–0.80g oil/g protein (Wang, Guldiken, et al., 2020).

3.3 Solubility

Protein solubility is measured by the amount of protein dissolved in water at specific testing condition. It is considered a critical functionality in the food industry and affects proteins' emulsifying, foaming, and water holding capacities (Ghribi et al., 2015; Lam et al., 2018). Protein solubility mainly depends on the proportion and distribution of hydrophilic and hydrophobic groups on the surface of protein molecules, which may be affected by intrinsic factors (amino acid composition and distribution, isoelectric point, molecular flexibility, and charge) and extrinsic factors (pH, temperature, and ionic strength) (Lam et al., 2017). When protein is dissolved in water, hydrophilic amino acids tend to orient toward the aqueous interface, while hydrophobic residues are in the interior of protein molecules to minimize free energy. Exposure of hydrophobic residues to the protein surface would reduce protein solubility (Lam et al., 2018). When pH value is below or above the isoelectric point, protein solubility would increase due to electrostatic repulsion by net negative or positive charge at protein surface. Protein exhibits the lowest solubility at the isoelectric point because it carries zero net charge, and no electrostatic repulsive force occurs. At this point, protein aggregation may be formed by hydrophobic interactions between protein molecules, which could result in protein precipitation and decrease protein solubility.

According to the study reported by Lam et al. (2017), pea protein solubility ranged from 62.5% to 75.2% at pH 7 among different varieties, with significant differences between cultivars and environments. Wang, Guldiken, et al. (2020) reported that alcohol-washed, air-classified pea protein had significantly lower solubility than untreated protein, with solubility decreasing from 85.4% to 42.6%–52.7% and 20.6%–38.0%, respectively, when washed with ethanol and isopropanol. This was because the lower polarity of solvents induced protein partial denaturation. During the washing stage, the ratio of albumin and globulin increased, while prolamin protein was washed away (Bader, Oviedo, Pickardt, & Eisner, 2011; Chang, Stone, Green, & Nickerson, 2019). Milad et al. (2019) found that pea protein had the lowest solubility at pH 4–5. Pea protein extracted by ultrafiltration-diafiltration precipitation exhibited relatively higher solubility compared with pea protein extracted by isoelectric precipitation and salt extraction, while isoelectric precipitation extraction led to the lowest solubility (Milad et al., 2019). This result could be explained by the higher surface charge and lower surface hydrophobicity of protein produced by ultrafiltration-diafiltration precipitation, and higher surface hydrophobicity of protein obtained by salt extraction. Therefore, different degrees of protein unfolding resulted in different protein solubility from various extraction methods. Drying methods also affect pea protein solubility. Freeze-dried pea protein was reported to exhibit better solubility than vacuum-oven-dried protein, indicating more availability of hydrophilic amino acid in the freeze-dried protein during hydration and more protein denaturation as a result of vacuum oven drying (Feyzi et al. (2018).

3.4 Emulsifying properties

Protein can naturally be an emulsifier due to its amphiphilicity. Emulsification property enhances proteins' role as food ingredient by controlling the quality and texture of products in many applications (Karaca, Low, & Nickerson, 2011). Two types of food emulsions are commonly seen in the food industry: oil in water (O/W) emulsions which create a creamy texture and water in oil (W/O) emulsions which exhibit greasy properties. Protein has been widely used as an emulsifier due to its ability to adsorb at the oil and water interface and form stabilized layers around oil or water droplets

(Dickinson, 2010; Kimura et al., 2008). During emulsion formation, proteins or their aggregates are adsorbed at the surface of droplets and reoriented at the interface, with their hydrophobic side chains moving to the oil phase while the hydrophilic moieties face the water phase, thus minimizing the interfacial tension to prevent coalescence or flocculation (Burger & Zhang, 2019; Stone, Avarmenko, Warkentin, & Nickerson, 2015). An emulsion is mostly stable at low ionic strength and when the pH is away from the isoelectric point of the protein. This is when the interactions between proteins are weak and protein is adsorbed into droplets, promoting it to form an interfacial film and avoid coalescence. On the other hand, an emulsion is the least stable at the isoelectric point and at high ionic strength, which is because the dispersed phase is close to each other so that electrostatic repulsion is weakened and the attractive force between the droplets is in turn relatively strengthened (Lam et al., 2018). The emulsifying properties of proteins are characterized by emulsifying capacity (EC) and emulsifying stability (ES). Emulsifying capacity is the ability of a protein to adsorb at the interfacial area of oil and water to form emulsions, and it depends on the shape, charge, and hydrophobicity of protein molecules; emulsifying stability is the stability of an emulsion over a certain time, which depends on the magnitude of these interactions such as electrostatic repulsion, van der Waals forces, etc. (Karaca et al., 2011; Ma et al., 2011).

Lam et al. (2017) reported that pea protein had an emulsion stability from 95.1% to 96.1% at pH 7 from different cultivars with no significant differences among cultivars and environmental factors. Wang, Guldiken, et al. (2020) demonstrated that the emulsion stability of pea protein significantly decreased from 40% to 28.13% with ethanol wash, especially with 80% ethanol, but isopropanol wash caused no difference in emulsion stability. Milad et al. (2019) found that pea protein extracted by ultrafiltration-diafiltration precipitation exhibited the highest emulsion capacity and stability compared with that extracted by isoelectric precipitation and salt extraction.

3.5 Foaming properties

Foaming is the ability of a protein dispersion to form a stable foam when air is beaten in. Because of the high free energy at the gas-liquid interface, foam is thermodynamically unstable and may undergo coalescence to reduce the interfacial area (Lam et al., 2018). Solubilized proteins diffuse and adsorb to the gas-liquid interface, with hydrophobic groups being oriented to the gas phase and hydrophilic regions to the liquid phase, thereby reducing surface tension and forming a cohesive film around the gas bubbles

(Wierenga & Gruppen, 2010). Characteristics of protein foaming properties include foaming capacity (FC) and foaming stability (FS). Foaming capacity refers to the amount of interfacial area that could be created by a protein, and it is positively related to protein hydrophobicity/hydrophilicity ratio. Foaming stability is the ability of a protein to stabilize a foam against stress during a certain time (Lam et al., 2018).

Lam et al. (2017) reported that the foaming capacity of pea proteins ranged from 167.4% to 243.7% and significantly differed among cultivars and with different environmental factors. In addition, compared with salt extraction and ultrafiltration-diafiltration precipitation, isoelectric precipitation was more effective at increasing pea proteins' foaming capacity and stability by increasing globulin molecules, which adsorb at the interface and reduce the interfacial tension (Milad et al., 2019). Makri et al. (2005) pointed out that to achieve good foaming capacity, proteins should adsorb quickly at the air-water interface after the rapid conformational change and rearrangement of protein at the interface. Moreover, Ribéreau et al. (2018) reported that foaming capacity and stability of pea flour treated by micronization and pre-germination were significantly decreased compared to those of untreated pea flour. The reduction may be attributed to the partial denaturation and aggregation of proteins in treated flours.

3.6 Gelation

Gelation is a critical functional property of globular proteins to modify the texture, quality, and sensory attributes of food products (Foegeding & Davis, 2011). A protein gel is defined as a three-dimensional network that entraps large amount of water. Protein gelation involve two stages: conformational changes or protein denaturation and the subsequent protein association or aggregation into the three-dimensional matrix structure. A protein's gel forming ability and viscoelastic property largely depend on the hydrogen and covalent bonds and electrostatic and hydrophobic interactions (Adebiyi & Aluko, 2011). Protein gelation can be affected by heat treatment, pH, salt, pressure, and other constituents (Lam et al., 2018).

Milad et al. (2019) investigated gelation concentrations of pea proteins that were prepared by different extraction methods. They found that the lowest least gelation concentration (10%, w/v) was achieved for the pea protein extracted by ultrafiltration-diafiltration precipitation, followed by that from isoelectric precipitation (12%, w/v) and salt extraction (14%, w/v). This was attributed to the intensity of intermolecular interaction from ultrafiltration-diafiltration precipitation being stronger than that from isoelectric precipitation and salt extraction (Milad et al., 2019). In addition, Adebiyi and Aluko (2011) compared the gelling properties of different pea protein fractions, including water-, salt-, alkaline-, and ethanol-soluble fractions. In their study, the ethanol-soluble fraction was not evaluated due to its complete insolubility in water. Water-soluble and salt-soluble fractions were found not be able to form a firm gel. However, the alkaline-soluble fraction was able to form a firm gel with the least gelation concentration of 10%. Since the alkaline-soluble fraction represents the majority of pea protein fractions, its gelling property could represent the gelation ability of pea protein (Adebiyi & Aluko, 2011). Agboola, Mofolasayo, Watts, and Aluko (2010) reported that pea protein isolates formed a paste texture rather than a cohesive gel in the room condition, which was possibly due to insufficient intensity of intermolecular interaction. Moreover, Shand, Ya, Pietrasik, and Wanasundara (2007) found that the addition of NaCl could enhance pea protein gel formation, presumably by increasing intermolecular hydrophobic interactions, decreasing electrostatic repulsion, and altering water structure around protein molecules to enhance protein hydration and gel network formation.

4. Protein modifications

Commonly used protein modification methods include protein-polysaccharide conjugation, enzymatic modification, acylation, deamidation, and physical modification, as summarized in Table 4.

4.1 Enzymatic modification

The most common method of enzymatic modification is to hydrolyze the protein with protease, which involves breaking down the polypeptide bonds with additional water molecules. Protease hydrolysis can affect protein functional properties, for example, increasing the solubility of proteins by hydrolyzing them to smaller peptides (Adler-Nissen, 1986; Ustunol, 2015). In general, hydrolyzed proteins have altered water binding, foaming, and emulsifying properties. Many studies (Mahdavi-Yekta, Nouri, & Azizi, 2019; Nongonierma, Le Maux, Dubrulle, Barre, & FitzGerald, 2015; Pownall et al., 2010; Shen, Hu, & Li, 2020) reported that bioactive peptides derived from controlled hydrolysis exhibited strong antioxidant activity and could inhibit lipid oxidation. In addition to protease hydrolysis, cross-linking of proteins with microbial transglutaminase is another commonly

Table 4 Functionality improvement of pea and other proteins through modifications.

Modification methods	Functional changes	References
Protein-polysaccharide conjugation/complexation		
Pea protein glycosylated with galactose, lactose, and galacturonic acid	Improved protein solubility and foaming properties	Baniel, Caer, Colas, and Gueguen (1992)
Pea protein with soluble soybean polysaccharide	Improved emulsifying and foaming properties	Zhan et al. (2019)
Pea protein with high methoxyl pectin	Improved thermal stability and solubility	Lan, Chen, and Rao (2018)
Pea protein with pectic polysaccharides	Improved solubility and coacervation	Warnakulasuriya, Pillai, Stone, and Nickerson (2018)
Pea protein with pectin	Improved emulsifying and encapsulation properties	Tamnak et al. (2016)
Pea protein with high methoxyl pectin	Improved emulsifying properties	Gharsallaoui, Yamauchi, Chambin, Cases, and Saurel (2010)
Pea protein with pectin	Improved emulsifying properties	Tamnak, Mirhosseini, Tan, Tabatabaee Amid, et al. (2016)
Pea protein with gum Arabic	Improved solubility and emulsifying properties; diminished beany flavor	Zha, Dong, et al. (2019a, 2019b), Zha, Yang, Rao, and Chen (2019)
Pea protein with gum Arabic	Improved emulsifying and foaming properties	Liu, Elmer, Low, and Nickerson (2010)
Pea protein with propylene glycol alginate	Improved solubility and emulsifying properties	Guo, Su, Yuan, Mao, and Gao (2019)
Pea protein with corn fiber gum/ high methoxyl pectin/ carboxymethyl cellulose/ konjac glucomannan	Improved emulsion physical stability	Wei et al. (2020)
Enzymatic modifications		
Pea protein hydrolyzed by trypsin	Improved solubility and emulsion stability	Klost and Drusch (2019)
Pea protein hydrolyzed by trypsin and Alcalase	Improved emulsifying properties	Tamm et al. (2016)

Table 4 Functionality improvement of pea and other proteins through modifications.—cont'd

Modification methods	Functional changes	References
Pea protein hydrolyzed by Protamex	Improved emulsifying properties, foam expansion, and solubility	Sijtsma et al. (1998)
Pea protein hydrolyzed with chymosin	Improved solubility and emulsifying and foaming properties	Barać et al. (2011)
Salt extracted pea protein catalyzed by transglutaminase	Improved gel strength	Sun and Arntfield (2011)
Quinoa protein catalyzed by transglutaminase	Reduced film surface roughness and improved physical stability	Escamilla-García et al. (2019)
Quinoa protein with high-intensity ultrasound combined with transglutaminase	Improved edible film thickness, decreased elongation percentage, and increased tensile strength	Vera, Tapia, and Abugoch (2020)
Pea protein deamidated by glutaminase	Improved solubility and reduced beany flavor	Fang et al. (2020)
Coconut protein deamidated by protein-glutaminase	Improved solubility, emulsifying stability, and foaming capacity	Kunarayakul, Thaiphanit, Anprung, and Suppavorasatit (2018)
Whey protein deamidated by protein-glutaminase	Decreased heat-induced aggregation; increased water holding capacity	Miwa, Yokoyama, Nio, and Sonomoto (2013)
Rice/ pea/ blend deamidated by glutaminase	Decreased solubility and increased aggregation	Marco, Pérez, Ribotta, and Rosell (2007)
Myofibrillar/pea protein mixture induced by transglutaminase crosslinking	Improved gelation	Sun and Arntfield (2012)
Chemical modification		
pea protein acetylation and succinylation	Improved solubility, water holding capacity, and emulsifying properties	Johnson and Brekke (1983)
Rapeseed protein acylation	Improved water holding capacity and gelation	Wang, Zhang, Zhang, Ju, and He (2018)

Continued

Table 4 Functionality improvement of pea and other proteins through modifications.—cont'd

Modification methods	Functional changes	References
Flax protein acylation	Improved emulsifying properties and solubility	Wanasundara and Shahidi (1997)
African Yam Bean acylation	Improved solubility, water holding capacity, and emulsifying properties	Arogundade et al. (2013)
Mung bean protein acylation	Improved water/oil holding capacity and foaming and emulsifying properties	El-Adawy (2000)
Oat protein succinylation	Improved solubility and emulsifying properties	Mirmoghtadaie, Kadivar, and Shahedi (2009)
Amyloid β protein deamidation	Improved aggregation; slower fibrillation formation	Osaki and Hiramatsu (2016)
Curli amyloid protein deamidation	Slower and less amyloid fibril formation	Wang et al. (2017)
Rice bran protein-alkaline deamidation	Improved solubility	Guan et al. (2017)
Physical modification		
Thermal treatment of pea protein	Increased protein aggregation and decreased solubility	Mession et al. (2013)
Ultrasound treatment of pea protein	Improved solubility, smaller particle size, and increased surface hydrophobicity	Jiang et al. (2017)
Ultrasound treatment of pea protein	Improved foaming properties	Xiong et al. (2018)
High pressure treatment of pea protein	Improved foaming stability at pH 7, but reduced foaming and emulsifying properties at pH 3	Chao, Jung, and Aluko (2018)
High pressure with supercritical carbon dioxide treatment of pea protein	Improved foam stability and surface properties	Saldanha Do Carmo et al. (2016)
High hydrostatic pressure of pea protein	Reduced solubility	Chao and Aluko (2018)

used enzymatic modification. This reaction involves the acyl transfer between the ε-amino group of lysine residues and the amide group of glutamine residues (Ustunol, 2015). Microbial transglutaminase is a polypeptide consisting of 331 amino acids with an isoelectric point of 8.9. It is active at pH 4–9 with an optimum pH range of 6–7 and an optimum temperature of 50 °C (Ustunol, 2015). It can induce cross-linking of proteins and enhance gelation, creating good gel stability at high temperatures. Such cross-linked proteins are used in restructured meat products for the purpose of binding meat pieces (Ustunol, 2015).

Klost and Drusch (2019) reported that trypsin hydrolysis significantly enhanced pea protein solubility at pH 4 to 6, which was due to the smaller protein size and increased amount of terminal carboxyl and amino groups. However, when pH was at 3 and 7, hydrolysis decreased solubility. This was because the increased exposure of hydrophobic residues caused aggregation and sedimentation of peptide fractions and impacted the electrostatic properties of hydrolysates (Klost & Drusch, 2019). When hydrolysis increased from degree of hydrolysis (DH) of 0 to 4%, there was a shift in the protein's isoelectric point from 4.91 to 4.37. From this, the authors deduced that the peptide fractions that aggregated and experienced lower solubility at pH conditions away from the isoelectric point may consist of more basic amino acids than acidic ones. Since trypsin hydrolysis occurred at pH 8 at the C-terminus of basic amino acids, small and basic peptides from pea protein were probably close to the isoelectric point (Klost & Drusch, 2019). This resulted in a larger fraction of acidic peptides in the soluble fractions, which required more protons to bind with dissociated carboxyl groups, thus lowering the isoelectric point (Klost & Drusch, 2019). In addition to protein solubility, enzymatic hydrolysis may affect protein emulsifying properties. Tamm et al. (2016) found that pea protein hydrolyzed by trypsin had improved emulsion stability, presenting smaller oil droplets compared with that hydrolyzed by Alcalase. Sijtsma et al. (1998) also showed that hydrolysis by Protamex significantly improved pea protein solubility, emulsifying properties, and foam expansion. Similar results were also found by Barać et al. (2011) using pea protein hydrolyzed by chymosin.

Transglutaminase cross-linking can improve protein gelation properties. Shand, Ya, Pietrasik, and Wanasundara (2008) and Sun and Arntfield (2011) demonstrated that pea protein treated with transglutaminase exhibited enhanced gel properties, evidenced by increased gel strength, elasticity, and lower least gelation capacity. Sun and Arntfield (2011) reported that addition of transglutaminase lowered the least gelation concentration of a salt-extracted pea protein from 5.5% to 3% (w/v). Increasing transglutaminase concentrations increased gel strength, which was confirmed by the increased magnitudes of both elastic (G′) and viscous (G″) modulus. In addition, slowing down heating and cooling rates allowed the formation of stronger gel by giving protein molecules more time to rearrange. Moreover, higher pea protein concentrations at the same level of transglutaminase resulted in higher G′ and G″ (Sun & Arntfield, 2011).

4.2 Protein-polysaccharide conjugation/complexation

Protein-polysaccharide conjugation is usually generated by heating the mixture of protein and carbohydrate below the denaturation temperature with controlled water activity via covalent bonds (Ustunol, 2015). It is a glycosylation reaction between the amine group of protein and carbonyl group of a reducing sugar, representing the early stage of the Maillard reaction. Glycosylation can affect the balance of hydrophilicity and hydrophobicity of proteins and enhance water-protein interactions (Baniel et al., 1992). Protein-polysaccharide complexation is a pH-dependent, electrostatic attractive interaction between the positively charged protein surface and negatively charged polysaccharide (Lan et al., 2018). Many studies reported that interactions between protein and polysaccharides improved protein functional properties, such as emulsifying and water holding properties (Gharsallaoui et al., 2010; Tamnak, Mirhosseini, Tan, Ghazali, & Muhammad, 2016; Tamnak, Mirhosseini, Tan, Tabatabaee Amid, et al., 2016; Zha, Dong, et al., 2019a).

Wei et al. (2020) studied the stabilization of pea protein dispersions using various polysaccharides, including corn fiber gum, high-methoxy pectin, carboxymethyl cellulose, and konjac glucomannan. They found that all these polysaccharides effectively improved the stability of pea protein dispersions due to electrostatic repulsions, steric hindrance, and high viscosity of the continuous phase. In addition, pH and concentrations of polysaccharides also affect the stability of protein dispersions (Wei et al., 2020). Guo et al. (2019) reported that pea protein complexed with propylene glycol alginate had improved solubility, emulsifying activity, and increased denaturation temperature compared with native pea protein or propylene glycol alginate alone. They also pointed out that the pea-protein-propylene complex exhibited the best functionalities at the mass ratio of 3: 1 (w/w) at pH4. Moreover, several studies have shown functional property improvements of pea-gum Arabic complexes (Liu, Elmer, et al., 2010; Zha, Dong, et al., 2019a, 2019b; Zha, Yang, et al., 2019). Zha, Dong, et al. (2019a) reported that pea protein-gum Arabic conjugates possessed better emulsifying properties, which was confirmed by their smaller droplet size, higher surface charge, and stronger steric hindrance. The emulsifying property was also affected by pH, temperature, and ionic strength. Similar findings were found by Liu et al. (2010), Liu, Elmer, et al. (2010). They reported that the emulsion stability and foaming property of pea protein were greatly improved by complexing with gum Arabic. Zha, Dong, et al. (2019b) and Zha, Yang, et al. (2019) further investigated the effect of pea-gum Arabic complexes on flavor attributes. They reported that the beany flavor was remarkably diminished after incubating the mixture at 60 °C with a longer incubation time, because key aromatic compounds were produced during the Maillard reaction process, such as pyrazines, thiophenes, ketones, and Strecker aldehydes (Zha, Dong, et al., 2019b). In addition, several studies reported that pea-pectin conjugates exhibited good rheological behavior and physicochemical properties in oil-in-water emulsions (Gharsallaoui et al., 2010; Lan et al., 2018; Tamnak, Mirhosseini, Tan, Ghazali, & Muhammad, 2016; Tamnak, Mirhosseini, Tan, Tabatabaee Amid, et al., 2016; Warnakulasuriya et al., 2018).

4.3 Acylation

Acylation is a chemical modification that covalently links an acyl group to the protein amino groups (Shen & Li, 2021). The ε-amino group of lysine is the most active and can be readily acylated (Ustunol, 2015). The two most common acylation methods are acetylation and succinylation, and they work by introducing acetic and succinic anhydride, respectively (Ustunol, 2015). Acetylation decreases proteins' positive charges by substituting the positively charged ε-- amino groups with hydrophobic acetyl groups; succinylation introduces anionic succinate residues, which are covalently linked to the ε-amino groups of lysine residues, to change proteins' net charge from positive to negative and also to change their electrostatic and conformational properties (El-Adawy, 2000). Acylation usually occurs at pH 8 in a protein slurry. It led to protein unfolding and exposure of hydrophilic residues of polypeptides, hence improving emulsifying properties (Johnson & Brekke, 1983). This modification has also been reported to greatly improve protein solubility and gelation (Ustunol, 2015).

Johnson and Brekke (1983) reported that acetylated and succinylated pea proteins had significantly improved water holding capacity, solubility, and emulsifying properties. In their study, acylation was found to lower the protein isoelectric point by increasing negative charges and replacing the ε-amino groups of lysine with neutral acetyl groups through acetylation or with negatively charged carboxyl groups through succinylation (Johnson & Brekke, 1983). El-Adawy (2000) found that succinylated mung bean protein had higher solubility than acetylated one. During succinylation, longer side chains were introduced, thus producing more electrostatic repulsions in the protein. This also alters protein conformation to diminish protein-protein interactions while enhancing protein–water interactions. The addition of carboxyl groups by succinylation enhanced the interaction between protein and the aqueous phase of emulsions. As a result, protein solubility increased, and a protein layer was formed around fat droplets to facilitate association with the aqueous phase, thereby enhancing interactions at the protein-oil interface (El-Adawy, 2000). Similar results were found in succinylated oat protein isolates (Mirmoghtadaie et al., 2009), acetylated and succinylated mung bean protein isolates (El-Adawy, 2000), African yam bean protein (Arogundade et al., 2013), flax protein isolates (Wanasundara & Shahidi, 1997), and rapeseed protein isolates (Wang et al., 2018).

4.4 Deamidation

Deamidation refers to the conversion of amide groups (glutamine/ asparagine) of protein to carboxyl groups (glutamic acid/ aspartic acid) by using acid, alkali, heat, or enzymes (Fang et al., 2020). Deamidation increases carboxyl groups, which shifts the isoelectric point to the acidic side and exposes hydrophobic regions (Ustunol, 2015). Deamidation can improve protein solubility and emulsifying and foaming properties, but excessive deamidation may also impair protein functional properties (Fang et al., 2020). The rate of deamidation depends on the ratio of asparagine to glutamine and amino acid composition (Hamada & Swanson, 1994). Among chemical, physical, and enzymatic methods for protein deamidation, enzymatic deamidation appears to be superior due to its high efficiency, safety, mild conditions, and fewer side reactions. Chemical deamidation may cause severe protein denaturation and hydrolysis of peptide bonds, which can induce undesirable changes of protein structures and affect functional properties (Kunarayakul et al., 2018). Amyloid-protein aggregation could be also affected by chemical deamidation (Osaki & Hiramatsu, 2016; Wang, Shu, Frieden, & Gross, 2017). The most common enzymes used in protein deamidation include transglutaminase, glutaminase, and protein-glutaminase.

Glutaminase is a food-grade commercial enzyme that can catalyze the glutamine and asparagine residues of protein with slight hydrolysis (Fang et al., 2020). Fang et al. (2020) reported that glutaminase deamidation improved solubility and reduced the beany flavor, grittiness, and lumpiness of pea proteins. Protein-glutaminase is an important deamidation enzyme that is isolated from soil bacterial *Chryseobacterium proteolyticum*. It has no side reaction and low susceptibility to hydrolysis compared with other enzymes (Kunarayakul et al., 2018). Functional changes with protein-glutaminase deamidation were reported for various food proteins, such as increased protein solubility and foaming properties in coconut proteins (Kunarayakul et al., 2018) and decreased gel formation in heat-induced whey protein aggregates (Miwa et al., 2013), which was resulted from the increase of the electrostatic repulsion between carboxylic acid groups and the reduction of disulfide bond formation.

4.5 Physical modification

Physical modifications play a critical role in altering protein structures and functionalities, with thermal treatment (Cerdán-Leal, López-Alarcón, Ortiz-Basurto, Luna-Solano, & Jiménez-Fernández, 2020; Chao & Aluko, 2018; Kaspchak et al., 2017; Mäkinen, Zannini, & Peter Koehler, 2016; Mession et al., 2013), ultrasound (Jiang et al., 2017; Li, Da, Li, Xue, & Zang, 2018; Mir, Riar, & Singh, 2019; Vera, Valenzuela, Yazdani-Pedram, Tapia, & Abugoch, 2019; Xiong et al., 2018), and high pressure (Chao et al., 2018) being the most widely used methods. Thermal treatment causes dissociation of protein quaternary structures, protein denaturation, and protein aggregation through disulfide, hydrophobic, and electrostatic interactions (Mirmoghtadaie et al., 2016; Shen, Tang, & Li, 2021). Chao and Aluko (2018) investigated the effect of thermal pretreatment on the emulsifying properties of pea protein at various pH conditions, finding an improvement in emulsion capacity at pH7.0 and deterioration at pH3.0.

The radicals and superoxides generated during the ultrasound process may promote cross-linking of protein molecules, while the hydrogen peroxide produced during cavitation may oxidize free sulfhydryl groups to undesirable sulfinic acid; therefore, the ultrasound technique can significantly alter protein structures and protein functional properties (Soria & Villamiel, 2010). Xiong et al. (2018) found that high-intensity ultrasound treatment increased the foaming ability and stability of pea protein from 145.6% to 200.0%, and 58.0% to 73.3%, respectively. The high-intensity ultrasound applied in the study was capable of inducing partial protein unfolding, leading to rapid protein adsorption at the freshly formed air-water interface, forming a viscoelastic film, and improving foam stability. However, the bubble size of foam increased with an extension of time, and the protein was desorbed from the interface with more hydrophobic groups being exposed to the surface. As a result, foam-induced protein aggregation may be formed.

High pressure treatment may disrupt protein hydrophobic and electrostatic interactions, while inducing new bond formation to enhance protein aggregation or precipitation as well as gelation properties (Mirmoghtadaie et al., 2016). Saldanha Do Carmo et al. (2016) illustrated that pea protein treated with supercritical carbon dioxide showed improved foam stability, which was attributed to the affinity of carbon dioxide to hydrophobic moieties enhancing the surface properties of pea protein while stabilizing air/water interfaces.

Wang, Zhang, et al. (2017) reported that pea protein hydrolysates irradiated by electron beams had improved protein emulsifying and foaming properties. This physical modification is considered a clean, safe, and efficient technique to improve protein functional properties.

5. Applications

Some applications of pea and other plant proteins, including meat analogs and meat extenders, bakery applications, sauces, beverages, dairy products and analogs, encapsulating wall materials, and Pickering particles, are summarized in Table 5.

5.1 Meat analogs and meat extenders

In recent years, plant-based meat analogs made using extrusion texturization techniques are becoming popular (Li, 2020). Texturization is the conversion of a protein to a fibrous structure. It alters protein textural properties through the induction of protein denaturation, cross-linking, and realignment of protein molecules (Ustunol, 2015). The pea-protein-based meat developed by Beyond Meat is popular among consumers, including vegetarians. Osen, Toelstede, Eisner, and Schweiggert-Weisz (2015) and Osen, Toelstede, Wild, Eisner, and Schweiggert-Weisz (2014) investigated using pea proteins as raw materials for fibrous meat-like alternatives and observed that the resulting fibrous structure was closely associated with different extrusion conditions. Since few published studies are available regarding the texture properties of plant-based meat products using pea proteins, we review

Table 5 Food applications of pea and other plant proteins and related functional improvement.

Plant proteins	Processes	Functional improvement	References
Meat/meat analogs			
Soy protein flour/protein	Added at various levels of flour/protein (0, 2, 3.5 and 5% kg/kg total mass) to beef patty	Improved water holding capacity and reduced cooking loss; softer and more cohesive texture of beef patty	Kassama, Ngadi, and Raghavan (2003)
Soy protein/texturized vegetable protein	Directly added or substitution of meat from 10% to 40%	Improved textural characteristics of burgers, beef sausage, and buffalo meat sausage	Ahmad, Rizawi, and Srivastava (2010), Akesowan (2010), Hidayat, Wea, and Andriati (2017)
Soy protein	Added at 2%–4% protein to meat batter	Improved meat batter emulsion stability	Li, Sukmanov, Kang, and Ma (2020)
Soy protein/wheat gluten	Extruded meat alternatives	Reduced hardness of sausage	Chiang, Hardacre, and Parker (2020)
Various plant proteins	Meat free sausage, and sausage with 20% and 60% meat	Minimized cooking loss and shrinkage; improved emulsion stability of meat sausage	Kamani, Meera, Bhaskar, and Modi (2019)
Vicia ervilia	100% plant based burger	Improved water holding capacity and extended shelf life	Hatamikia, Elhamirad, Heydari, Sharayei, and Azarpazhooh (2019)
Bakery products			
Oat flour and Pea protein	80% oat flour and 20% pea protein were used	Improved nutritional quality and reduced cracker hardness	Morales-Polanco, Campos-Vega, Gaytán-Martínez, Enriquez, and Loarca-Piña (2017)

Continued

Table 5 Food applications of pea and other plant proteins and related functional improvement.—cont'd

Plant proteins	Processes	Functional improvement	References
Rice/pea protein	Adding 5% and 10%	Less crispy texture of bread crust; higher water activity	Pico, Reguilón, Bernal, and Gómez (2019)
Succinylated pea protein	Adding 3.45% modified pea protein	Improved viscosity, cell count, and rise and softness of eggless cake	Shah, Umesh, and Singhal (2019)
Cowpea protein	Cowpea protein mixed with wheat flour at 2%, 4% and 6% in wheat bread and sponge cake	Improved water holding capacity; softer texture	Campbell, Euston, and Ahmed (2016)
Quinoa protein	Adding 0%–30% quinoa protein	Increased firmness of cupcake and water activity; decreased growth of microorganisms and extended shelf life	López-Alarcón et al. (2019)
Soy protein	Adding 4% soy protein to rice flour	Lower gluten-free bread volume with inhomogeneous crumb structure	Masure, Wouters, Fierens, and Delcour (2019)
Soy protein hydrolysate	Adding 1% soy protein hydrolysate	Improved dough fermentation ability and acidification of steam bread	Liu et al. (2016)
Extruded soy protein	1000 g soy protein or extruded soy protein added with 400 g water	Improved volume and interior structure of Chinese steam bread	Du et al. (2016)
Sauces and dressings			
Lupin seed globulin protein	50:50 oil/water ratio, with protein concentration from 1.5% to 4%	Improved emulsion stability and rheological behavior of mayonnaise	Papalamprou, Doxastakis, and Kiosseoglou (2006)

Hydrolyzed canola protein	Prepared with 70% oil, and replaced 10%–50% egg yolk with the hydrolyzed protein	Improved emulsion stability	Aluko and McIntosh (2005)
Faba bean protein	Prepared with 40% oil, 0%–0.75% protein	Finest mean particle diameter and better texture of mayonnaise	Ouraji, Alimi, Motamedzadegan, and Shokoohi (2020)
Broad bean protein/ chickpea protein	Prepared with 75%–76% oil, and 3% protein	Improved lightness of mayonnaise and antioxidant activity	Alu'datt et al. (2017)
Soy protein	Prepared with 37%–43% oil, and 4%–7% soy protein concentrate	Optimized sensory attributes	Garcia, Sriwattana, No, Corredor, and Prinyawiwatkul (2009)
Soy protein-dextran conjugates	50:50 oil/water ratio, and 1% protein	Improved viscoelastic properties and creaming behavior of salad dressing emulsion	Diftis, Biliaderis, and Kiosseoglou (2005)
Beverages			
Pea protein	1:1 milk and pea protein solution	Viscous liquid behavior; more rapidly hydrolyzed under gastrointestinal conditions	Štreimikytė et al. (2020)
Pea protein with ultra-high temperature	Prepared by adding 3% pea protein isolate under ultra-high temperature	Reduced beany off-flavor of beverage	Trikusuma, Paravisini, and Peterson (2020)
Pea, lentil, soy protein	Hydrolyzed protein fining agent in plastic-stoppered 100 mL cylinders, filled with 120 mL wine, minimize air	Effective wine clarification	Granato, Ferranti, Iametti, and Bonomi (2018), Granato, Nasi, Ferranti, Iametti, and Bonomi (2014), Marangon, Vincenzi, and Curioni (2019)

Continued

Table 5 Food applications of pea and other plant proteins and related functional improvement.—cont'd

Plant proteins	Processes	Functional improvement	References
Rice protein	Adding glutelin (19 or 38 g), albumin (1.2 or 2.4 g)	Improved sensory properties of Chinese rice wine	Chen et al. (2020)
Soy protein with high hydrostatic pressure	Adding 25 g soy protein in 1 L peach beverage	Improved physical stability of peach juice	Manassero, Speroni, and Vaudagna (2019)
Protein-carbohydrate extracts from soybean meal	Prepared by adding 0.6% whey powder and soybean meal extracts obtained through different electro-activation conditions	Better solubility, foaming properties, and water holding capacity	Gerliani, Hammami, and Aïder (2019)
Soy protein-dextran conjugates	5% modified protein added to the drink	Increased solubility, wettability, and dispersibility	Boostani, Aminlari, Moosavi-nasab, Niakosari, and Mesbahi (2017)
Dairy products or analogs			
Soy protein	Low fat yogurt fortified with 0%–5% soy protein concentrate	Higher viscosity; darker color of yogurt	Drake, Chen, Tamarapu, and Leenanon (2000)
Soy protein	Yogurt enriched with 10 mg/mL soy protein isolate	Increased water holding capacity of yogurt	Soleymanpuori, Madadlou, Zeynali, and Khosrowshahi (2014)
Soy protein	Yogurt was prepared by adding 4% soy protein isolate	Increased biologically activity of yogurt	Pham and Shah (2009)
Soy protein	Prepared by 4 g oil, and 8 g total solids/100 g soy protein nanoparticles emulsion	Improved emulsion stability	Sengupta, Bhattacharyya, Goswami, and Bhowal (2019)

Oat protein	Prepared by adding 150 g oat protein concentrate in 1 L distilled water	Increased gel properties of yogurt	Brückner-Gühmann et al. (2019)
Encapsulating wall materials			
Soy protein	Prepared by ionic gelation method. Soy protein ratio was from 0% to 100%	Increased hygroscopicity of carrot powder beads	Kamaldeen, Ariahu, and Yusufu (2020)
Pumpkin protein	Prepared by spray drying method. Protein suspended in beet root juice in 1:20 core-wall ratio	Higher encapsulation efficiency in freeze-dried beetroot juice and higher antioxidant activity	Čakarević et al. (2020)
Acylated soy protein	Prepared by spray drying method. 8% protein solution/core ratio is 2:1	Reduced oil droplet size of microencapsulation	Nesterenko, Alric, Silvestre, and Durrieu (2014)
Soy protein-gum acacia conjugates	Protein: oleoresin from 2:1 to 6:1	Improved water vapor permeability and emulsion stability	Li et al. (2015)
Cross-linked soy protein by transglutaminase	Prepared by inulin/soy protein isolate ratio from 0.5 to 1.5	Improved oil holding capacity and thermal stability	Rios-Mera et al. (2019)
Pickering particles			
Pea protein	Protein dispersion at 0.25–3.0 g/100 mL. Oil at ø = 0.2, pH 3	Improved emulsion stability	Liang and Tang (2014)

Continued

Table 5 Food applications of pea and other plant proteins and related functional improvement.—cont'd

Plant proteins	Processes	Functional improvement	References
Pea protein	Protein dispersion at 6%. Oil at ø = 0.2–0.6, pH 3.	Acted as release delivery system for beta-carotene	Shao and Tang (2016)
Pea protein microgel particles	Oil: pea protein microgel = 20:80	Responsiveness to pH and ionic strength	Zhang, Holmes, Ettelaie, and Sarkar (2020)
Soy protein-anthocyanin	6% Soy protein solution with 0%–0.15% anthocyanins	Improved emulsion stability, oxidative stability, and resistance to in vitro digestion	Ju et al. (2020)
Soy protein-Jackfruit filum pectin (SPP)	SPP solution (2%, pH 7): oil = 9:1.	Improved wettability, thermal stability, and freeze-thaw stability	Jin et al. (2019)
Quinoa protein nanoparticles	Quinoa protein nanoparticle dispersions (0.25%–6%), oil at ø = 0.2–0.7	Improved wettability and emulsification efficiency	Qin, Luo, and Peng (2018)
Flaxseed protein and mucilage nano-assemblies	Flaxseed protein:flaxseed mucilage = 7:30 or 50:50, total concentration = 0.45%, pH 3	Improved emulsion stability	Nikbakht Nasrabadi et al. (2019)
Rapeseed protein nanogel	Acylated rapeseed protein concentration = 0.1%–1%, oil at ø = 0.1–0.5	Improved emulsion stability	Wang, Zhang, Chen, He, and Ju (2020)

plant-based meat using other plant proteins in the following paragraphs, which can still provide useful information for developing pea protein based meat analogs and products.

Chiang et al. (2020) compared the textures of sausages made from soy protein/wheat gluten with different moisture content and found that when moisture content increased, hardness of the sausages decreased, suggesting that higher moisture content creates softer texture. This could be attributed to the addition of water in the formulation and manufacture process disrupting the sausage intact structure and reducing hardness and chewiness. Sausage made from soy protein/wheat gluten showed a higher adhesiveness than control chicken breast due to the fact that wheat gluten creates sticky consistency and higher adhesion. In addition, sausage with soy protein/wheat gluten was more susceptible to oxidation compared with the control sausage, which could be caused by the longer storage time after the extrusion and frozen process and higher level of oxygen in the package that decreased the product's ability to maintain its storage stability (Zhang, Xiao, & Ahn, 2013). Kamani et al. (2019) reported that the full substitution with plant protein in sausage minimized cooking loss and shrinkage and greatly improved emulsion stability, while control meat sausage had better elasticity and gel strength. This was because the soy protein used in the study is hydrophilic and could strongly retain moisture and fat to form an adhesive gel matrix and consequently stabilize the emulsion. It was acting as a fat-encapsulating agent that prevented oil separation during cooking, thus reducing cooking loss. Hatamikia et al. (2019) conducted a study to investigate the functional properties of plant-based protein burgers based on various *Vicia ervilia* (L.) Willd. protein isolates produced by different extraction methods. They found that the protein produced by salt-dialysis had higher water holding capacity in comparation to that produced by alkaline solubilization/acid precipitation, because of a higher degree of denaturation during the extraction and the reduced purity and solubility of the protein. Moreover, the plant-based burger did not undergo tangible changes after 6 months of storage, which confirmed that the protein isolate of *V. ervilia* could be used to produce novel products with a longer shelf life.

Plant proteins have also been widely used as extenders in meat products because of their oil and water holding capacities and emulsifying properties and relatively lower cost (Dakhili, Abdolalizadeh, Hosseini, Shojaee-Aliabadi, & Mirmoghtadaie, 2019; Owusu-Ansah & McCurdy, 1991). Kassama et al. (2003) investigated the textural properties of beef patties containing soy protein flour (SPF) and texturized soy protein (TSP), and found that increasing soy protein concentration in both SPF and TSP enhanced the water holding capacity of meat patties and reduced cooking loss. Besides, patties with SPF and TSP exhibited softer and more cohesive textures than the control beef patty (Kassama et al., 2003). Akesowan (2010) studied light pork burgers fortified with 2% soy protein isolates, which showed significantly improved textural characteristics, including cohesiveness, springiness, and chewiness. Similar findings were observed in beef sausage with texturized vegetable protein (Hidayat et al., 2017) and buffalo meat sausage with soy protein (Ahmad et al., 2010).

Saturated fats have a harder texture at room temperature than unsaturated fats. Replacing meat with texturized proteins reduces fat content and results in softer texture of the final products. The rehydration effect of soy protein during meat formulation as well as its water holding capacity could also contribute to the softer texture and juiciness of meat products. In addition, texturized protein produced by extrusion cooking of defatted protein can give an elastic structure and imitate spongy chewy texture (Wild et al., 2014). Zhang, Lin, Leng, Huang, and Zhou (2013) indicated that the increased hardness of meat patties might be associated with emulsion destabilization caused by the separation of fat and water. Protein oxidation may also impact hardness through the formation of carbonyls and the induction of protein cross-linking via disulfide bonding. Li et al. (2020) reported that treating low-sodium pork meat with high pressure and 2% soy protein significantly improved the emulsion stability of the meat batter due to the ability of the added protein to form gels with good water and oil holding capacities. In addition, adding soy protein to pork meat batters and treating them with high pressure delayed thermal denaturation, reduced pre-gel formation generated by the denaturation of myosin tail, and significantly changed the protein secondary structure by decreasing the α-helix structure and increasing the β-sheet, β-turn, and random coil structures. Despite these advantages of plant-based meat products, pea protein based meat products and their physicochemical, functional, and sensory properties need to be further studied.

5.2 Bakery applications

The addition of pea protein or other modified proteins in bakery products improves nutritional value, textural properties, and sensory properties. Morales-Polanco et al. (2017) reported that crackers baked with dehulled oat flour and pea protein had higher nutritional content and lower hardness.

Shah et al. (2019) found that eggless cake batter with succinic anhydride modified pea protein possessed good viscosity, lower cake bulk density, along with improved cell number and softness. These improvements were attributed to the alteration of emulsifying and foaming properties due to the hydrophobic modification of the protein. Pico et al. (2019) conducted a study to investigate effects of plant proteins (rice and pea proteins) and animal proteins (egg white and whey proteins) on the crust quality of gluten-free bread. They found that bread containing 10% of pea protein and whey protein exhibited a less crispy texture. They also mentioned that higher moisture content or water activity reduced the jaggedness of the deformation curve and that high water activity caused a transition from the glassy to the rubbery state, both of which led to a less crispy texture.

Campbell et al. (2016) investigated the textural properties of wheat bread and sponge cake fortified with cowpea protein that had been thermally denatured and glycated. It was concluded that the bread dough with the denatured protein had significantly increased water holding capacity, leading to softer texture of the bread. Wheat bread with unmodified cowpea protein showed higher crumb hardness compared with bread containing other modified proteins (Campbell et al., 2016). This is because unmodified cowpea protein was more susceptible to thermal denaturation and loss of solubility than glycated protein during heat treatment and the water holding capacity of protein was highly related to its solubility. In addition, replacing 20% of whole egg with glycated cowpea protein (3.5% w/w in dry bases) in sponge cake did not affect sensory acceptability, whereas using unmodified and thermally denatured cowpea protein resulted in significantly lower acceptance (Campbell et al., 2016).

López-Alarcón et al. (2019) reported that cupcakes containing heat-denatured and lyophilized quinoa proteins exhibited greater firmness and water activity and an extended shelf life than cupcakes containing unmodified protein. The modified proteins were found to have a more porous surface to retain water and consequently led to less free water for microorganism growth and strong interactions between protein-carbohydrates (López-Alarcón et al., 2019). In addition, lyophilization could modify the nanostructure of protein, improving the distribution of water molecules, causing lower water mobility with high viscosity, and thus decreasing the growth of microorganisms on cupcakes and extending their shelf life (López-Alarcón et al., 2019). Masure et al. (2019) illustrated that gluten-free batter containing soy protein had lower surface activity and stability and resulted in a lower bread volume with inhomogeneous crumb structure compared with batter containing egg white protein. The volume and crumb structure of gluten-free bread were mainly determined by the balance between gas cell opening and crumb structure setting. The rate and total carbon dioxide release of batter with soy were higher than those of bread with egg white, which created less efficient gas cell stabilization of bread containing soy protein.

Apart from traditional bakery products, plant proteins could also be applied in steamed bread to improve nutrition and quality. Liu et al. (2016) added 1% of soy protein hydrolysates to improve dough fermentation and acidification, which was confirmed by the dough's increased gas holding capacity. In addition, soy protein hydrolysates reduced the time needed for steam dough to develop and stabilize, as less energy was required for the formation of the gluten dough network. Du et al. (2016) found that extruded soy protein improved the volume and interior structure of Chinese steam bread. The extruded soy protein formed a larger polymer network with the gluten protein, which acted as the nitrogen source of yeast and increased the S—S bonds of the gluten network.

5.3 Dressings

Mayonnaise is one of the oldest dressings or sauces, normally used in sandwiches or salad in North America. It is a semi-solid oil-in-water emulsion made from egg yolk, vinegar, and oil. The stability of mayonnaise depends on many factors, including the amount of oil and egg yolks, viscosity, mixing, water quality, and temperature. For health reasons, replacing egg yolks with cholesterol-free ingredients or plant proteins has been widely investigated. Very few studies have reported the usage of pea proteins in mayonnaise or other dressings or sauces, but other comparable analogs have been used. Papalamprou et al. (2006) observed higher stability and improved rheological behavior in salad dressing emulsions containing lupin seed globulins protein, compared with that containing lupin seed albumin or the 1:1 mixture of the two isolates. This was attributed to the "bridging" effect of the globulin aggregates strengthening the droplet-droplet interactions of the emulsion gel network. Creaming properties depended mainly on the concentration of lupin seed proteins, regardless of the isolate type. Aluko and McIntosh (2005) found that mayonnaise made with hydrolyzed canola protein exhibited a reddish-brown color as the concentration of the protein increased, which may be caused by the oxidization of phenolic compounds during protein extraction. This result implied that the addition of ascorbic acid or bisulfite reducing agents during extraction may help to prevent the oxidation of canola polyphenols. In addition, larger amounts of egg yolk (up to 50% w/w) can be replaced with hydrolyzed protein than native protein (up to 15% w/w) without disrupting the stable emulsion, confirming that limited enzymatic hydrolysis can unfold the protein's globular structure and increase hydrophobicity, which resulted in intensified interactions among oil droplets and the enhancement of emulsion properties.

Ouraji et al. (2020) reported that mayonnaise made with equal amounts of faba bean protein and egg yolk (0.375%) had the finest mean particle diameters and better texture properties, which was attributed to the reduction of surface tension and the formation of a flexible protein film around the dispersed oil droplets to prevent coalescence and flocculation. As a result, the finer dispersed emulsion presented higher physical stability, monodispersity, and elasticity. Moreover, Alu'datt et al. (2017) found that mayonnaise made with broad bean protein or the mixture of either broad bean/chickpea protein or broad bean protein/lupin flour was superior in lightness. In addition to functional properties, Garcia et al. (2009) optimized the taste, mouthfeel, and overall liking of mayonnaise using the formulation of 37% rice bran oil, 6% soy protein, and 57% water.

The inclusion of hydrocolloids in the mayonnaise formulation containing plant protein could enhance the stability of the interfacial film around droplets and consequently prevent coalescence. Diftis et al. (2005) reported that the conjugation of soy protein with dextran improved the viscoelastic properties and creaming behavior of salad dressing emulsions because the steric repulsion effects between droplets were enhanced by conjugation. Similar findings were observed by Ghoush, Samhouri, Al-Holy, and Herald (2008), who reported that the combination of egg yolk and protein (0.1% iota-carrageenan + 4% wheat protein) at 25:75 mass ratio significantly increased emulsion stability compared with whole egg yolk. The stability of mayonnaise emulsion decreased as the storage temperature elevated, which was because of the rapid flocculation or coalescence of small droplets at higher temperatures.

5.4 Beverages

Beverages fully or partially made with plant-based protein can help prevent cardiovascular disease, cancer, and osteoporosis (Gerliani et al., 2019) as well as animal-protein allergies such as lactose intolerance and milk protein allergies. A study conducted by Štreimikytė et al. (2020) investigated the formulation of protein-based beverages for the elderly who had dysphagia. Mixtures of milk and pea protein showed viscous liquid behaviors and could be more rapidly hydrolyzed under gastrointestinal conditions than milk protein alone. This finding was attributed to the aggregation of pea protein in the stomach and its better intestinal bioavailability than casein. The coagulation properties were also affected by different proteins during digestion.

The beany off-flavor of legume proteins is poorly accepted by consumers, and it is associated with lipid oxidation. Trikusuma et al. (2020) demonstrated that ultra-high temperature processing greatly changed the volatile aroma composition of the pea-protein-based beverage. The dominant pathway of pea protein oxidation was through lipoxygenase, which is the most abundant enzyme in legume seeds. Lipoxygenase converted polyunsaturated fatty acids into unsaturated fatty acid hydroperoxides, which were unstable and could be further degraded to carbonyls such as aldehydes and ketones. These chemicals were further oxidized to alcohols and acids, which were aroma active and could alter the flavor of pea protein beverages (Trikusuma et al., 2020). Similarly, Chen et al. (2020) demonstrated that rice protein could induce the formation of alcohol esters, contributing to the improvement of sensory properties of Chinese rice wine.

Proteins used in beverages should satisfy specific requirements, such as high solubility, good emulsifying properties, and low viscosity. Protein modifications help to achieve these properties. Physical modification and protein-polysaccharide conjugation are the most common protein modification approaches used in foods and beverages for their chemical-free and "clean labels" claim. Manassero et al. (2019) reported that the high hydrostatic pressure (HHP) treatment improved the physical stability of peach juice made with soy protein and $20\,\text{mmol}\,\text{L}^{-1}$ $CaCl_2$ compared with unpressurized juice, while the addition of calcium in the unpressurized juice destabilized the dispersion and induced protein aggregation. The HHP treatment caused parts of protein aggregates to remain in the suspension and significantly increased the viscosity of calcium added juice. This result was attributed to the interactions between soy protein, pectin, and calcium. HHP was also reported to enhance the activity of pectin methylesterase but decrease that of polygalacturonase to form a three-dimensional gel structure. Gerliani et al. (2019) investigated the protein-carbohydrate extract obtained from soybean meal by electro-activation and found that different protein extracts resulted in significantly different beverage functionalities. Alteration of treatment parameters, such as alkaline concentration, time, and protein slurry concentration could improve protein solubility, foaming property, and/or water holding capacity. Protein contents and fractions in the extracted materials were responsible for the differences in the foaming properties, and beverage with higher protein content showed better foaming ability. Boostani et al. (2017) reported that beverages made with spray-dried, soy-protein-dextran conjugation had higher solubility, wettability, and dispersibility than beverages made with unmodified protein, due to increased thermal stability and solubility of glycated proteins. In addition, the conjugated protein had significantly higher DPPH antioxidant radical scavenging activity (20.5%) than unmodified protein (4.7%), which was attributed to the products of the Maillard reaction affecting the antioxidant potential.

Legume proteins have been investigated for their wine fining efficiency in comparison to commercial gelatins. The fining treatment, carried out for wine clarification and stabilization without disturbing the color of the wine, involved the addition of substance that could bind target compounds and form insoluble aggregates that were later removed from the wine (Marangon et al., 2019). A few studies reported using the insoluble proteins from pea, lentil, and soybean as fining agents, because these proteins effectively sped up clarification (Granato et al., 2018; Granato et al., 2014; Marangon et al., 2019). Lentil protein was the best at removing monomeric and dimeric flavonols, while pea and soy proteins were similarly effective at reducing aroma compounds compared to commercial agents. Generally, the dose of legume-based fining agents ranged from 5 to 30 g/hL, though it was dependent on the vinification stage and treatment conditions (Marangon et al., 2019).

5.5 Dairy products and analogs

Low-fat and nonfat dairy products are growing in popularity. Low-fat yogurt usually exhibits poor textural properties such as weak structure and whey separation (Drake et al., 2000). The use of plant proteins or combinations of plant and milk proteins bring added nutritional value to yogurt, while improving its emulsion stability, functional properties, and biological activity. To date, few studies have reported using pea proteins in non-dairy products; however, other plant-based proteins have been successfully applied. Drake et al. (2000) reported that dairy yogurt fortified with soy protein exhibited higher viscosity than the control with similar sensory thickness. The fortified yogurt showed a darker color because of Maillard browning during processing and storage. Yogurt with soy protein maintained the fermented dairy aroma and flavor, but such sensory attributes decreased as the concentration of soy protein increased. The elevated content of soy protein along with the decreased concentration of nonfat dried milk resulted in a decline of lactose, which may reduce the sweetness of yogurt. Soleymanpuori et al. (2014) showed that yogurt enriched with soy protein possessed improved water holding capacity compared with the control during extended storage. They explained that soy protein was acting as an inert filler in the yogurt gel matrix on day 1, but experienced a gradual hydration on the following days that enhanced the product's water holding capacity. Milk treatment with transglutaminase had higher water holding capacity at day 1. This was attributed to the formation of covalent cross-linkages with the gel structure achieved by the enzyme treatment, which intensified the 3-D network of yogurt gel to prevent the liquid from flowing out. In addition, enzymatically treated yogurt had higher viability of probiotics, which was attributed to the gel formed by transglutaminase creating a good microenvironment for starter bacteria growth. Similar findings were observed in the study of Pham and Shah (2009), who reported that 4% of soy protein supplemented to yogurt increased the biological activity of starter bacteria.

Sengupta et al. (2019) investigated the emulsion stabilized by soy protein nanoparticles as potential ingredients in non-dairy yogurt. They reported that with increased particle concentrations, the fortified emulsion had a smaller droplet size and higher stability against coalescence and creaming. With an increased number of repeated cycles during solubilization, crystallization, and ultrasonication, more soy protein nanoparticles aggregated and adsorbed at the interface of oil droplets to form a gel-like network of oil droplets that inhibited creaming. In addition, the viscosity of yogurt increased as soy protein nanoparticles decreased in size, which was because of the intermolecular attraction between various soy protein nanoparticles and lipids of soy yogurts. The yogurt had excellent water holding capacity to retain a large quantity of water and enable a slow release of water with decreased syneresis. In another study conducted by Brückner-Gühmann et al. (2019), yogurt developed by the fermentation of oat protein concentrate underwent acidification and exhibited increased gel properties due to the production of lactic acid during fermentation. The porous network structure of fermented oat protein concentrate confirmed by scanning electron microscopy suggested the capability of oat protein as an alternative source for yogurt-type products.

5.6 Encapsulating wall materials

Microencapsulation has been widely used to entrap biological compounds in a carrier and protect food against oxidation and degradation during storage. Microencapsulation protect sensitive compounds, control the release of core agent, mask unpleasant taste and odor, and protect biological ingredients from undesirable light, moisture, and oxygen, thus extending products' shelf life (Li et al., 2015). The most common techniques used for encapsulation include spray drying, ionic gelation, and complex coacervation (Rios-Mera et al., 2019). The spray drying method is inexpensive and fast but utilizes high temperature, which may cause thermal degradation of active encapsulated compounds. Ionic gelation is the gelation of biopolymers during an extrusion process and does not involve heat treatment. This method renders the longest shelf life of encapsulated compounds due to its provision of an impermeable barrier against oxygen. Complex coacervation occurs between oppositely charged biopolymers (proteins and polysaccharides) through electrostatic interactions and results in smaller particle sizes (Rios-Mera et al., 2019).

Kamaldeen et al. (2020) reported that different ratios of soy protein isolates to cassava starch contributed to the different hygroscopicity of ionic encapsulated carrot powder beads, with higher proportion of cassava starch in the soy protein isolate film solution resulting in a higher hygroscopicity. This may be attributed to the differences in the molecular weights of cassava starch and soy protein as they form a matrix to encapsulate carrot powder beads. The addition of plasticization to the film was expected to increase hygroscopicity due to an increase in molecular mobility within the film. Čakarević et al. (2020) investigated the encapsulation efficiency of phenols from beetroot juice encapsulated by freeze- and spray-dried pumpkin protein. Higher encapsulation efficiency was found in freeze-dried than spray-dried samples, suggesting that drying techniques played a role in encapsulation efficiency. Furthermore, they found that beetroot juice encapsulated in pumpkin protein had higher antioxidant activity than the proteins alone after in vitro digestion, which indicated that pumpkin protein could be a good carrier agent in the encapsulation of bioactive compounds in the gastrointestinal tract.

Apart from native proteins, modified proteins also exhibit excellent properties in encapsulation applications. Nesterenko et al. (2014) found that in comparison to microencapsulation with unmodified proteins, microencapsulation with acylated soy protein brought about reduced oil droplet size due to increased surface activity of soy protein, while acylation modification enhanced the amphiphilic characteristic of the protein. On the other hand, encapsulation with protein modified by cationization resulted in decreased viscosity, which may be attributed to the enhanced hydrophilic properties of protein during cationization that increased the mobility of proteinic chains in the aqueous media, thus decreasing the viscosity of encapsulated samples. Xue, Gu, Wang, Li, and Adhikari (2019) reported that the encapsulation of grapefruit essential oil in emulsion-based edible films prepared by soy protein-gum acacia conjugates showed lower water vapor permeability compared with the control, because of the smaller droplet size of emulsions that resulted in a uniform distribution of the oil phase in the film structure. The reduced water vapor permeability was also related to the surface hydrophobicity and compact microstructure. Higher tensile strength was also achieved because of thick or multiple layers of conjugates at the interface that led to the formation of a stronger film network. Furthermore, films containing the grapefruit essential oil showed better water-repelling ability and a higher glass transition temperature than those containing oregano, lemon, fruit of *Amomum tsaoko* Crevost et Lemaire, indicating that the grapefruit essential oil had stronger molecular interactions with the soy protein-gum acacia matrix during encapsulation. A similar result was found by Li et al. (2015) in the encapsulation of tomato oleoresin using soy protein-gum acacia conjugates as carrier agents. Besides functional properties, the soy protein-gum acacia conjugate also exhibited better biocompatibility compared with the control emulsifier (Tween 80). Moreover, Rios-Mera et al. (2019) reported that the encapsulation of fish oil by complex coacervation formed by transglutaminase induced cross-linked soy protein and that inulin enhanced the resistance of microparticles and increased oil holding capacity due to improved thermal stability.

5.7 Pickering particles

Pickering emulsions are emulsions stabilized by solid particles. These surface-active particles remain stable in the water and oil system and have a contact angle (wettability). There is an increasing interest in Pickering emulsions because they are "surfactant-free," flocculation recovery and suitable for the environmental responsiveness in high internal phase emulsions (Qin et al., 2018). Particle wettability, particle size, and surface charge are the most important characteristics governing the formation of Pickering emulsions.

Particle wettability is characterized by the three-phase contact angle (θ), which measures the distribution of particles at the oil-water interface. A particle could stabilize oil-in-water emulsions if the contact angle is below 90° (hydrophilic particles), while water-in-oil emulsions may form when the contact angle is above 90° (hydrophobic particles). For Pickering emulsions, particles should have a contact angle close to 90°, so that they could be equally immersed in both the dispersed and continuous phases to form a stable film at the oil-water interface (Albert et al., 2019; Huang et al., 2019). A smaller particle size is preferred in Pickering emulsions because they can adsorb better at the oil-water interface (Zhu, 2019). Surface charge is the distribution of ions on the particle surface, which may affect the interfacial region of a solution, and it could be induced by both dispersed and continuous phases (Hu et al., 2016).

Zhang et al. (2020) developed pea protein microgel particles as a Pickering stabilizer and investigated the emulsion properties that were affected by pH and ionic strength. They found that salt enhanced emulsion viscosity and shear-thinning properties. The highest degree of particle aggregation of microgel formed at pH 5 due to the activation energy barrier in particle-particle interaction, which resulted in higher adsorption efficiency and higher viscosity. Liang and Tang (2014) reported that pea protein isolates exhibited a good Pickering stabilization effect for oil-in-water emulsions at pH 3. Furthermore, Shao and Tang (2016) demonstrated that gel-like pea protein Pickering emulsions at pH 3 could be used as a potential intestine-targeted and sustained-release delivery system for β-carotene. They found that the release of β-carotene during the intestinal digestion of the emulsion could be altered by oil fractions. The emulsion at the oil fraction of 0.6 showed a much slower release of β-carotene and higher stability against degradation during digestion than the emulsion at the oil fraction of 0.3. This might be due to the change of interfacial layers resulted from the different actions of pepsin during gastric digestion. Similar findings were reported by Wang, Zhang, et al. (2020) and Nikbakht Nasrabadi et al. (2019), who demonstrated that rapeseed and flaxseed proteins, respectively, could be used as stabilizers in Pickering emulsions.

Other modified proteins also exhibit excellent properties in Pickering emulsion applications. Qin et al. (2018) reported that quinoa protein nanoparticles treated with ultrasound improved the wettability and surface hydrophobicity of Pickering emulsions, thus enhancing emulsification efficiency, especially at higher quinoa protein concentrations. The ultrasound treatment altered disulfide bonds and hydrophobic interactions. In another study, soy protein conjugated with jackfruit filum pectin by photocatalysis exhibited good wettability and led to exhibit good freeze-thaw stability in stabilized emulsions (Jin et al., 2019). This was because the photocatalysis treatment caused the exposure of hydrophobic residues on the protein surface and strengthened the affinity of soy protein to jackfruit filum pectin. Improvement in freeze-thaw stability was caused by interactions between proteins and polysaccharides, which enabled the formation of thick interfacial membranes that were difficult to penetrate by fat crystals and created steric repulsions between emulsion droplets. In addition, Ju et al. (2020) developed a novel Pickering emulsion stabilized by soy protein-anthocyanin complex nanoparticles. The fabricated Picking emulsion showed better emulsion oxidative stability, which was attributed to beneficial biological activities of anthocyanins, including anti-inflammation, anti-oxidation, and inhibition of lipid peroxidation.

6. Conclusions

Plant proteins are gaining increasing attention due to the growing world population, rising protein deficiency, and their versatile environmental, functional, nutritional, and health benefits. Pea is one of the most widely consumed and cultivated legumes. Different from soybean and wheat, pea involves low allergic effects or few GMO concerns. Current utilization of commercial pea protein as a food ingredient is limited partly due to its less desirable techno-functional properties. Various modification methods (enzymatic, chemical, and physical) have been used to alter pea protein chemical structures and enhance functional properties for different food applications. Pea proteins, along with other types of plant proteins have been investigated in different applications, such as meat products and analogs bakery products, dressings, beverages, dairy products and analogs, encapsulating wall materials, and Pickering emulsions, although research studies utilizing pea protein are still relatively limited. How to apply pea proteins without affecting foods' textural and sensory properties needs to be further investigated. Future studies are also needed to explore how different pea protein fractions and modifications affect functional properties for specific applications. Comprehensive understandings of protein characteristics would provide meaningful insights to create and develop better functional ingredients for food applications.

Acknowledgments

This is contribution no. 22-097-B from the Kansas Agricultural Experimental Station. This project is in part supported by the USDA Pulse Crop Health Initiative (Award# 58-3060-0-046).

References

Adebiyi, A. P., & Aluko, R. E. (2011). Functional properties of protein fractions obtained from commercial yellow field pea (*Pisum sativum* L.) seed protein isolate. *Food Chemistry*, *128*(4), 902–908. https://doi.org/10.1016/J.FOODCHEM.2011.03.116.

Adler-Nissen, J. (1986). *Enzymic hydrolysis of food proteins*. Elsevier Applied Science Publishers. Retrieved from https://openlibrary.org/books/OL2535282M/Enzymic_hydrolysis_of_food_proteins.

Agboola, S. O., Mofolasayo, O. A., Watts, B. M., & Aluko, R. E. (2010). Functional properties of yellow field pea (*Pisum sativum* L.) seed flours and the in vitro bioactive properties of their polyphenols. *Food Research International*, *43*(2), 582–588. https://doi.org/10.1016/J.FOODRES.2009.07.013.

Ahmad, S., Rizawi, J. A., & Srivastava, P. K. (2010). Effect of soy protein isolate incorporation on quality characteristics and shelf-life of buffalo meat emulsion sausage. *Journal of Food Science and Technology*, *47*(3), 290. https://doi.org/10.1007/S13197-010-0045-X.

Akesowan, A. (2010). Quality characteristics of light pork burgers fortified with soy protein isolate. *Food Science and Biotechnology*, *19*(5), 1143–1149. https://doi.org/10.1007/s10068-010-0163-2.

Albert, C., Beladjine, M., Tsapis, N., Fattal, E., Agnely, F., & Huang, N. (2019). Pickering emulsions: Preparation processes, key parameters governing their properties and potential for pharmaceutical applications. *Journal of Controlled Release*, *309*, 302–332. https://doi.org/10.1016/J.JCONREL.2019.07.003.

Alu'datt, M. H., Rababah, T., Alhamad, M. N., Ereifej, K., Gammoh, S., Kubow, S., et al. (2017). Preparation of mayonnaise from extracted plant protein isolates of chickpea, broad bean and lupin flour: Chemical, physiochemical, nutritional and therapeutic properties. *Journal of Food Science and Technology*, *54*(6), 1395. https://doi.org/10.1007/S13197-017-2551-6.

Aluko, R. E., & McIntosh, T. (2005). Limited enzymatic proteolysis increases the level of incorporation of canola proteins into mayonnaise. *Innovative Food Science & Emerging Technologies*, *6*(2), 195–202. https://doi.org/10.1016/J.IFSET.2004.11.003.

Arogundade, L. A., Eromosele, C. O., Eromosele, I. C., Ademuyiwa, O., Etafo, N. O., & Adesokan, A. O. (2013). Effect of glycosylation via Maillard reaction and acylation on African yam bean (*Sphenostylis stenocarpa*) protein functionality. *Food Science and Biotechnology*, *22*(4), 951–960. https://doi.org/10.1007/s10068-013-0169-7.

Asgar, M. A., Fazilah, A., Huda, N., Bhat, R., & Karim, A. A. (2010). Nonmeat protein alternatives as meat extenders and meat analogs. *Comprehensive Reviews in Food Science and Food Safety*, *9*(5), 513–529. https://doi.org/10.1111/j.1541-4337.2010.00124.x.

Bader, S., Oviedo, J. P., Pickardt, C., & Eisner, P. (2011). Influence of different organic solvents on the functional and sensory properties of lupin (*Lupinus angustifolius* L.) proteins. *LWT—Food Science and Technology*, *44*(6), 1396–1404. https://doi.org/10.1016/J.LWT.2011.01.007.

Baniel, A., Caer, D., Colas, B., & Gueguen, J. (1992). Functional properties of glycosylated derivatives of the 11S storage protein from pea (*Pisum sativum* L.). *Journal of Agricultural and Food Chemistry*, *40*(2), 200–205. https://doi.org/10.1021/jf00014a007.

Barać, M., Čabrilo, S., Pešić, M., Stanojević, S., Pavlićević, M., Maćej, O., et al. (2011). Functional properties of pea (*Pisum sativum*, L.) protein isolates modified with chymosin. *International Journal of Molecular Sciences, 12*(12), 8372. https://doi.org/10.3390/IJMS12128372.

Barac, M., Stanojevic, S., Jovanovic, S., & Pesic, M. (2004). Soy protein modification: A review. *Acta Periodica Technologica, 35*, 3–16. https://doi.org/10.2298/APT0435003B.

Batista, A. P., Portugal, C. A. M., Sousa, I., Crespo, J. G., & Raymundo, A. (2005). Accessing gelling ability of vegetable proteins using rheological and fluorescence techniques. *International Journal of Biological Macromolecules, 36*(3), 135–143. https://doi.org/10.1016/J.IJBIOMAC.2005.04.003.

Boostani, S., Aminlari, M., Moosavi-nasab, M., Niakosari, M., & Mesbahi, G. (2017). Fabrication and characterisation of soy protein isolate-grafted dextran biopolymer: A novel ingredient in spray-dried soy beverage formulation. *International Journal of Biological Macromolecules, 102*, 297–307. https://doi.org/10.1016/J.IJBIOMAC.2017.04.019.

Boye, J. I., Aksay, S., Roufik, S., Ribéreau, S., Mondor, M., Farnworth, E., et al. (2010). Comparison of the functional properties of pea, chickpea and lentil protein concentrates processed using ultrafiltration and isoelectric precipitation techniques. *Food Research International, 43*, 537–546. https://doi.org/10.1016/j.foodres.2009.07.021.

Boye, J., Zare, F., & Pletch, A. (2010). Pulse proteins: Processing, characterization, functional properties and applications in food and feed. *Food Research International, 43*, 414–431.

Brückner-Gühmann, M., Vasil'eva, E., Culetu, A., Duta, D., Sozer, N., & Drusch, S. (2019). Oat protein concentrate as alternative ingredient for non-dairy yoghurt-type product. *Journal of the Science of Food and Agriculture, 99*(13), 5852–5857. https://doi.org/10.1002/jsfa.9858.

Burger, T. G., & Zhang, Y. (2019). Recent progress in the utilization of pea protein as an emulsifier for food applications. *Trends in Food Science and Technology*. https://doi.org/10.1016/j.tifs.2019.02.007.

Čakarević, J., Šeregelj, V., Tumbas Šaponjac, V., Ćetković, G., Čanadanović Brunet, J., Popović, S., et al. (2020). Encapsulation of beetroot juice: A study on the application of pumpkin oil cake protein as new carrier agent. *Journal of Microencapsulation, 37*(2), 121–133. https://doi.org/10.1080/02652048.2019.1705408.

Campbell, L., Euston, S. R., & Ahmed, M. A. (2016). Effect of addition of thermally modified cowpea protein on sensory acceptability and textural properties of wheat bread and sponge cake. *Food Chemistry, 194*, 1230–1237. https://doi.org/10.1016/J.FOODCHEM.2015.09.002.

Casey, R. (1982). The genetics of pea seed storage proteins. *Qualitas Plantarum Plant Foods for Human Nutrition, 31*(3), 281–295. https://doi.org/10.1007/BF01108636.

Cerdán-Leal, M. A., López-Alarcón, C. A., Ortiz-Basurto, R. I., Luna-Solano, G., & Jiménez-Fernández, M. (2020). Influence of heat denaturation and freezing–lyophilization on physicochemical and functional properties of quinoa protein isolate. *Cereal Chemistry, 97*(2), 373–381. https://doi.org/10.1002/cche.10253.

Chang, C., Stone, A. K., Green, R., & Nickerson, M. T. (2019). Reduction of off-flavours and the impact on the functionalities of lentil protein isolate by acetone, ethanol, and isopropanol treatments. *Food Chemistry, 277*, 84–95. https://doi.org/10.1016/J.FOODCHEM.2018.10.022.

Chao, D., & Aluko, R. E. (2018). Modification of the structural, emulsifying, and foaming properties of an isolated pea protein by thermal pretreatment. *CyTA—Journal of Food, 16*(1), 357–366. https://doi.org/10.1080/19476337.2017.1406536.

Chao, D., Jung, S., & Aluko, R. E. (2018). Physicochemical and functional properties of high pressure-treated isolated pea protein. *Innovative Food Science & Emerging Technologies, 45*, 179–185. https://doi.org/10.1016/J.IFSET.2017.10.014.

Chapleau, N., & de Lamballerie-Anton, M. (2003). Improvement of emulsifying properties of lupin proteins by high pressure induced aggregation. *Food Hydrocolloids*, *17*(3), 273–280. https://doi.org/10.1016/S0268-005X(02)00077-2.

Chen, T., Wu, F., Guo, J., Ye, M., Hu, H., Guo, J., et al. (2020). Effects of glutinous rice protein components on the volatile substances and sensory properties of Chinese rice wine. *Journal of the Science of Food and Agriculture*, *100*(8), 3297–3307. https://doi.org/10.1002/jsfa.10343.

Chiang, J. H., Hardacre, A. K., & Parker, M. E. (2020). Extruded meat alternatives made from Maillard-reacted beef bone hydrolysate and plant proteins. Part II—Application in sausages. *International Journal of Food Science & Technology*, *55*(3), 1207–1217. https://doi.org/10.1111/ijfs.14362.

Croy, R. R., Hoque, M. S., Gatehouse, J. A., & Boulter, D. (1984). The major albumin proteins from pea (*Pisum sativum* L). Purification and some properties. *Biochemical Journal*, *218*(3), 795. https://doi.org/10.1042/BJ2180795.

Dakhili, S., Abdolalizadeh, L., Hosseini, S. M., Shojaee-Aliabadi, S., & Mirmoghtadaie, L. (2019). Quinoa protein: Composition, structure and functional properties. *Food Chemistry*, *299*, 125161. https://doi.org/10.1016/J.FOODCHEM.2019.125161.

Dickinson, E. (2010). Food emulsions and foams: Stabilization by particles. *Current Opinion in Colloid and Interface Science*. https://doi.org/10.1016/j.cocis.2009.11.001.

Diftis, N. G., Biliaderis, C. G., & Kiosseoglou, V. D. (2005). Rheological properties and stability of model salad dressing emulsions prepared with a dry-heated soybean protein isolate–dextran mixture. *Food Hydrocolloids*, *19*(6), 1025–1031. https://doi.org/10.1016/J.FOODHYD.2005.01.003.

Drake, M. A., Chen, X. Q., Tamarapu, S., & Leenanon, B. (2000). Soy protein fortification affects sensory, chemical, and microbiological properties of dairy yogurts. *Journal of Food Science*, *65*(7), 1244–1247. https://doi.org/10.1111/j.1365-2621.2000.tb10272.x.

Du, Z., Chen, F., Liu, K., Lai, S., Zhang, L., Bu, G., et al. (2016). Effects of extruded soy protein on the quality of Chinese steamed bread. *Journal of Chemistry*, *2016*, 1–8. https://doi.org/10.1155/2016/3691523.

Egbert, W. R., & Payne, C. T. (2009). Plant proteins. In *Ingredients in meat products* (pp. 111–129). New York, NY: Springer New York. https://doi.org/10.1007/978-0-387-71327-4_5.

El-Adawy, T. A. (2000). Functional properties and nutritional quality of acetylated and succinylated mung bean protein isolate. *Food Chemistry*, *70*, 83–91. https://doi.org/10.1016/S0308-8146(00)00079-0.

Escamilla García, M., Delgado Sánchez, L. F., Ríos Romo, R. A., García Almendárez, B. E., Calderón-Domínguez, G., Méndez-Méndez, J. V., et al. (2019). Effect of transglutaminase cross-linking in protein isolates from a mixture of two quinoa varieties with chitosan on the physicochemical properties of edible films. *Coatings*, *9*(11), 736. https://doi.org/10.3390/coatings9110736.

Fang, L., Xiang, H., Sun-Waterhouse, D., Cui, C., & Lin, J. (2020). Enhancing the usability of pea protein isolate in food applications through modifying its structural and sensory properties via deamidation by glutaminase. *Journal of Agricultural and Food Chemistry*, *68*(6), 1691–1697. https://doi.org/10.1021/acs.jafc.9b06046.

Fehrenbach, K. S., Righter, A. C., & Santo, R. E. (2016). A critical examination of the available data sources for estimating meat and protein consumption in the USA. *Public Health Nutrition*, *19*(8), 1358–1367. https://doi.org/10.1017/S1368980015003055.

Feyzi, S., Milani, E., & Golimovahhed, Q. A. (2018). Grass pea (*Lathyrus sativus* L.) protein isolate: The effect of extraction optimization and drying methods on the structure and functional properties. *Food Hydrocolloids*, *74*, 187–196. https://doi.org/10.1016/J.FOODHYD.2017.07.031.

Fleming, S. E., & Sosulski, F. W. (1977). Breadmaking properties of four concentrated plant proteins. *Cereal Chemistry*, *54*, 1124–1140. Retrieved from https://www.cerealsgrains.org/publications/cc/backissues/1977/Documents/CC1977a143.html.

Foegeding, E. A., & Davis, J. P. (2011). Food protein functionality: A comprehensive approach. *Food Hydrocolloids*. https://doi.org/10.1016/j.foodhyd.2011.05.008.

Food and Agriculture Organization of the United Nations. (2015). *FAOSTAT data, Food balance sheets*. Retrieved April 29, 2020, from http://www.fao.org/faostat/en/#data.

Föste, M., Elgeti, D., Brunner, A.-K., Jekle, M., & Becker, T. (2015). Isolation of quinoa protein by milling fractionation and solvent extraction. *Food and Bioproducts Processing*, *96*, 20–26. https://doi.org/10.1016/J.FBP.2015.06.003.

Gao, Z., Shen, P., Lan, Y., Cui, L., Ohm, J.-B., Chen, B., et al. (2020). Effect of alkaline extraction pH on structure properties, solubility, and beany flavor of yellow pea protein isolate. *Food Research International*, *131*, 109045. https://doi.org/10.1016/J.FOODRES.2020.109045.

Garcia, K., Sriwattana, S., No, H. K., Corredor, J. A. H., & Prinyawiwatkul, W. (2009). Sensory optimization of a mayonnaise-type spread made with rice bran oil and soy protein. *Journal of Food Science*, *74*(6), S248–S254. https://doi.org/10.1111/j.1750-3841.2009.01203.x.

Ge, J., Sun, C. X., Corke, H., Gul, K., Gan, R. Y., & Fang, Y. (2020). The health benefits, functional properties, modifications, and applications of pea (*Pisum sativum* L.) protein: Current status, challenges, and perspectives. *Comprehensive Reviews in Food Science and Food Safety*, *19*(4), 1835–1876. https://doi.org/10.1111/1541-4337.12573.

Gerliani, N., Hammami, R., & Aïder, M. (2019). Production of functional beverage by using protein-carbohydrate extract obtained from soybean meal by electro-activation. *LWT*, *113*, 108259. https://doi.org/10.1016/J.LWT.2019.108259.

Ghaly, A. E., & Alkoaik, F. N. (2010). Extraction of protein from common plant leaves for use as human food. *American Journal of Applied Sciences*, 7(3), 331–342. Retrieved from https://thescipub.com/PDF/ajassp.2010.331.342.pdf.

Gharsallaoui, A., Saurel, R., Chambin, O., & Voilley, A. (2012). Pea (*Pisum sativum*, L.) protein isolate stabilized emulsions: A novel system for microencapsulation of lipophilic ingredients by spray drying. *Food and Bioprocess Technology*, *5*(6), 2211–2221. https://doi.org/10.1007/s11947-010-0497-z.

Gharsallaoui, A., Yamauchi, K., Chambin, O., Cases, E., & Saurel, R. (2010). Effect of high methoxyl pectin on pea protein in aqueous solution and at oil/water interface. *Carbohydrate Polymers*, *80*(3), 817–827. https://doi.org/10.1016/J.CARBPOL.2009.12.038.

Ghoush, M. A., Samhouri, M., Al-Holy, M., & Herald, T. (2008). Formulation and fuzzy modeling of emulsion stability and viscosity of a gum–protein emulsifier in a model mayonnaise system. *Journal of Food Engineering*, *84*(2), 348–357. https://doi.org/10.1016/J.JFOODENG.2007.05.025.

Ghribi, A. M., Gafsi, I. M., Blecker, C., Danthine, S., Attia, H., & Besbes, S. (2015). Effect of drying methods on physico-chemical and functional properties of chickpea protein concentrates. *Journal of Food Engineering*, *165*, 179–188. https://doi.org/10.1016/j.jfoodeng.2015.06.021.

Granato, T. M., Ferranti, P., Iametti, S., & Bonomi, F. (2018). Affinity and selectivity of plant proteins for red wine components relevant to color and aroma traits. *Food Chemistry*, *256*, 235–243. https://doi.org/10.1016/J.FOODCHEM.2018.02.085.

Granato, T. M., Nasi, A., Ferranti, P., Iametti, S., & Bonomi, F. (2014). Fining white wine with plant proteins: Effects of fining on proanthocyanidins and aroma components. *European Food Research and Technology*, *238*(2), 265–274. https://doi.org/10.1007/s00217-013-2108-5.

Guan, J., Takai, R., Toraya, K., Ogawa, T., Muramoto, K., Mohri, S., et al. (2017). Effects of alkaline deamidation on the chemical properties of rice bran protein. *Food Science and Technology Research*, *23*(5), 697–704. https://doi.org/10.3136/fstr.23.697.

Gueguen, J. (1983). Legume seed protein extraction, processing, and end product characteristics. *Qualitas Plantarum Plant Foods for Human Nutrition, 32*(3–4), 267–303. https://doi.org/10.1007/BF01091191.

Gueguen, J., & Cerletti, P. (1994). Proteins of some legume seeds: Soybean, pea, fababean and lupin. In *New and developing sources of food proteins* (pp. 145–193). Boston, MA: Springer US. https://doi.org/10.1007/978-1-4615-2652-0_6.

Gueguen, J., Chevalier, M., And, J. B., & Schaeffer, F. (1988). Dissociation and aggregation of pea legumin induced by pH and ionic strength. *Journal of the Science of Food and Agriculture, 44*(2), 167–182. https://doi.org/10.1002/jsfa.2740440208.

Guo, Q., Su, J., Yuan, F., Mao, L., & Gao, Y. (2019). Preparation, characterization and stability of pea protein isolate and propylene glycol alginate soluble complexes. *LWT, 101*, 476–482. https://doi.org/10.1016/j.lwt.2018.11.057.

Hamada, J. S., & Swanson, B. (1994). Deamidation of food proteins to improve functionality. *Critical Reviews in Food Science and Nutrition, 34*(3), 283–292. https://doi.org/10.1080/10408399409527664.

Hatamikia, M., Elhamirad, A. H., Heydari, R., Sharayei, P., & Azarpazhooh, E. (2019). Investigating the physicochemical, sensory and microbial properties of plant-based protein products (meat-free burgers) formulated using various *Vicia ervilia* (L.) Willd. protein isolates. *Plant Science Today, 6*(3), 346. https://doi.org/10.14719/pst.2019.6.3.563.

Hidayat, B. T., Wea, A., & Andriati, N. (2017). Physicochemical, sensory attributes and protein profile by SDS-PAGE of beef sausage substituted with texturized vegetable protein. *Food Research, 2*(1), 20–31. https://doi.org/10.26656/fr.2017.2(1).106.

Hu, Y.-Q., Yin, S.-W., Zhu, J.-H., Qi, J.-R., Guo, J., Wu, L.-Y., et al. (2016). Fabrication and characterization of novel Pickering emulsions and Pickering high internal emulsions stabilized by gliadin colloidal particles. *Food Hydrocolloids, 61*, 300–310. https://doi.org/10.1016/J.FOODHYD.2016.05.028.

Huang, X.-N., Zhou, F.-Z., Yang, T., Yin, S.-W., Tang, C.-H., & Yang, X.-Q. (2019). Fabrication and characterization of Pickering High Internal Phase Emulsions (HIPEs) stabilized by chitosan-caseinophosphopeptides nanocomplexes as oral delivery vehicles. *Food Hydrocolloids, 93*, 34–45. https://doi.org/10.1016/j.foodhyd.2019.02.005.

Jiang, S., Ding, J., Andrade, J., Rababah, T. M., Almajwal, A., Abulmeaty, M. M., et al. (2017). Modifying the physicochemical properties of pea protein by pH-shifting and ultrasound combined treatments. *Ultrasonics Sonochemistry, 38*, 835–842. https://doi.org/10.1016/j.ultsonch.2017.03.046.

Jin, B., Zhou, X., Guan, J., Yan, S., Xu, J., & Chen, J. (2019). Elucidation of stabilizing Pickering emulsion with jackfruit filum pectin-soy protein nanoparticles obtained by photocatalysis. *Journal of Dispersion Science and Technology, 40*(6), 909–917. https://doi.org/10.1080/01932691.2018.1489277.

Johnson, E. A., & Brekke, C. J. (1983). Functional properties of acylated pea protein isolates. *Journal of Food Science, 48*(3), 722–725. https://doi.org/10.1111/j.1365-2621.1983.tb14883.x.

Jones, O. G. (2016). Recent advances in the functionality of non-animal-sourced proteins contributing to their use in meat analogs. *Current Opinion in Food Science, 7*, 7–13. https://doi.org/10.1016/J.COFS.2015.08.002.

Ju, M., Zhu, G., Huang, G., Shen, X., Zhang, Y., Jiang, L., et al. (2020). A novel Pickering emulsion produced using soy protein-anthocyanin complex nanoparticles. *Food Hydrocolloids, 99*, 105329. https://doi.org/10.1016/J.FOODHYD.2019.105329.

Kamaldeen, O. S., Ariahu, C. C., & Yusufu, M. I. (2020). Application of soy protein isolate and cassava starch based film solutions as matrix for ionic encapsulation of carrot powders. *Journal of Food Science and Technology*, 1–11. https://doi.org/10.1007/s13197-020-04455-w.

Kamani, M. H., Meera, M. S., Bhaskar, N., & Modi, V. K. (2019). Partial and total replacement of meat by plant-based proteins in chicken sausage: Evaluation of mechanical, physico-chemical and sensory characteristics. *Journal of Food Science and Technology*, *56*(5), 2660–2669. https://doi.org/10.1007/s13197-019-03754-1.

Karaca, A. C., Low, N., & Nickerson, M. (2011). Emulsifying properties of chickpea, faba bean, lentil and pea proteins produced by isoelectric precipitation and salt extraction. *Food Research International*, *44*(9), 2742–2750. https://doi.org/10.1016/j.foodres.2011.06.012.

Kaspchak, E., de Oliveira, M. A. S., Simas, F. F., Franco, C. R. C., Silveira, J. L. M., Mafra, M. R., et al. (2017). Determination of heat-set gelation capacity of a quinoa protein isolate (*Chenopodium quinoa*) by dynamic oscillatory rheological analysis. *Food Chemistry*, *232*, 263–271. https://doi.org/10.1016/j.foodchem.2017.04.014.

Kassama, L. S., Ngadi, M. O., & Raghavan, G. S. V. (2003). Structural and instrumental textural properties of meat patties containing soy protein. *International Journal of Food Properties*, *6*(3), 519–529. https://doi.org/10.1081/JFP-120021456.

Kimura, A., Takako, F., Meili, Z., Shiori, M., Maruyama, N., & Utsumi, S. (2008). Comparison of physicochemical properties of 7S and 11S globulins from pea, fava bean, cowpea, and French bean with those of soybean-French bean 7S globulin exhibits excellent properties. *Journal of Agricultural and Food Chemistry*, *56*, 10273–10279. https://doi.org/10.1021/jf801721b.

Klompong, V., Benjakul, S., Kantachote, D., & Shahidi, F. (2007). Antioxidative activity and functional properties of protein hydrolysate of yellow stripe trevally (*Selaroides leptolepis*) as influenced by the degree of hydrolysis and enzyme type. *Food Chemistry*, *102*(4), 1317–1327. https://doi.org/10.1016/J.FOODCHEM.2006.07.016.

Klost, M., & Drusch, S. (2019). Functionalisation of pea protein by tryptic hydrolysis—Characterisation of interfacial and functional properties. *Food Hydrocolloids*, *86*, 134–140. https://doi.org/10.1016/J.FOODHYD.2018.03.013.

Krause, J.-P., Schultz, M., & Dudek, S. (2002). Effect of extraction conditions on composition, surface activity and rheological properties of protein isolates from flaxseed (*Linum usitativissimum* L). *Journal of the Science of Food and Agriculture*, *82*(9), 970–976. https://doi.org/10.1002/jsfa.1140.

Kunarayakul, S., Thaiphanit, S., Anprung, P., & Suppavorasatit, I. (2018). Optimization of coconut protein deamidation using protein-glutaminase and its effect on solubility, emulsification, and foaming properties of the proteins. *Food Hydrocolloids*, *79*, 197–207. https://doi.org/10.1016/J.FOODHYD.2017.12.031.

Lam, A. C. Y., Can Karaca, A., Tyler, R. T., & Nickerson, M. T. (2018). Pea protein isolates: Structure, extraction, and functionality. *Food Reviews International*, *34*(2), 126–147. https://doi.org/10.1080/87559129.2016.1242135.

Lam, A. C. Y., Warkentin, T. D., Tyler, R. T., & Nickerson, M. T. (2017). Physicochemical and functional properties of protein isolates obtained from several pea cultivars. *Cereal Chemistry Journal*, *94*(1), 89–97. https://doi.org/10.1094/CCHEM-04-16-0097-FI.

Lan, Y., Chen, B., & Rao, J. (2018). Pea protein isolate–high methoxyl pectin soluble complexes for improving pea protein functionality: Effect of pH, biopolymer ratio and concentrations. *Food Hydrocolloids*, *80*, 245–253. https://doi.org/10.1016/J.FOODHYD.2018.02.021.

Li, Y. (2020). Feeding the future: Plant-based meat for global food security and environmental sustainability. *Cereal Foods World*, *65*, 4. https://doi.org/10.1094/CFW-65-4-0042.

Li, X., Da, S., Li, C., Xue, F., & Zang, T. (2018). Effects of high-intensity ultrasound pretreatment with different levels of power output on the antioxidant properties of alcalase hydrolyzates from Quinoa (*Chenopodium quinoa* Willd.) protein isolate. *Cereal Chemistry*, *95*(4), 518–526. https://doi.org/10.1002/cche.10055.

Li, Y., Sukmanov, V., Kang, Z., & Ma, H. (2020). Effect of soy protein isolate on the techno-functional properties and protein conformation of low-sodium pork meat batters treated by high pressure. *Journal of Food Process Engineering*, *43*(2). https://doi.org/10.1111/jfpe.13343.

Li, C., Wang, J., Shi, J., Huang, X., Peng, Q., & Xue, F. (2015). Encapsulation of tomato oleoresin using soy protein isolate-gum aracia conjugates as emulsifier and coating materials. *Food Hydrocolloids*, *45*, 301–308. https://doi.org/10.1016/J.FOODHYD.2014.11.022.

Liang, H.-N., & Tang, C. (2014). Pea protein exhibits a novel Pickering stabilization for oil-in-water emulsions at pH 3.0. *LWT—Food Science and Technology*, *58*(2), 463–469. https://doi.org/10.1016/j.lwt.2014.03.023.

Liu, C., Chen, Y., Wang, X., Huang, J., Chang, P. R., & Anderson, D. P. (2010). Improvement in physical properties and cytocompatibility of zein by incorporation of pea protein isolate. *Journal of Materials Science*, *45*(24), 6775–6785. https://doi.org/10.1007/s10853-010-4774-z.

Liu, S., Elmer, C., Low, N. H., & Nickerson, M. T. (2010). Effect of pH on the functional behaviour of pea protein isolate–gum Arabic complexes. *Food Research International*, *43*(2), 489–495. https://doi.org/10.1016/J.FOODRES.2009.07.022.

Liu, X., Li, T., Liu, B., Zhao, H., Zhou, F., & Zhang, B. (2016). An external addition of soy protein isolate hydrolysate to sourdough as a new strategy to improve the quality of Chinese steamed bread. *Journal of Food Quality*, *39*(1), 3–12. https://doi.org/10.1111/jfq.12172.

López, D. N., Galante, M., Robson, M., Boeris, V., & Spelzini, D. (2018). Amaranth, quinoa and chia protein isolates: Physicochemical and structural properties. *International Journal of Biological Macromolecules*, *109*, 152–159. https://doi.org/10.1016/J.IJBIOMAC.2017.12.080.

López-Alarcón, C. A., Cerdán-Leal, M. A., Beristain, C. I., Pascual-Pineda, L. A., Azuara, E., & Jiménez-Fernández, M. (2019). The potential use of modified quinoa protein isolates in cupcakes: Physicochemical properties, structure and stability of cupcakes. *Food & Function*, *10*(7), 4432–4439. https://doi.org/10.1039/C9FO00852G.

López-Castejón, M. L., Bengoechea, C., Díaz-Franco, J., & Carrera, C. (2020). Interfacial and emulsifying properties of quinoa protein concentrates. *Food Biophysics*, *15*(1), 122–132. https://doi.org/10.1007/s11483-019-09603-0.

Lu, Z. X., He, J. F., Zhang, Y. C., & Bing, D. J. (2020). Composition, physicochemical properties of pea protein and its application in functional foods. In *Critical reviews in food science and nutrition* Taylor and Francis Inc. https://doi.org/10.1080/10408398.2019.1651248.

Ma, Z., Boye, J. I., Simpson, B. K., Prasher, S. O., Monpetit, D., & Malcolmson, L. (2011). Thermal processing effects on the functional properties and microstructure of lentil, chickpea, and pea flours. *Food Research International*, *44*(8), 2534 2541. https://doi.org/10.1016/j.foodres.2010.12.017.

Mahdavi-Yekta, M., Nouri, L., & Azizi, M. H. (2019). The effects of hydrolysis condition on antioxidant activity of protein hydrolyzate from *quinoa*. *Food Science & Nutrition*, 7(3), 930–936. https://doi.org/10.1002/fsn3.871.

Mäkinen, O. E., Zannini, E., & Peter Koehler, E. K. A. (2016). Heat-denaturation and aggregation of quinoa (*Chenopodium Quinoa*) globulins as affected by the pH value. *Food Chemistry*, *196*. https://doi.org/10.1016/J.FOODCHEM.2015.08.069.

Makri, E., Papalamprou, E., & Doxastakis, G. (2005). Study of functional properties of seed storage proteins from indigenous European legume crops (lupin, pea, broad bean) in admixture with polysaccharides. *Food Hydrocolloids*, *19*(3), 583–594. https://doi.org/10.1016/J.FOODHYD.2004.10.028.

Manassero, C. A., Speroni, F., & Vaudagna, S. R. (2019). High hydrostatic pressure treatment improves physicochemical properties of calcium- and soybean protein-added peach juice. *LWT*, *101*, 54–60. https://doi.org/10.1016/J.LWT.2018.11.021.

Marangon, M., Vincenzi, S., & Curioni, A. (2019). Wine fining with plant proteins. *Molecules*, *24*(11), 2186. https://doi.org/10.3390/molecules24112186.

Marco, C., Pérez, G., Ribotta, P., & Rosell, C. M. (2007). Effect of microbial transglutaminase on the protein fractions of rice, pea and their blends. *Journal of the Science of Food and Agriculture*, *87*(14). https://doi.org/10.1002/JSFA.3006.
Martínez, K. D., Sánchez, C. C., Ruíz-Henestrosa, V. P., Rodríguez Patino, J. M., & Pilosof, A. M. R. (2007). Effect of limited hydrolysis of soy protein on the interactions with polysaccharides at the air–water interface. *Food Hydrocolloids*, *21*(5–6), 813–822. https://doi.org/10.1016/J.FOODHYD.2006.09.008.
Masure, H. G., Wouters, A. G. B., Fierens, E., & Delcour, J. A. (2019). Impact of egg white and soy proteins on structure formation and crumb firming in gluten-free breads. *Food Hydrocolloids*, *95*, 406–417. https://doi.org/10.1016/j.foodhyd.2019.04.062.
McWatters, K. H. (1980). Replacement of Milk protein with protein from cowpea and field pea flours in baking powder biscuits. *Cereal Chemistry*, *57*, 223–226. Retrieved from https://www.cerealsgrains.org/publications/cc/backissues/1980/Documents/CC1980a57.html.
Mession, J.-L., Sok, N., Assifaoui, A., & Saurel, R. (2013). Thermal denaturation of pea globulins (*Pisum sativum* L.)—Molecular interactions leading to heat-induced protein aggregation. *Journal of Agricultural and Food Chemistry*, *61*(6), 1196–1204. https://doi.org/10.1021/jf303739n.
Micha, R., Wallace, S. K., & Mozaffarian, D. (2010). Red and processed meat consumption and risk of incident coronary heart disease, stroke, and diabetes mellitus. *Circulation*, *121*(21), 2271–2283. https://doi.org/10.1161/CIRCULATIONAHA.109.924977.
Milad, H. Z., Reihaneh, A. G., Mehdi, G. D., & Akram, A. (2019). Functional properties of Grass pea protein concentrates prepared using various precipitation methods. *Journal of Food Science and Technology*, *56*(11), 4799–4808. https://doi.org/10.1007/s13197-019-03930-3.
Mir, N. A., Riar, C. S., & Singh, S. (2019). Structural modification of quinoa seed protein isolates (QPIs) by variable time sonification for improving its physicochemical and functional characteristics. *Ultrasonics Sonochemistry*, *58*, 104700. https://doi.org/10.1016/J.ULTSONCH.2019.104700.
Mirmoghtadaie, L., Kadivar, M., & Shahedi, M. (2009). Effects of succinylation and deamidation on functional properties of oat protein isolate. *Food Chemistry*, *114*(1), 127–131. https://doi.org/10.1016/j.foodchem.2008.09.025.
Mirmoghtadaie, L., Shojaee Aliabadi, S., & Hosseini, S. M. (2016). Recent approaches in physical modification of protein functionality. *Food Chemistry*, *199*, 619–627. https://doi.org/10.1016/j.foodchem.2015.12.067.
Miwa, N., Yokoyama, K., Nio, N., & Sonomoto, K. (2013). Effect of enzymatic deamidation on the heat-induced conformational changes in whey protein isolate and its relation to gel properties. *Journal of Agricultural and Food Chemistry*, *61*(9), 2205–2212. https://doi.org/10.1021/jf3047626.
Morales-Polanco, E., Campos-Vega, R., Gaytán-Martínez, M., Enriquez, L. G., & Loarca-Piña, G. (2017). Functional and textural properties of a dehulled oat (*Avena sativa* L) and pea (*Pisum sativum*) protein isolate cracker. *LWT*, *86*, 418–423. https://doi.org/10.1016/j.lwt.2017.08.015.
Nesterenko, A., Alric, I., Silvestre, F., & Durrieu, V. (2014). Comparative study of encapsulation of vitamins with native and modified soy protein. *Food Hydrocolloids*, *38*, 172–179. https://doi.org/10.1016/J.FOODHYD.2013.12.011.
Nielsen, M. A., Sumner, A. K., & Whalley, L. L. (1980). Fortification of pasta with pea flour and air-classified pea protein concentrate. *Cereal Chemistry*, *57*, 203–206. Retrieved from https://www.cerealsgrains.org/publications/cc/backissues/1980/Documents/CC1980a52.html.
Nikbakht Nasrabadi, M., Goli, S. A. H., Sedaghat Doost, A., Roman, B., Dewettinck, K., Stevens, C. V., et al. (2019). Plant based Pickering stabilization of emulsions using soluble

flaxseed protein and mucilage nano-assemblies. *Colloids and Surfaces A: Physicochemical and Engineering Aspects*, *563*, 170–182. https://doi.org/10.1016/J.COLSURFA.2018.12.004.

Nongonierma, A. B., Le Maux, S., Dubrulle, C., Barre, C., & FitzGerald, R. J. (2015). Quinoa (*Chenopodium quinoa* Willd.) protein hydrolysates with in vitro dipeptidyl peptidase IV (DPP-IV) inhibitory and antioxidant properties. *Journal of Cereal Science*, *65*, 112–118. https://doi.org/10.1016/J.JCS.2015.07.004.

Nylander, T., Arnebrant, T., Cárdenas, M., Bos, M., & Wilde, P. (2019). Protein/emulsifier interactions. In *Food emulsifiers and their applications* (pp. 101–192). Cham: Springer International Publishing. https://doi.org/10.1007/978-3-030-29187-7_5.

Osaki, D., & Hiramatsu, H. (2016). Citrullination and deamidation affect aggregation properties of amyloid β-proteins. *Amyloid*, *23*(4), 234–241. https://doi.org/10.1080/13506129.2016.1240076.

Osen, R., Toelstede, S., Eisner, P., & Schweiggert-Weisz, U. (2015). Effect of high moisture extrusion cooking on protein-protein interactions of pea (*Pisum sativum* L.) protein isolates. *International Journal of Food Science & Technology*, *50*(6), 1390–1396. https://doi.org/10.1111/ijfs.12783.

Osen, R., Toelstede, S., Wild, F., Eisner, P., & Schweiggert-Weisz, U. (2014). High moisture extrusion cooking of pea protein isolates: Raw material characteristics, extruder responses, and texture properties. *Journal of Food Engineering*, *127*, 67–74. https://doi.org/10.1016/j.jfoodeng.2013.11.023.

Ouraji, M., Alimi, M., Motamedzadegan, A., & Shokoohi, S. (2020). Faba bean protein in reduced fat/cholesterol mayonnaise: Extraction and physico-chemical modification process. *Journal of Food Science and Technology*, *57*(5), 1774–1785. https://doi.org/10.1007/s13197-019-04211-9.

Owusu-Ansah, Y. J., & McCurdy, S. M. (1991). Pea proteins: A review of chemistry, technology of production, and utilization. *Food Reviews International*, 7(1), 103–134. https://doi.org/10.1080/87559129109540903.

Pan, A., Sun, Q., Bernstein, A. M., Schulze, M. B., Manson, J. E., Stampfer, M. J., et al. (2012). Red meat consumption and mortality: Results from two prospective cohort studies. *Archives of Internal Medicine*, *172*(7), 555. https://doi.org/10.1001/ARCHINTERNMED.2011.2287.

Papalamprou, E., Doxastakis, G., & Kiosseoglou, V. (2006). Model salad dressing emulsion stability as affected by the type of the lupin seed protein isolate. *Journal of the Science of Food and Agriculture*, *86*(12), 1932–1937. https://doi.org/10.1002/jsfa.2566.

Pham, T. T., & Shah, N. P. (2009). Performance of starter in yogurt supplemented with soy protein isolate and biotransformation of isoflavones during storage period. *Journal of Food Science*, *74*(4), M190–M195. https://doi.org/10.1111/j.1750-3841.2009.01141.x.

Pico, J., Reguilón, M. P., Bernal, J., & Gómez, M. (2019). Effect of rice, pea, egg white and whey proteins on crust quality of rice flour-corn starch based gluten-free breads. *Journal of Cereal Science*, *86*, 92–101. https://doi.org/10.1016/J.JCS.2019.01.014.

Pownall, T. L., Udenigwe, C. C., & Aluko, R. E. (2010). Amino acid composition and antioxidant properties of pea seed (*Pisum sativum* L.) enzymatic protein hydrolysate fractions. *Journal of Agricultural and Food Chemistry*, *58*(8), 4712–4718. https://doi.org/10.1021/jf904456r.

Qin, X.-S., Luo, Z.-G., & Peng, X.-C. (2018). Fabrication and characterization of quinoa protein nanoparticle-stabilized food-grade Pickering emulsions with ultrasound treatment: Interfacial adsorption/arrangement properties. *Journal of Agricultural and Food Chemistry*, *66*(17), 4449–4457. https://doi.org/10.1021/acs.jafc.8b00225.

Rawal, V., & Navarro, D. K. (Eds.). (2019). *The global economy of pulses*. Rome: FAO. http://www.fao.org/3/i7108en/i7108en.pdf.

Ribéreau, S., Aryee, A. N. A., Tanvier, S., Han, J., & Boye, J. I. (2018). Composition, digestibility, and functional properties of yellow pea as affected by processing. *Journal of Food Processing and Preservation*, *42*(1), e13375. https://doi.org/10.1111/jfpp.13375.

Rios-Mera, J. D., Saldaña, E., Ramírez, Y., Auquiñivín, E. A., Alvim, I. D., & Contreras-Castillo, C. J. (2019). Encapsulation optimization and pH- and temperature-stability of the complex coacervation between soy protein isolate and inulin entrapping fish oil. *LWT*, *116*, 108555. https://doi.org/10.1016/J.LWT.2019.108555.

Rohrmann, S., Overvad, K., Bueno-de-Mesquita, H. B., Jakobsen, M. U., Egeberg, R., Tjønneland, A., et al. (2013). Meat consumption and mortality—Results from the European Prospective Investigation into Cancer and Nutrition. *BMC Medicine*, *11*(1), 63. https://doi.org/10.1186/1741-7015-11-63.

Rubio, L., Pérez, A., Ruiz, R., Guzmán, M., Aranda-Olmedo, I., & Clemente, A. (2014). Characterization of pea (*Pisum sativum*) seed protein fractions. *Journal of the Science of Food and Agriculture*, *94*(2), 280–287. https://doi.org/10.1002/JSFA.6250.

Sadler, M. J. (2004). Meat alternatives—Market developments and health benefits. *Trends in Food Science & Technology*, *15*(5), 250–260. https://doi.org/10.1016/J.TIFS.2003.09.003.

Saldanha Do Carmo, C., Nunes, A. N., Silva, I., Maia, C., Poejo, J., Ferreira-Dias, S., et al. (2016). Formulation of pea protein for increased satiety and improved foaming properties. *RSC Advances*, *6*(8), 6048–6057. https://doi.org/10.1039/c5ra22452g.

Sandberg, A.-S. (2011). Developing functional ingredients: A case study of pea protein. *Functional Foods*, 358–382. https://doi.org/10.1533/9780857092557.3.358.

Sengupta, S., Bhattacharyya, D. K., Goswami, R., & Bhowal, J. (2019). Emulsions stabilized by soy protein nanoparticles as potential functional non-dairy yogurts. *Journal of the Science of Food and Agriculture*, *99*(13), 5808–5818. https://doi.org/10.1002/jsfa.9851.

Shah, N. N., Umesh, K. V., & Singhal, R. S. (2019). Hydrophobically modified pea proteins: Synthesis, characterization and evaluation as emulsifiers in eggless cake. *Journal of Food Engineering*, *255*, 15–23. https://doi.org/10.1016/J.JFOODENG.2019.03.005.

Shand, P. J., Ya, H., Pietrasik, Z., & Wanasundara, P. K. J. P. D. (2007). Physicochemical and textural properties of heat-induced pea protein isolate gels. *Food Chemistry*, *102*(4), 1119–1130. https://doi.org/10.1016/J.FOODCHEM.2006.06.060.

Shand, P. J., Ya, H., Pietrasik, Z., & Wanasundara, P. K. J. P. D. (2008). Transglutaminase treatment of pea proteins: Effect on physicochemical and rheological properties of heat-induced protein gels. *Food Chemistry*, *107*(2), 692–699. https://doi.org/10.1016/j.foodchem.2007.08.095.

Shao, Y., & Tang, C. H. (2016). Gel-like pea protein Pickering emulsions at pH3.0 as a potential intestine-targeted and sustained-release delivery system for β-carotene. *Food Research International*, *79*, 64–72. https://doi.org/10.1016/j.foodres.2015.11.025.

Shen, Y., Hu, R., & Li, Y. (2020). Antioxidant and emulsifying activities of corn gluten meal hydrolysates in oil-in-water emulsions. *Journal of the American Oil Chemists' Society*, *97*(2), 175–185. https://doi.org/10.1002/aocs.12286.

Shen, Y., & Li, Y. (2021). Acylation modification and/or guar gum conjugation enhanced functional properties of pea protein isolate. *Food Hydrocolloids*, *117*, 106686. https://doi.org/10.1016/j.foodhyd.2021.106686.

Shen, Y., Tang, X., & Li, Y. (2021). Drying methods affect physicochemical and functional properties of quinoa protein isolate. *Food Chemistry*, *339*, 127823. https://doi.org/10.1016/j.foodchem.2020.127823.

Shevkani, K., Singh, N., Kaur, A., & Rana, J. C. (2015). Structural and functional characterization of kidney bean and field pea protein isolates: A comparative study. *Food Hydrocolloids*, *43*, 679–689.

Sijtsma, L., Tezera, D., Hustinx, J., & Vereijken, J. M. (1998). Improvement of pea protein quality by enzymatic modification. *Food/Nahrung*, *42*(03–04), 215–216. https://doi.org/10.1002/(SICI)1521-3803(199808)42:03/04<215::AID-FOOD215>3.0.CO;2-A.

Soleymanpuori, R., Madadlou, A., Zeynali, F., & Khosrowshahi, A. (2014). Enzymatic cross-linking of soy proteins within non-fat set yogurt gel. *The Journal of Dairy Research*, *81*(3), 378–384. https://doi.org/10.1017/S0022029914000302.

Soria, A. C., & Villamiel, M. (2010). Effect of ultrasound on the technological properties and bioactivity of food: A review. *Trends in Food Science & Technology*, *21*(7), 323–331. https://doi.org/10.1016/J.TIFS.2010.04.003.

Speroni, F., Beaumal, V., de Lamballerie, M., Anton, M., Añón, M. C., & Puppo, M. C. (2009). Gelation of soybean proteins induced by sequential high-pressure and thermal treatments. *Food Hydrocolloids*, *23*(5), 1433–1442. https://doi.org/10.1016/J.FOODHYD.2008.11.008.

Steffolani, M. E., Villacorta, P., Morales-Soriano, E. R., Repo-Carrasco, R., León, A. E., & Pérez, G. T. (2016). Physicochemical and functional characterization of protein isolated from different quinoa varieties (*Chenopodium quinoa* Willd.). *Cereal Chemistry Journal*, *93*(3), 275–281. https://doi.org/10.1094/CCHEM-04-15-0083-R.

Stone, A. K., Avarmenko, N. A., Warkentin, T. D., & Nickerson, M. T. (2015). Functional properties of protein isolates from different pea cultivars. *Food Science and Biotechnology*, *24*, 827–833. https://doi.org/10.1007/s10068-015-0107-y.

Štreimikytė, P., Keršienė, M., Eisinaitė, V., Jasutienė, I., Lesauskaitė, V., Damulevičienė, G., et al. (2020). Formulating protein-based beverages for the dysphagia diets of the elderly: Viscosity, protein quality, *in vitro* digestion, and consumers acceptability. *Journal of the Science of Food and Agriculture*. https://doi.org/10.1002/jsfa.10431.

Sun, X. D., & Arntfield, S. D. (2011). Gelation properties of salt-extracted pea protein isolate catalyzed by microbial transglutaminase cross-linking. *Food Hydrocolloids*, *25*(1), 25–31. https://doi.org/10.1016/J.FOODHYD.2010.05.002.

Sun, X. D., & Arntfield, S. D. (2012). Gelation properties of myofibrillar/pea protein mixtures induced by transglutaminase crosslinking. *Food Hydrocolloids*, *27*(2), 394–400. https://doi.org/10.1016/J.FOODHYD.2011.11.001.

Swanson, B. G. (1990). Pea and lentil protein extraction and functionality. *Journal of the American Oil Chemists' Society*, *67*(5), 276–280. https://doi.org/10.1007/BF02539676.

Tamm, F., Herbst, S., Brodkorb, A., & Drusch, S. (2016). Functional properties of pea protein hydrolysates in emulsions and spray-dried microcapsules. *Food Hydrocolloids*, *58*, 204–214. https://doi.org/10.1016/J.FOODHYD.2016.02.032.

Tamnak, S., Mirhosseini, H., Tan, C. P., Ghazali, H. M., & Muhammad, K. (2016). Physicochemical properties, rheological behavior and morphology of pectin-pea protein isolate mixtures and conjugates in aqueous system and oil in water emulsion. *Food Hydrocolloids*, *56*, 405–416. https://doi.org/10.1016/J.FOODHYD.2015.12.033.

Tamnak, S., Mirhosseini, H., Tan, C. P., Tabatabaee Amid, B., Kazemi, M., & Hedayatnia, S. (2016). Encapsulation properties, release behavior and physicochemical characteristics of water in oil in water (W/O/W) emulsion stabilized with pectin pea protein isolate conjugate and Tween 80. *Food Hydrocolloids*, *61*, 599–608. https://doi.org/10.1016/j.foodhyd.2016.06.023.

Timilsena, Y. P., Adhikari, R., Barrow, C. J., & Adhikari, B. (2016). Physicochemical and functional properties of protein isolate produced from Australian chia seeds. *Food Chemistry*, *212*, 648–656. https://doi.org/10.1016/j.foodchem.2016.06.017.

Trikusuma, M., Paravisini, L., & Peterson, D. G. (2020). Identification of aroma compounds in pea protein UHT beverages. *Food Chemistry*, *312*, 126082. https://doi.org/10.1016/J.FOODCHEM.2019.126082.

Ustunol, Z. (2015). Physical, chemical, and processing-induced changes in proteins. In *Applied food protein chemistry* (pp. 23–46). Chichester, UK: John Wiley & Sons, Ltd. https://doi.org/10.1002/9781118860588.ch4.

Vera, A., Tapia, C., & Abugoch, L. (2020). Effect of high-intensity ultrasound treatment in combination with transglutaminase and nanoparticles on structural, mechanical, and physicochemical properties of quinoa proteins/chitosan edible films. *International Journal of Biological Macromolecules*, *144*, 536–543. https://doi.org/10.1016/J.IJBIOMAC.2019.12.120.

Vera, A., Valenzuela, M. A., Yazdani-Pedram, M., Tapia, C., & Abugoch, L. (2019). Conformational and physicochemical properties of quinoa proteins affected by different conditions of high-intensity ultrasound treatments. *Ultrasonics Sonochemistry, 51*, 186–196. https://doi.org/10.1016/J.ULTSONCH.2018.10.026.
Wanasundara, K. J. P. D., & Shahidi, F. (1997). Functional properties of acylated flax protein isolates. *Journal of Agricultural and Food Chemistry, 45*(7), 2431–2441. https://doi.org/10.1021/JF9607829.
Wang, Y., Guldiken, B., Tulbek, M., House, J. D., & Nickerson, M. (2020). Impact of alcohol washing on the flavour profiles, functionality and protein quality of air classified pea protein enriched flour. *Food Research International, 132*, 109085. https://doi.org/10.1016/J.FOODRES.2020.109085.
Wang, H., Shu, Q., Frieden, C., & Gross, M. L. (2017). Deamidation slows curli amyloid-protein aggregation. *Biochemistry, 56*(23), 2865. https://doi.org/10.1021/ACS.BIOCHEM.7B00241.
Wang, Z., Zhang, N., Chen, C., He, R., & Ju, X. (2020). Rapeseed protein nanogels as novel Pickering stabilizers for oil-in-water emulsions. *Journal of Agricultural and Food Chemistry, 68*(11), 3607–3614. https://doi.org/10.1021/acs.jafc.0c00128.
Wang, L., Zhang, X., Liu, F., Ding, Y., Wang, R., Luo, X., et al. (2017). Study of the functional properties and anti-oxidant activity of pea protein irradiated by electron beam. *Innovative Food Science & Emerging Technologies, 41*, 124–129. https://doi.org/10.1016/J.IFSET.2017.01.005.
Wang, Z., Zhang, C., Zhang, T., Ju, X., & He, R. (2018). Effects of acylation and glycation treatments on physicochemical and gelation properties of rapeseed protein isolate. *RSC Advances, 8*(70), 40395–40406. https://doi.org/10.1039/C8RA07912A.
Warnakulasuriya, S., Pillai, P. K. S., Stone, A. K., & Nickerson, M. T. (2018). Effect of the degree of esterification and blockiness on the complex coacervation of pea protein isolate and commercial pectic polysaccharides. *Food Chemistry, 264*, 180–188. https://doi.org/10.1016/J.FOODCHEM.2018.05.036.
Wei, Y., Cai, Z., Wu, M., Guo, Y., Tao, R., Li, R., et al. (2020). Comparative studies on the stabilization of pea protein dispersions by using various polysaccharides. *Food Hydrocolloids, 98*, 105233. https://doi.org/10.1016/J.FOODHYD.2019.105233.
WHO. (2002). Childhood nutrition and progress. In *Implementing the international code of marketing of breast-milk substitutes*. Geneva: World Health Organization.
Wierenga, P. A., & Gruppen, H. (2010). New views on foams from protein solutions. *Current Opinion in Colloid & Interface Science, 15*(5), 365–373. https://doi.org/10.1016/J.COCIS.2010.05.017.
Wild, F., Czerny, M. F., Janssen, A. M., Kole, A., Žunabović, M., & Domig, K. J. (2014). The evolution of a plant-based alternative to meat. From niche markets to widely accepted meat alternatives. *Agro FOOD Industry Hi Tech, 25*(1), 45–49. Retrieved from https://www.semanticscholar.org/paper/The-evolution-of-a-plant-based-alternative-to-meat.-Wild-Czerny/cf90d287aa226b483aed430ff4f0432081bfd3d7.
Wu, G. (2016). Dietary protein intake and human health. *Food & Function*, 7(3), 1251–1265. https://doi.org/10.1039/C5FO01530H.
Xiong, Y. L. (2014). Protein functionality. In *Encyclopedia of meat sciences*. https://doi.org/10.1016/B978-0-12-384731-7.00088-X.
Xiong, T., Xiong, W., Ge, M., Xia, J., Li, B., & Chen, Y. (2018). Effect of high intensity ultrasound on structure and foaming properties of pea protein isolate. *Food Research International, 109*, 260–267. https://doi.org/10.1016/j.foodres.2018.04.044.
Xue, F., Gu, Y., Wang, Y., Li, C., & Adhikari, B. (2019). Encapsulation of essential oil in emulsion based edible films prepared by soy protein isolate-gum acacia conjugates. *Food Hydrocolloids, 96*, 178–189. https://doi.org/10.1016/J.FOODHYD.2019.05.014.

Zha, F., Dong, S., Rao, J., & Chen, B. (2019a). Pea protein isolate-gum Arabic Maillard conjugates improves physical and oxidative stability of oil-in-water emulsions. *Food Chemistry*, *285*, 130–138. https://doi.org/10.1016/j.foodchem.2019.01.151.

Zha, F., Dong, S., Rao, J., & Chen, B. (2019b). The structural modification of pea protein concentrate with gum Arabic by controlled Maillard reaction enhances its functional properties and flavor attributes. *Food Hydrocolloids*, *92*, 30–40. https://doi.org/10.1016/J.FOODHYD.2019.01.046.

Zha, F., Yang, Z., Rao, J., & Chen, B. (2019). Gum Arabic-mediated synthesis of Glyco-pea protein hydrolysate via Maillard reaction improves solubility, flavor profile, and functionality of plant protein. *Journal of Agricultural and Food Chemistry*, *67*(36), 10195–10206. https://doi.org/10.1021/acs.jafc.9b04099.

Zhan, F., Shi, M., Wang, Y., Li, B., & Chen, Y. (2019). Effect of freeze-drying on interaction and functional properties of pea protein isolate/soy soluble polysaccharides complexes. *Journal of Molecular Liquids*, *285*, 658–667. https://doi.org/10.1016/J.MOLLIQ.2019.04.126.

Zhang, S., Holmes, M., Ettelaie, R., & Sarkar, A. (2020). Pea protein microgel particles as Pickering stabilisers of oil-in-water emulsions: Responsiveness to pH and ionic strength. *Food Hydrocolloids*, *102*, 105583. https://doi.org/10.1016/J.FOODHYD.2019.105583.

Zhang, L., Lin, Y. H., Leng, X. J., Huang, M., & Zhou, G. H. (2013). Effect of sage (*Salvia officinalis*) on the oxidative stability of Chinese-style sausage during refrigerated storage. *Meat Science*, *95*(2), 145–150. https://doi.org/10.1016/J.MEATSCI.2013.05.005.

Zhang, W., Xiao, S., & Ahn, D. U. (2013). Protein oxidation: Basic principles and implications for meat quality. *Critical Reviews in Food Science and Nutrition*, *53*(11), 1191–1201. https://doi.org/10.1080/10408398.2011.577540.

Zhu, F. (2019). Starch based Pickering emulsions: Fabrication, properties, and applications. *Trends in Food Science & Technology*, *85*, 129–137. https://doi.org/10.1016/J.TIFS.2019.01.012.

CHAPTER FOUR

Edible insects, a valuable protein source from ancient to modern times

Andrea M. Liceaga*
Protein Chemistry and Bioactive Peptides Laboratory, Department of Food Science, Purdue University, West Lafayette, IN, United States
*Corresponding author: e-mail address: aliceaga@purdue.edu

Contents

Abstract

The current COVID-19 pandemic has demonstrated that we are not prepared to deal with food security amid unexpected situations; the FAO (Food and Agriculture Organization) has stipulated that the future of our food & agriculture looks challenging toward the year 2050; primarily in response to the fact that global population is expected to increase by 9 billion people by 2050. Although entomophagy has been practiced by humans for thousands of years, until recently, edible insects have gained special attention due to their high nutritional value (particularly their high protein and essential amino acid content) and lower environmental impact that could help alleviate the global food demand. Edible insects are classified into eight main orders belonging to Blattodea (cockroaches and termites), Coleoptera (beetles), Diptera (flies), Hemiptera (cicadas, stink bugs), Hymenoptera (bees, wasps, ants), Lepidoptera (butterflies, moths), Odonata (dragonflies), and Orthoptera (crickets, grasshoppers, locusts). Several traditional cooking (e.g., boiling, roasting, sun-drying) and processing technologies (e.g., pasteurization, enzymatic proteolysis, high pressure processing) have shown that it is feasible to prepare safe and nutritious insects and/or foods with insects. Nevertheless, challenges associated with consumers acceptance to eat insects, as well as potential presence of anti-nutritive factors and allergens, need to be carefully evaluated as the industry grows in the coming years. Foreseeing such food shortages during pandemics and future food security concerns, consumers, scientists, and the food industry need to consider the value of farming insects as promising protein sources.

Advances in Food and Nutrition Research, Volume 101
ISSN 1043-4526
https://doi.org/10.1016/bs.afnr.2022.04.002

1. Humand entomophagy: Historical, nutritional and sustainability perspective

Historical evidence shows that insects have been used as a food source, a term known as *entomophagy*. Although entomophagy is defined as the dietary consumption of insects by any organism, it is most likely used to refer to the human consumption of insects (Costa-Neto & Dunkel, 2016). The fact is that entomophagy has been practiced since early hominids like *Paranthropus* (or *Australopithecus*) *robustus* (Late Pliocene and Early Pleistocene) in South Africa, who used bone tools do dig into termite mounds. In Northern Spain, dental plaque from an early hominid revealed microfossils of insect fragments, while coprolites found in the Lakeside cave (Utah, USA), suggest that migratory grasshoppers (*Melanoplus sanguinipes*) were consumed by humans; other prehistoric human coprolites containing chitinous exoskeletons from insects were also found in different States of the USA as well as in Mexico and Peru (Van Huis, 2017). Based on these findings, paleoanthropologists believe that insects could have indeed played an important role in the diet of early humans.

As we go through historical records, we have evidence from Aristotle in the *Historia Animalium* that cicadas were harvested and considered a delicacy in ancient Greece, while Pliny the Elder (AD 23/24-79) wrote that Romans consumed "cossus," a highly coveted dish consisting of larvae from the long-horn beetle (Van Huis et al., 2013). The Old Testament in the Bible describes the four kinds of locusts which the Hebrews were allowed to eat (Leviticus, XI: 21–22), and the New Testament (Mark 1:6) describes John the Baptist's food as consisting of "locusts and wild honey" (Costa-Neto & Dunkel, 2016). Moving into our modern-day cultures, we know that pre-Hispanic Mesoamerican cultures also practiced entomophagy. For example, documents from the New Spain (Nueva España) describe that the Aztecs consumed a variety of insects, including winged ants, grasshoppers, mosquito eggs, and worms from the maguey plant (Novo, 1997). Interestingly, many of these insects are still consumed today in public markets in Mexico City (Fig. 1) (Liceaga, 2021). Indigenous peoples from the United States and Canada were also known to eat insects such as grasshoppers, crickets, caterpillars, flies, cicadas, beetles, ants, bees and yellowjackets (Capinera, 2008). For example, the Tlicho from the Northwest Territories of Canada ate warble fly larvae (*Oedemagena* or *Hypoderma*) collected during the butchering of caribou carcasses (Lesnik, 2018). Some cultures also give

Fig. 1 Degustation plate of insects traditionally eaten by the Aztecs and still available in pubic markets in Mexico City. Insects, clockwise from the top, are grasshoppers (plain), chicatana ants, jumiles (stink bugs), chinicuiles (red maguey worms), cocopache (leaf-footed bug), grasshoppers (salted), grasshoppers (adobo), ahuautles or "mosco de río" (water fly), and acociles (crayfish). Scorpions are in the center of the plate. *Photo by Andrea M. Liceaga.*

spiritual attributes to certain insect species, such as consuming wasps for bringing prosperity, protection and abundance (Ramos-Elorduy, 2009).

Although through history several plant and animal species have been domesticated and become part of our modern-day staple diet, we know that nearly 2.5 million people worldwide who continue to eat insects and have remained as an important aspect of their culture (van Huis, Dicke, & Van Loon, 2015). In countries such as Thailand, China, Africa, Mexico and Colombia, insects are well known for their nutritional benefits and considered dietary staples (Pal & Roy, 2014), while in southern Ghana, palm weevil larvae (locally known as "Gbamedo") are one of the most widely consumed insects considered a delicacy (Agbemafle, Hadzi, Amagloh, Zotor, & Reddy, 2020). There are close to 2000 edible insect species catalogued to this day, classified within 8 main orders belonging to Blattodea (cockroaches and termites), Coleoptera (beetles), Diptera (flies), Hemiptera (cicadas, stink bugs), Hymenoptera (bees, wasps, ants), Lepidoptera (butterflies, moths), Odonata (dragonflies), and Orthoptera (crickets, grasshoppers, locusts) (Liceaga, Aguilar-Toalá, Vallejo-Cordoba, González-Córdova, & Hernández-Mendoza, 2021). Worldwide, more families of Lepidoptera are reared by humans than any other insect. In the United States, the waxworm (*G. mellonella*) is a lepidopteran commonly mass-reared for the animal feed industry as well as for fish bait (Dossey, Tatum, & McGill, 2016). However, worldwide the house cricket (order: Orthoptera) and yellow mealworm (order: Coleoptera) are the most popular farmed insects exclusively for human consumption (Melgar-Lalanne, Hernández-Álvarez, & Salinas-Castro, 2019). Most edible insect species have nutritional yields comparable to conventional meat on a per-gram basis (Table 1). A compilation of the nutrient compositions of over 200 edible insect species showed that these insects are composed primarily of protein and fat, followed by fiber, nitrogen free-extract, non-fiber carbohydrates and ash (Rumpold & Schlüter, 2013). Insects can primarily be an excellent source of protein, as they contain all essential amino acids. The protein content will vary by insect species and their life-cycle stage, with crickets, grasshoppers and locusts (order: Orthoptera) having overall the highest protein content (61% dry basis), followed by dragonflies and damselflies (order: Odonata) with 55% protein (dry basis); cockroaches and termites (order: Blattodea) have the lowest overall protein content (35% dry basis) (Fig. 2) (Liceaga et al., 2021).

Table 1 Comparison of nutritional composition found on traditional (raw) protein sources and two domesticated insects (crickets and yellow mealworms) used for human consumption.

Nutrien	Salmon[a]	Chicken[b]	Beef[c]	Pork[d]	Crickets[e]	Yellow mealworms[e]
Protein	22.2%	22.2%	22.5%	21.0%	21.3%%	20.3%
Fat	4.7%	2.6%	8.7%	2.2%	7.3%	13.8%
Carbohydrates	0.0%	0.0%	0.0%	0.0%	4.1%	3.1%
Fiber	0.0%	0.0%	0.0%	0.0%	3.2%	1.7%

[a]National Nutrient Database for Standard reference (ndb.nal.usda.gov): report 173691 Salmon, sockeye, raw.
[b]National Nutrient Database for Standard reference (ndb.nal.usda.gov): report 05062 Chicken, broiler or fryers, breast, skinless, boneless, meat only, raw.
[c]National Nutrient Database for Standard reference (ndb.nal.usda.gov): report 1390 Beef Round, prime, raw.
[d]National Nutrient Database for Standard reference (ndb.nal.usda.gov): report 10060, Pork, fresh, lion, tenderloin, separable lean only, raw.
[e]Whole, raw crickets and mealworms (wet basis with 69.07% and 62.44% moisture content, respectively).
Data adapted from Finke, M. D. (2004). Nutrient content of insects: Springer.; Liceaga, A. M., Aguilar-Toalá, J. E., Vallejo-Cordoba, B., González-Córdova, A. F., & Hernández-Mendoza, A. (2021). Insects as an Alternative Protein Source. Annual review of food science and technology, 13, 19–34; Rumpold, B. A., & Schlüter, O. K. (2013). Nutritional composition and safety aspects of edible insects. Molecular nutrition & food research, 57(5), 802-823.

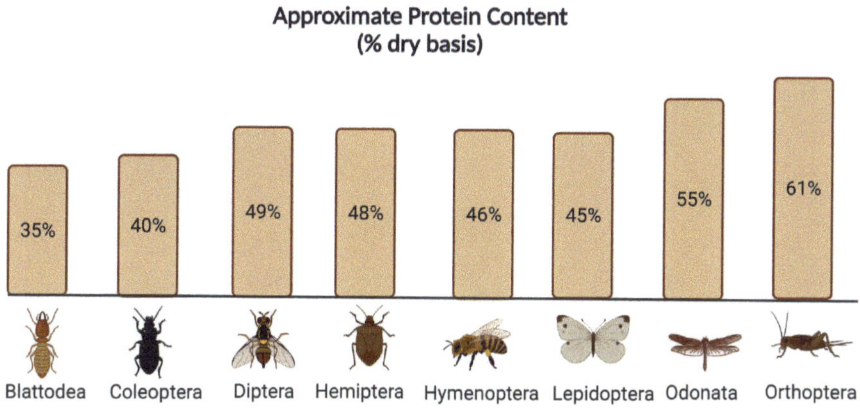

Fig. 2 Approximate protein content for the eight most common edible insect orders. *Figure created with BioRender.com*

In most western cultures, particularly North American and European, entomophagy is exercised sparingly and insects are mainly considered a food novelty rather than a source of nutrients (Raubenheimer & Rothman, 2013). In fact, in several western countries, human consumption of insects is regarded as a cultural taboo and opinions on eating insects are associated with feelings of disgust and reluctance (Pal & Roy, 2014). However, this perception has slowly begun to change as reports from the Food and Agriculture Organization (FAO) for the United Nations indicate that the global population is likely to increase to 9 billion by 2050 (Van Huis et al., 2013). Currently, nearly 1 billion people go hungry; therefore, as the population continues to rise, so will the amount people that go without food. It is well documented that agricultural land is already pressured by the food demand of the current population. Thus, by the year 2050, world food demand will need to increase by at least 50% with farmers producing 60% more crop calories (7400 trillion calories), and increasing land use by 593 million hectares (twice the size of India) (Searchinger, Walte, Hanson, & Ranganathan, 2019). Exploring other sources of protein that are also more sustainable should alleviate pressures on current food sources such as livestock. Domesticated (farmed) insects are highlighted among such alternatives because they already are integrated in many food cultures around the globe. At present, the FAO encourages entomophagy due to the high economic opportunity that insect farming represents and lower environmental impact (Hall, Jones, O'Haire, & Liceaga, 2017). Compared to other domesticated animals, farm-raised insects require less resources to produce

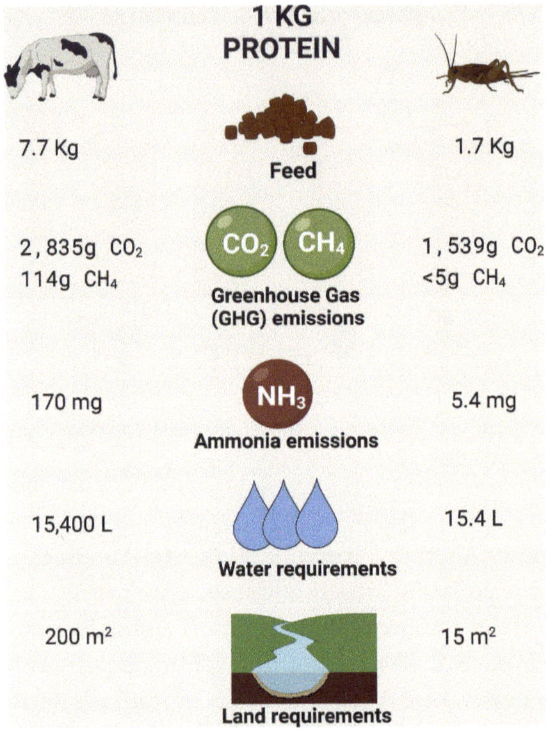

Fig. 3 Comparison of estimated resources needed to produce 1 kg of protein from livestock and farmed (domesticated) insects, respectively. *Figure created using data from Goodland, R., & Anhang, J. (2009). Livestock and climate change: What if the key actors in climate change are... cows, pigs, and chickens? World Watch; Van Huis, A., Van Itterbeeck, J., Klunder, H., Mertens, E., Halloran, A., & Vantomme, P. (2013). Edible Insects: Future Prospects for Food and Feed Security. Food and Agriculture Organization of the United Nations. FAO Forestry Paper, FAO, Rome (187 pp). Figure created with BioRender.com.*

the same amount of protein. Overall, insect breeding requires much less food and land relative to livestock production. Furthermore, reared (farmed) insects inflict a smaller climate impact in terms of greenhouse gases (GHG) and ammonia emissions. Insects can be reared on less land for short periods, due to their short life cycle, requiring significantly less water and energy with a higher intrinsic growth rate than traditional livestock (Fig. 3).

For example, reared crickets use 15 m^2 of land to produce 1 kg of protein, while cattle use 200 m^2, they emit lower GHG emissions and ammonia per unit of protein, use less water and feed. In perspective, livestock can produce up to 100 times more GHG than mealworms (*Tenebrio molitor*) and emit 8–12 times more ammonia compared to house crickets (*Acheta domesticus*)

(Goodland & Anhang, 2009; Oonincx et al., 2010). Another important consideration when selecting sustainable protein sources relates to the efficiency conversion of ingested food (ECI), which estimates the ability of the animal to convert feed into body mass. In some insects, the ECI can be up to 44%; crickets in particular are twice as efficient as pigs and broiler chickens, four times greater than sheep, and six times higher than a steer (Nakagaki & Defoliart, 1991). These data confirm the need to develop an alternative agriculture system beyond conventional food sources that considers the rising global population. It has been proposed that substituting at least 25% of protein from livestock with other more sustainable proteins, would allow the reforestation of agrarian land and reduce 4% or more of agricultural greenhouse gases (GHG), equivalent to 23 million metric tons per year (EPA, U. S. E. P. A, 2017; Searchinger et al., 2019; Steinfeld, Gerber, Wassenaar, Castel, & de Haan, 2006). In this sense, insect farming (rearing) has gained attention as a nutritional and sustainable approach to this problem. In fact, insect farming has been labeled as an emerging "mini-livestock production system" (Abbasi & Abbasi, 2016). Currently, the majority of industrial-scale edible insect farms are located in Europe (e.g., France and the Netherlands) and North America (e.g., Canada and USA) (Fig. 4). These vertical, sustainable farms rear mainly crickets (e.g., *Acheta domesticus*) and yellow mealworms (*Tenebrio molitor*) and rely on their own core breeding stock to ensure a great production of insect biomass, thus limiting the possibility of introducing diseases into the system (Baiano, 2020). Another major advantage of farmed insects is that current farming practices do not utilize chemical agents, for example, antibiotics, steroids, hormones, pesticides, or other synthetic chemical components that are often used in vertebrate livestock operations (Dossey et al., 2016).

2. Traditional methods and commercial processing technologies used for insects

Edible insects are customarily prepared using traditional methods such as sun-drying, roasting, boiling, steaming, baking, frying, and stewing, among others (Table 2). Nowadays, they are typically consumed as whole insects (raw or cooked), processed (non-recognizable form), and in the form of extracts (Liceaga et al., 2021). The food industry is showing interest in this novel protein source, as evidenced by several startup companies and number of scientific publications in the last decade, with market trends leading toward a global edible insect market of approximately USD 8 billion in

A

B

Fig. 4 (A) Aspire Food Group's 100,000 sq ft fully-automated cricket production and processing facility located in London, Ontario, Canada. Once operational, this landmark plant is expected to produce 10,000 tons of crickets/year. (B) An early iteration of robotic watering technology at Aspire Food Group Research & Development facility in Austin, Texas, USA. *Photographs reprinted with permission of Aspire Food Group.*

the next 10 years (Liceaga et al., 2021). As a result of this market increase aimed toward Western cultures, other approaches to preparing insects must rely in processing methods that render insects into non-recognizable forms, like flours or powders, protein hydrolysates, fermentable substrates, etc. (Table 2) (Liceaga, 2021; Melgar-Lalanne et al., 2019). The use of different drying technologies seems to be the most commonly used approach for

Table 2 List of some traditional cooking and commercial processing methods reported in literature for edible insects.

Insect species	Cooking or processing method used	References
Alphitobius diaperinus (beetle) *Ruspolia differens* (grasshopper) *Tenebrio molitor* (yellow mealworm) *Archea domesticus* (house cricket)	Boiling (submerged in boiling water), blanching, steaming, sautéed	Wynants et al. (2018), Grabowski and Klein (2017), Fombong, Van Der Borght, and Vanden Broeck (2017), Purschke, Brüggen, Scheibelberger, and Jäger (2018), Kamau et al. (2018), Ssepuuya, Aringo, Mukisa, and Nakimbugwe (2016), and Nyangena et al. (2020)
Ruspolia nitidula (grasshopper) *Rhynchophorus phoenicis* (palm weevil) *Ruspolia differens* (longhorn grasshopper) *Tenebrio molitor* (yellow mealworm) *Polyrhachis vicina* Roger (Black ant) *Nauphoeta cinerea* (speckled cockroach)	Drying (sun/solar, oven, freeze-drying, pan-fried, fluidized bed, microwave-assisted drying)	Tiencheu et al. (2013), Fombong et al. (2017), Purschke et al. (2018), Alves, Sanjinez-Argandoña, Linzmeier, Cardoso, and Macedo (2016), Bußler et al. (2016), Wynants et al. (2018), Vandeweyer, Lenaerts, Callens, and Van Campenhout (2017), Kröncke, Böschen, Woyzichovski, Demtröder, and Benning (2018), de Oliveira, da Silva Lucas, Cadaval, and Mellado (2017), and Hernández-Álvarez, Mondor, Piña-Domínguez, Sánchez-Velázquez, and Melgar Lalanne (2021)
Acheta domesticus (house cricket) *Schistocerca gregaria* (desert locust) *Spodoptera littoralis* (leaf worm), *Gryllodes sigillatus* (tropical banded cricket) *Tenebrio molitor* (yellow mealworm)	Enzymatic proteolysis; sonication, fermentation, ultrasound-, pasteurized-liquid, and microwave-assisted extractions, extrusion	Hall et al. (2017), Zhao, Vázquez-Gutiérrez, Johansson, Landberg, and Langton (2016), Mishyna, Martinez, Chen, and Benjamin (2019), Zielińska, Karaś, and Baraniak (2018), Mendoza-Salazar et al. (2021), Carcea (2020), Otero, Gutierrez-Docio, Del Hierro, Reglero, and Martin (2020), Del Hierro, Gutiérrez-Docio, Otero, Reglero, and Martin (2020)

preserving and processing edible insects. However, each drying method used will have different effects on the insects' nutritional composition and stability. For example, (Kröncke et al., 2018) reported that drying techniques caused minor changes in protein, fat, and fiber content of yellow mealworms (*Tenebrio molitor*). However, oven drying, microwave drying, fluidized bed drying, and drying with a vacuum decreased ($P < 0.05$) the protein solubility, while freeze dried mealworms exhibited the highest lipid oxidation compared to the other drying methods. Overall, vacuum oven and microwave drying technologies were reported to be an alternative to conventional oven drying and freeze drying. In contrast, Lenaerts, Van Der Borght, Callens, and Van Campenhout (2018) showed that freeze drying *Tenebrio molitor* increased lipid oxidation compared to microwave drying, which displayed minor changes in protein, fat, and ash content of the mealworms; the application of a vacuum during the microwave drying process did not add an advantage, since the proximate and fatty acid composition of the mealworms were not significantly affected. Strategies for process-optimized drying of edible insects are still needed to ensure nutrient quality and product functionality.

There are clear indications that Western consumers are more inclined to eat insect protein when insects are in non-visible or unrecognizable form in the food or masked by a familiar flavor (e.g., chocolate covered). Studies report that showing a full image of the insect as a marketing strategy for insect-based food products in a retail setting, significantly decreased consumers' willingness to buy that particular food product (Baker, Shin, & Kim, 2016). This demonstrates the frail acceptance by consumers toward edible insects and/or entomophagy. Additionally, consumers have shown a more positive emotional response to food products that were formulated with the incorporation of non-recognizable insects (i.e., in the form of a flour or powder) compared to those foods that were formulated with insects that remained in a recognizable (i.e., visible) form (Gmuer, Nuessli Guth, Hartmann, & Siegrist, 2016). Sensory evaluation studies also indicate that meals containing visible insects were rated much more negatively in terms of attractiveness and likelihood of eating it, compared to foods formulated with insect protein or insect flours (i.e., pulverized insects into a fine powder) (Caparros Megido et al., 2014; Schösler, De Boer, & Boersema, 2012; Tranter, 2013; Tucker, 2014). Dossey et al. (2016) indicates that caution should be used when using terms such as "insect flour" as this may cause confusion in consumers who might think that the insect flour will have the same properties for cooking and baking as those found in products like grain flours. The authors explain that while insects are composed primarily of protein, followed by fat and fiber (chitin), true flours (e.g., wheat) are made primarily of starches and fiber, followed by protein. Nevertheless, food extrusion, typically used in the production of cereal-based foods using flours, is a good example of a processing method that has allowed for the incorporation of insects as enrichment ingredients, into baked goods and pasta (Carcea, 2020). Other methods reported in literature for processing edible insects rely on separating the protein from the insect exoskeleton (high in chitin) by means of controlled, enzymatic proteolysis with commercial food-grade proteases such as alcalase (Liceaga, 2019). The resulting protein hydrolysates or protein powders tend to have an overall improvement on the protein's functional properties (e.g., solubility, emulsifications, foaming) by effectively separating the insoluble chitin from the protein; these highly-soluble protein hydrolysates can be used in food formulation as protein supplements, emulsifiers and stabilizers, and flavor enhancers, among others (Liceaga, 2019). In one study, corn tortillas were successfully formulated with 20% cricket protein powder, which increased the limiting amino acid lysine from 0.2 g to 1.0 g/100g and also received positive acceptability scores (degree of liking >6.5) for aroma and flavor, despite panelists ($n = 112$) knowing that the tortillas contained cricket protein (Calzada-Luna, Martin-Gonzalez, Mauer, & Liceaga, 2021). This demonstrates the potential to develop familiar or staple food products that have some of their traditional protein replaced by insect protein derived from a processing method that transforms the insect into a non-recognizable form.

As with other traditional protein sources (e.g., dairy, meat, etc.), processing methods that involve heat treatments like pasteurization and commercial sterilization, are known to effectively decrease microbial loads, inactivate enzymes as well as increase the nutritional quality, and digestibility of insects (Agbemafle et al., 2020; Liceaga, 2021). There is limited scientific literature available on commercial thermal processing methods applied to edible insects and their effect on the safety, nutritional quality, and protein functionality. This is because the *insects-as-food* industry remains in the early stages of production, processing, and commercialization in comparison to the other long-established food industries (e.g., dairy, meat, poultry) (Cho, Zhao, Kim, Kim, & Chung, 2018). Meyer-Rochow, Gahukar, Ghosh, and Jung (2021) provide a detailed description on the effects of different traditional cooking and processing methods, grouped by insect order, on final product quality.

In terms of microbial safety, studies on raw and heat-treated yellow mealworm (*Tenebrio molitor*) and house crickets (*Acheta domesticus*) indicate that thermal processing was effective at eliminating pathogenic bacteria (Klunder, Wolkers-Rooijackers, Korpela, & Nout, 2012). Grabowski and Klein (2017) evaluated the microbial quality of a variety of processed edible insect species (e.g., deep fried, seasoned, cooked, dried, powdered, and frozen). Their results showed that dried and seasoned insects contained higher microbial counts than those that were cooked or deep fried. All samples tested negative for *Salmonella, L. monocytogenes, E. coli* and *Stapyhlococcus aureus*; however, dried and powdered insects contained *B. cereus*, coliforms, *Listeria ivanovii, Aspergillus* spp., *Penicillium* spp., and *Cryptococcus neoformans* (Grabowski & Klein, 2017). In another study, Nyangena et al. (2020) examined the effects of different traditional processing techniques (i.e., boiling, toasting, solar-drying, and sun-drying, etc.) on the proximate composition and microbiological quality of different edible insect species (*Acheta domesticus, Ruspolia differens, Hermetia illucens* and *Spodoptera littoralis*) relative to the raw and/or un-processed insects. Authors reported that boiling and roasting or toasting were the most effective methods for increasing the protein content and decreasing or eliminating aerobic mesophilic bacteria, *Staphylococcus aureus, Salmonella*, yeasts, and molds. Bußler et al. also reported a 3-log microbial reduction was achieved for *Tenebrio molitor* flour following exposure to cold atmospheric pressure plasma for 15 min; whereas equally long thermal treatments at 120°C and 140°C were found to completely inactivate the native microorganism flora (Bußler et al., 2016). Lastly, other studies have reported that fermentation of soy sauce-analog using *Tenebrio molitor, Bacillus licheniformis*, and *Aspergillus oryzaep*, resulted in the amino-nitrogen and aromatic compound content increasing indicating protein degradation that did not affect the nutritional and sensory quality of the fermented sauce (Cho et al., 2018; Mouritsen, Duelund, Calleja, & Frøst, 2017; Yi, Van Boekel, Boeren, & Lakemond, 2016).

Another important consideration is that applying processing technologies insects can lead to development of insect-based food ingredients beyond proteins, including fiber, lipids, and other insect components (e.g., polyphenols) that could serve multiple functions in the food and beverage industries. In addition to fermentation, methods like enzymatic proteolysis, microwave-extraction, ultrasonication, and high-pressure processing are capable of releasing bioactive compounds such as peptides, phenolic compounds, and chitin that can have potential benefits to human health in the prevention or control of diseases such as hypertension, inflammation, and type-II diabetes (Hall & Liceaga, 2020). For example, edible cricket chitosan obtained from the chitin, a by-product of the protein extraction using microwave-assisted enzymatic proteolysis, was shown to have antimicrobial and hypolipidemic activity comparable to that of commercial shellfish chitosan (Malm & Liceaga, 2021); whereas some phenolic compounds with antioxidant and antimicrobial activity were successfully extracted and characterized from farmed edible crickets (Nino, Reddivari, Ferruzzi, & Liceaga, 2021).

3. Applications of insect protein in food and beverage formulations

Several applications using edible insects have been considered for the food and beverage industries. The most common application of edible insects has been in bakery and cereal-based products such as cookies, bread, tortillas, and pasta. For example, roasted speckled cockroach (*Nauphoeta cinerea*) powder ranging from 5% to 15% (w/w) was used as protein enrichment of wheat flour to formulate bread. Results showed that the 10% enrichment formulation was the most similar to the whole what bread control, and also presented the best nutritional characteristics including higher protein (22.6% vs 9.7%, dry basis) and fiber (2.3% vs 2.0%, dry basis) and an acceptability index above 75% (de Oliveira et al., 2017).

Replacing wheat flour with 5% insect flour from house cricket (*A. domesticus*), mealworm, and black soldier fly (*Hermetia illucens*), resulted in bread loafs with decreased water absorption and increased dough formation and stability (González, Garzón, & Rosell, 2018). Cricket powder (10–30%) was also used to enrich wheat bread. Compared to control breads (using only wheat flour), breads containing cricket powder showed a higher nutritional profile in terms of fatty acid composition, high protein content and also showed a significant enrichment in the essential amino acids lysine, tyrosine, valine, and methionine (Osimani et al., 2018). In another study, microwave-dried yellow mealworm powder (20%) was used to enrich wheat flour and produce different cereal-based snacks using 3D printing technology (Severini, Azzollini, Albenzio, & Derossi, 2018). Corn tortillas is another example of a baked cereal product fortified with edible insects. In this study, 20% cricket (*Acheta domesticus*) protein powders were used to formulate corn tortillas. The improved nutritional quality (including essential amino acids) as well as comparable physico-chemical and sensory acceptability scores of the fortified products, compared to control tortillas, demonstrates the high potential of using edible insects to fortify staple food products that have limiting essential amino acids without compromising the palatability (Calzada-Luna et al., 2021).

Reports indicate that companies are currently working on extraction and restructuration of insect proteins into versatile food ingredients, like soluble protein powders for beverages and textured insect proteins for meat analogues, and egg or dairy replacements in baking and food processing applications (Shockley, Lesnik, Allen, & Muñoz, 2018). Minced cooked insects are also being used to formulate meat analogue foods like hamburgers, meatballs, and sausages (Elhassan, Wendin, Olsson, & Langton, 2019; Fraqueza & Patarata, 2017). Efforts to produce protein supplements, beverages and energy bars based on insect powder are also documented (Mutungi et al., 2019). One study looked at the application of insects to formulate animal-sourced foods as a strategy to achieve protein and micronutrient density of infants and young children in developing countries. In this study, crickets and palm weevil larvae were blended with a sweet potato porridge and compared to the maize-peanut-soybean blend (Weanimix). The results showed that the edible insect foods had several advantages over the mainstream blend, including ease of preparation, improved nutritional composition (e.g., meeting protein requirements), and lower risk of aflatoxin contamination (Agbemafle et al., 2020). Other recent applications include the incorporation of edible insects for space food applications such as the Mars mission. In this context, the required food to support life in a space mission and/or closed ecological environments could be harvested from enclosed agriculture systems. Because protein from animal origin will be difficult to produce due to its constraints related to the extraterrestrial environment, efficiency in the use of biomass energy, such as reared edible insects, should be considered (Katayama, Yamashita, Wada, & Mitsuhashi, 2005).

In addition to the use of insects and their protein to formulate foods, some studies have investigated the application of other insect components for food formulation. For example, lipids derived from insect biomass of two species (*Hermetia illucens* and *Tenebrio molitor*) were applied as an alternative for plant and animal lipids in spreadable products like margarine or butter (Smetana, Leonhardt, Kauppi, Pajic, & Heinz, 2020). Authors reported that it was possible to substitute up to 75% of the lipids with insect-derived fats without negative effects on the spreading ability or color of the products. In another study, cookies prepared with insect oils had higher omega-3 fatty acids, flavonoids, and vitamin E than the cookies formulated with plant oils. Consumers' acceptance was also high for those cookies prepared with *Ruspolia differens* and sesame oils, respectively, compared to those formulated with olive and *Schistocerca gregaria* oils (Cheseto, Baleba, Tanga, Kelemu, & Torto, 2020).

Finally, the effort to develop novel processing methods and foods that incorporate edible insects can also be evidenced by the vast number of patent applications. Baiano (2020) lists the most recent patents containing edible insects or insect powders. These include cookies, rice cakes, energy bars, soup, tea, rice noodles, jerky, coffee, salad dressing, and tofu, among others. These commercial patent applications highlight the broad interest by scientists and the private sector to include edible insects in food and beverage formulations.

4. Challenges and future prospects of insect protein

One of the major challenges facing edible insects is the acceptance by consumers particularly in Western society, where food neophobic factors have resulted in edible insects being regarded with a feeling of disgust or viewed as a cultural taboo (Gmuer et al., 2016). Food neophobia is regarded as the fear to eating new foods and can occur in all type of consumers; however, the level of neophobia response will vary amongst consumers depending on their age, gender, education, social status, among others (Tuorila, Lähteenmäki, Pohjalainen, & Lotti, 2001). In this sense, studies have shown that consumers categorized as neophobics have a low willingness to purchase food products containing edible insects, compared to consumers who have a positive disposition to eat new foods (neophilics) (Lombardi, Vecchio, Borrello, Caracciolo, & Cembalo, 2019). However, there is no evidence confirming that neophobics will not accept edible insects. As previously discussed in this chapter, research suggests that incorporating insects as part of an ingredient within a familiar food will help alleviate some of those neophobic, psychological constrains. Moving forward, education and industry efforts will need to find pathways to promote edible insects in order to eventually normalize their consumption just as it has been done over decades with other "novel" foods like sushi, plant-based meat-analogues, etc.

Although insects are consumed by many people all over the world, safety aspects remain important challenges. This chapter has already discussed some of the microbial safety concerns associated with edible insects and the processing methods that have shown to decrease their microbial load. Other safety concerns are related to the anti-nutrient content in some edible insect species. Due to insects' herbivore feeding behavior, farmed insects are primarily fed plant-based diets rich in allelochemicals such as phenolic compounds (Nino et al., 2021). These allelochemicals can be a good source of antioxidants, but some can also have anti-nutritive effects. For example, crickets are reported to have 3159 mg/100 g and 900 mg/100 g of phytate and tannins, respectively; while grasshoppers have 1100 mg/100 g and

1050 mg/100 g of phytate and tannins, respectively (Meyer-Rochow et al., 2021). Nevertheless, similar to antinutritive compounds found in other foods, most processing methods (e.g., boiling, drying) can decrease their content (Liceaga et al., 2021). Furthermore, the advantage of farming edible insects will allow for selecting carefully-designed plant diets that can minimize the concentration of these compounds.

Other insect compounds, like chitin and allergens, can also present challenges. Excessive consumption of chitin, found primarily in the insect's exoskeleton, is speculated to increase risks of urinary stone formation and chronic degenerative disease (Yhoung-aree, 2008). Potential allergic responses from insect chitin ingestion have also been reported (Bush, 2008). However, there is no clear link to these effects with chitin consumption. The presence of protein allergens, on the other hand, remains a major safety concern surrounding edible insects. Ongoing studies have identified antigens and IgE-binding proteins from several insect species that are correlated to an allergic reaction upon exposure or consumption (Feng et al., 2018; Pali-Schöll et al., 2019; Ribeiro, Cunha, Sousa-Pinto, & Fonseca, 2018). For instance, immunoreactions have been associated with silkworm (*Bombyx mori*) (Liu, Tian, & Chen, 2001), teak caterpillar cocoons *(Hyblaea puera)* (Lukiwati, 2010), grasshoppers (Srivastava, Babu, & Pandey, 2009; Vetter, 1995), house crickets (*Acheta* domesticus) (Abdelmoteleb et al., 2018) and farmed tropical banded crickets (*Gryllodes sigillatus*) (Hall, Johnson, & Liceaga, 2018). The major shrimp allergen, tropomyosin, is known to be a cross-sensitizing allergen in several edible insects due to its reported immunological relationships between crustaceans and arthropods (which include insects) (Abdelmoteleb et al., 2018; Ayuso, Reese, Leong-Kee, Plante, & Lehrer, 2002; Hall & Liceaga, 2021; Wong, Huang, & Lee, 2016). This is because tropomyosin, a highly-conserved protein that can exist in different isoforms, is present in all vertebrate species as well as among invertebrate's (e.g., insects) muscle and non-muscle cells (Leung et al., 1996). Therefore, there is a high degree of cross-reactivity between homologous proteins found in crustaceans (shellfish) and other arthropods (Leoni, Volpicella, Dileo, Gattulli, & Ceci, 2019; Pali-Schöll et al., 2019; Volpicella, Leoni, Dileo, & Ceci, 2019), suggesting that individuals with a shellfish allergy should avoid eating insects (MacEvilly, 2000). Table 3 shows the immunoinformatics results of shared sequence homology (>60% identity) for cricket tropomyosin and allergens from various species of shellfish, insects, and nematodes. It can be observed that the top matches were for tropomyosin from Lep s 1 silverfish (*Lepisma saccharina*) and Pan b 1.0101 northern shrimp

Table 3 Cricket tropomyosin predicted sequence homology with reported allergens derived from insects, shellfish and nematodes.

Species	Allergen	Sequence link in SwissProt/NCBI	Full alignment[a]	
			E-val	%ID
Lepisma saccharina	Lep s 1	CAC84590	1.7e-050	81.50%
Pandalus borealis	Pan b 1.0101	CBY17558	4.8e-049	78.50%
Penaeus monodon	Pen m 1	AAX37288	2.2e-040	67.30%
Penaeus aztecus	Pen a 1	11893851	1.2e-039	65.40%
Homarus americanus	Hom a 1.0102	AAC48288	9.5e-042	69.30%
Litopenaeus vannamei	Lit v 1.0101	EU410072	1.1e-040	67.30%
Homarus americanus	Hom a 1.0101	O44119	3.3e-041	67.80%
Periplaneta americana	Per a 7.0102	AAD19606	2.9e-041	68.30%
Blattella germanica	Bla g 7.0101	AAF72534	4.5e-041	68.30%
Dermatophagoides farinae	Der f 10.0101	BAA04557	3.7e-041	67.80%
Chironomus kiiensis	Chi k 10	CAA09938	7.4e-042	68.80%
Tyrophagus putrescentiae	Tyr p 10.0101	AAT40866	9.8e-038	65.60%
Blomia tropicalis	Blo t 10.0101	ABU97466	1.4e-041	68.30%
Metapenaeus ensis	Met e 1	Q25456	3.6e-039	66.80%
Panulirus stimpsoni	Pan s 1	O61379	4.7e-041	67.30%
Lepidoglyphus destructor	Lep d 10	Q9NFZ4	5.3e-038	66.20%
Dermatophagoides farinae	Der p 10	O18416	6.5e-042	68.30%
Charybdis feriatus	Cha f 1	Q9N2R3	2.1e-040	67.30%
Ascaris lumbricoides	Asc l 3.0101	ACN32322	3.5e-042	68.80%
Anisakis simplex	Ani s 3	Q9NAS5	5.8e-042	69.30%
Helix aspersa	Hel as 1	CAB38044	3.7e-041	67.80%
Haliotis diversicolor	Hal d 1	AAG08987	1e-039	65.40%
Mimachlamys nobilis	Mim n 1	AAG08989	6e-041	67.80%
Perna viridis	Per v 1	AAG08988	2e-041	68.30%
Crassostrea gigas	Cra g 1	AAK96889	1.1e-040	67.30%
Dermatophagoides farinae	Der p 11	AAO73464	3.7e-041	67.80%
Blomia tropicalis	Blo t 11	AAM83103	3.5e-042	69.30%
Dermatophagoides farina	Der f 11.0101	AAK39511	4.2e-041	68.30%

[a]Parameters assessed are % Identity and E-score. Significance is assumed when the expected score is below 1.0 or a >50% identity match (www.allermatch.org). Duplicates were not included on the list.
Table reprinted from Hall, F., & Liceaga, A. (2021). Isolation and proteomic characterization of tropomyosin extracted from edible insect protein. *Food Chemistry: Molecular Sciences*, 100049, with permission from Elsevier.

(*Pandalus borealis*). Matches were also obtained with other insect allergens such as house dust mite (*Dermatophagoides farinae*) paramyosin (Der f 11), American cockroach (*Periplaneta americana*) tropomyosin (Per a 7), and German cockroach (*Blattella germanica*) tropomyosin (Bla g 7) (Hall & Liceaga, 2021).

Further research is needed in order to establish the effect of the different food processing technologies on edible insect allergens. For example, studies report that treating insects with heat (e.g., baking, boiling), frying, and high-pressure processing did not have much effect on lowering an immuno-response (i.e., IgE-binding) (Broekman et al., 2015; Jeong et al., 2016; Mills & Mackie, 2008; Phiriyangkul, Srinroch, Srisomsap, Chokchaichamnankit, & Punyarit, 2015; Van Broekhoven, Bastiaan-Net, de Jong, & Wichers, 2016). Conversely, Hall and Liceaga (2021) demonstrated that microwave-heated and protease-treated cricket tropomyosin had lower IgE and IgG reactivity compared to cricket tropomyosin treated by convection heating. Based on immunoinformatics and proteomics analyses, the decreased allergenicity was associated with increased cleavage of the epitope region of the tropomyosin that had been protease-treated with microwave heating. This agrees with other reports by El Mecherfi et al. (2011) and Ketnawa and Liceaga (2017) that microwave heating can increase the rate of an allergenic protein's unfolding, enhancing the exposure of its epitope region, to proteases that would have not been otherwise accessible under convection heating. As more edible insects become part of formulated foods and beverages, it will be crucial to continue the assessment for the presence of potential allergens, toxicants, and anti-nutritive factors.

5. Conclusion

Edible insects are gaining attention as potential protein sources that could help alleviate the predicted protein demand by the year 2050. The lower environmental impact of insect farming places them as leaders in the future development of more sustainable foods worldwide. The incorporation of these novel protein sources as viable ingredients will largely depend on consumers' perception and acceptance of products containing edible insects. Decades of research on processing technologies and product development for the plant, meat and dairy industries have created the necessary knowledge to overcome food processing and safety challenges to produce safe and palatable foods for consumers. Like with traditional protein sources, the incorporation of insect protein into food and beverage formulations will present its challenges and limitations that will require extensive research to ensure that processing technologies and formulation strategies work in the same form as they have done for traditional proteins. Overall, edible insects are a highly nutritious source of protein, fiber, and lipids. Early education efforts and mechanisms to process them into non-recognizable forms, will allow for the normalized attitude towards eating insects by modern-day society. Studies show that insects can be processed using similar technologies to those applied for traditional proteins; therefore, the possibilities of developing convenient, safe, palatable, or even shelf-stable, insect-based food products is vast. Foreseeing food shortages during pandemics like the 2020 COVID-19 pandemic, and food security concerns towards the year 2050, consumers, scientists, and the food industry need to consider the value of farming insects as promising protein sources.

References

Abbasi, T., & Abbasi, S. (2016). Reducing the global environmental impact of livestock production: the minilivestock option. *Journal of Cleaner Production, 112*, 1754–1766.

Abdelmoteleb, M., Palmer, L. K., Pavlovikj, N., Marsh, J. T., Johnson, P. E., & Goodman, R. E. (2018). Bioinformatics and Proteomics Evaluations to Consider IgE Binding Assays for Potential Cross-Reactivity Between House-Cricket (Aceta domesticus) Used in Food, Crustaceans and Cockroaches. *Journal of Allergy and Clinical Immunology, 141*(2), AB263.

Agbemafle, I., Hadzi, D., Amagloh, F. K., Zotor, F. B., & Reddy, M. B. (2020). Nutritional, microbial, and sensory evaluation of complementary foods made from blends of orange-fleshed sweet potato and edible insects. *Food, 9*(9), 1225.

Alves, A. V., Sanjinez-Argandoña, E. J., Linzmeier, A. M., Cardoso, C. A. L., & Macedo, M. L. R. (2016). Food value of mealworm grown on Acrocomia aculeata pulp flour. *PLoS One, 11*(3), e0151275.

Ayuso, R., Reese, G., Leong-Kee, S., Plante, M., & Lehrer, S. B. (2002). Molecular basis of arthropod cross-reactivity: IgE-binding cross-reactive epitopes of shrimp, house dust mite and cockroach tropomyosins. *International Archives of Allergy and Immunology, 129*(1), 38–48.

Baiano, A. (2020). Edible insects: An overview on nutritional characteristics, safety, farming, production technologies, regulatory framework, and socio-economic and ethical implications. *Trends in Food Science & Technology, 100*, 35–50.

Baker, M. A., Shin, J. T., & Kim, Y. W. (2016). An exploration and investigation of edible insect consumption: the impacts of image and description on risk perceptions and purchase intent. *Psychology & Marketing, 33*(2), 94–112.

Broekman, H., Knulst, A., den Hartog Jager, S., Monteleone, F., Gaspari, M., De Jong, G., et al. (2015). Effect of thermal processing on mealworm allergenicity. *Molecular Nutrition & Food Research, 59*(9), 1855–1864.

Bush, R. K. (2008). Indoor allergens, environmental avoidance, and allergic respiratory disease. In *Paper presented at the Allergy and Asthma Proceedings*.

Bußler, S., Rumpold, B. A., Fröhling, A., Jander, E., Rawel, H. M., & Schlüter, O. K. (2016). Cold atmospheric pressure plasma processing of insect flour from Tenebrio molitor: Impact on microbial load and quality attributes in comparison to dry heat treatment. *Innovative Food Science & Emerging Technologies, 36*, 277–286.

Calzada-Luna, G., Martin-Gonzalez, F. S., Mauer, L., & Liceaga, A. (2021). Cricket (Acheta domesticus) protein hydrolysates' impact on the physicochemical, structural and sensory properties of tortillas and tortilla chips. *Journal of Insects as Food and Feed*, 7(1), 109–120.

Caparros Megido, R., Sablon, L., Geuens, M., Brostaux, Y., Alabi, T., Blecker, C., et al. (2014). Edible insects acceptance by Belgian consumers: Promising attitude for entomophagy development. *Journal of Sensory Studies*, *29*(1), 14–20.

Capinera, J. L. (2008). Native American Culture and Insects. In J. L. Capinera (Ed.), *Encyclopedia of Entomology* (pp. 2546–2550). Dordrecht, The Netherlands: Springer.

Carcea, M. (2020). *Quality and Nutritional/Textural Properties of Durum Wheat Pasta Enriched with Cricket Powder*. Multidisciplinary Digital Publishing Institute.

Cheseto, X., Baleba, S., Tanga, C. M., Kelemu, S., & Torto, B. (2020). Chemistry and sensory characterization of a bakery product prepared with oils from African edible insects. *Food*, *9*(6), 800.

Cho, J.-H., Zhao, H.-L., Kim, J.-S., Kim, S.-H., & Chung, C.-H. (2018). Characteristics of fermented seasoning sauces using Tenebrio molitor larvae. *Innovative Food Science & Emerging Technologies*, *45*, 186–195.

Costa-Neto, E. M., & Dunkel, F. (2016). Insects as food: history, culture, and modern use around the world. In *Insects as sustainable food ingredients* (pp. 29–60). Elsevier.

de Oliveira, L. M., da Silva Lucas, A. J., Cadaval, C. L., & Mellado, M. S. (2017). Bread enriched with flour from cinereous cockroach (Nauphoeta cinerea). *Innovative Food Science & Emerging Technologies*, *44*, 30–35.

Del Hierro, J. N., Gutiérrez-Docio, A., Otero, P., Reglero, G., & Martin, D. (2020). Characterization, antioxidant activity, and inhibitory effect on pancreatic lipase of extracts from the edible insects Acheta domesticus and Tenebrio molitor. *Food Chemistry*, *309*, 125742.

Dossey, A., Tatum, J., & McGill, W. (2016). Modern insect-based food industry: current status, insect processing technology, and recommendations moving forward. In *Insects as sustainable food ingredients* (pp. 113–152). Elsevier.

El Mecherfi, K. E., Saidi, D., Kheroua, O., Boudraa, G., Touhami, M., Rouaud, O., et al. (2011). Combined microwave and enzymatic treatments for β-lactoglobulin and bovine whey proteins and their effect on the IgE immunoreactivity. *European Food Research and Technology*, *233*(5), 859–867.

Elhassan, M., Wendin, K., Olsson, V., & Langton, M. (2019). Quality aspects of insects as food—nutritional, sensory, and related concepts. *Food*, *8*(3), 95.

EPA, U. S. E. P. A. (2017). Sources of Greenhouse Gas Emissions. Retrieved from https://www.epa.gov/ghgemissions/sources-greenhouse-gas-emissions.

Feng, Y., Chen, X. M., Zhao, M., He, Z., Sun, L., Wang, C. Y., et al. (2018). Edible insects in China: Utilization and prospects. *Insect Science*, *25*(2), 184–198.

Fombong, F. T., Van Der Borght, M., & Vanden Broeck, J. (2017). Influence of freeze-drying and oven-drying post blanching on the nutrient composition of the edible insect Ruspolia differens. *Insects*, *8*(3), 102.

Fraqueza, M., & Patarata, L. (2017). *Constraints of HACCP application on edible insect for food and feed* (pp. 89–113). London: Future Foods. InTech.

Gmuer, A., Nuessli Guth, J., Hartmann, C., & Siegrist, M. (2016). Effects of the degree of processing of insect ingredients in snacks on expected emotional experiences and willingness to eat. *Food Quality and Preference*, *54*, 117–127. https://doi.org/10.1016/j.foodqual.2016.07.003.

González, C. M., Garzón, R., & Rosell, C. M. (2018). Insects as ingredients for bakery goods. A comparison study of H. illucens, A. domestica and T. molitor flours. *Innovative Food Science & Emerging Technologies*.

Goodland, R., & Anhang, J. (2009). *Livestock and climate change: What if the key actors in climate change are.. cows, pigs, and chickens?* World Watch.

Grabowski, N. T., & Klein, G. (2017). Microbiology of cooked and dried edible Mediterranean field crickets (Gryllus bimaculatus) and superworms (Zophobas atratus) submitted to four different heating treatments. *Food Science and Technology International, 23*(1), 17–23.

Hall, F., Johnson, P. E., & Liceaga, A. (2018). Effect of enzymatic hydrolysis on bioactive properties and allergenicity of cricket (Gryllodes sigillatus) protein. *Food Chemistry, 262*, 39–47. Retrieved from https://doi.org/10.1016/j.foodchem.2018.04.058.

Hall, F., Jones, O. G., O'Haire, M. E., & Liceaga, A. M. (2017). Functional properties of tropical banded cricket (Gryllodes sigillatus) protein hydrolysates. *Food Chemistry, 224*, 414–422. Retrieved from https://doi.org/10.1016/j.foodchem.2016.11.138.

Hall, F., & Liceaga, A. (2020). Effect of microwave-assisted enzymatic hydrolysis of cricket (Gryllodes sigillatus) protein on ACE and DPP-IV inhibition and tropomyosin-IgG binding. *Journal of Functional Foods, 64*, 103634. Retrieved from https://doi.org/10.1016/j.jff.2019.103634.

Hall, F., & Liceaga, A. (2021). Isolation and proteomic characterization of tropomyosin extracted from edible insect protein. *Food Chemistry: Molecular Sciences, 100049*.

Hernández-Álvarez, A.-J., Mondor, M., Piña-Domínguez, I.-A., Sánchez-Velázquez, O.-A., & Melgar Lalanne, G. (2021). Drying technologies for edible insects and their derived ingredients. *Drying Technology, 39*(13), 1991–2009.

Jeong, K. Y., Son, M., Lee, J. Y., Park, K. H., Lee, J.-H., & Park, J.-W. (2016). Allergenic characterization of 27-kDa glycoprotein, a novel heat stable allergen, from the pupa of silkworm, Bombyx mori. *Journal of Korean Medical Science, 31*(1), 18–24.

Kamau, E., Mutungi, C., Kinyuru, J., Imathiu, S., Tanga, C., Affognon, H., et al. (2018). Moisture adsorption properties and shelf-life estimation of dried and pulverised edible house cricket Acheta domesticus (L.) and black soldier fly larvae Hermetia illucens (L.). *Food Research International, 106*, 420–427.

Katayama, N., Yamashita, M., Wada, H., & Mitsuhashi, J. (2005). Entomophagy as part of a space diet for habitation on Mars. *The Journal of Space Technology and Science, 21*(2), 2_27-22_38.

Ketnawa, S., & Liceaga, A. M. (2017). Effect of Microwave Treatments on Antioxidant Activity and Antigenicity of Fish Frame Protein Hydrolysates. *Food and Bioprocess Technology, 10*, 582–591.

Klunder, H., Wolkers-Rooijackers, J., Korpela, J., & Nout, M. (2012). Microbiological aspects of processing and storage of edible insects. *Food Control, 26*(2), 628–631.

Kröncke, N., Böschen, V., Woyzichovski, J., Demtröder, S., & Benning, R. (2018). Comparison of suitable drying processes for mealworms (Tenebrio molitor). *Innovative Food Science & Emerging Technologies, 50*, 20–25.

Lenaerts, S., Van Der Borght, M., Callens, A., & Van Campenhout, L. (2018). Suitability of microwave drying for mealworms (Tenebrio molitor) as alternative to freeze drying: Impact on nutritional quality and colour. *Food Chemistry, 254*, 129–136.

Leoni, C., Volpicella, M., Dileo, M. C., Gattulli, B. A., & Ceci, L. R. (2019). Chitinases as food allergens. *Molecules, 24*(11), 2087.

Lesnik, J. J. (2018). *Edible insects and human evolution*. University Press of Florida.

Leung, P. S., Chow, W. K., Duffey, S., Kwan, H. S., Gershwin, M. E., & Chu, K. H. (1996). IgE reactivity against a cross-reactive allergen in crustacea and mollusca: evidence for tropomyosin as the common allergen. *Journal of Allergy and Clinical Immunology, 98*(5), 954–961.

Liceaga, A. M. (2019). Approaches for Utilizing Insect Protein for Human Consumption: Effect of Enzymatic Hydrolysis on Protein Quality and Functionality. *Annals of the Entomological Society of America, 112*(6), 529–532.

Liceaga, A. M. (2021). Processing insects for use in the food and feed industry. *Current opinion in insect science, 48*, 32–36.

Liceaga, A. M., Aguilar-Toalá, J. E., Vallejo-Cordoba, B., González-Córdova, A. F., & Hernández-Mendoza, A. (2021). Insects as an Alternative Protein Source. *Annual Review of Food Science and Technology*, *13*, 19–34.

Liu, X., Tian, J., & Chen, D. (2001). An allergic shock case that resulted from consuming silkworm pupae. *Chinese Recuperative Medicine*, *10*(2), 80.

Lombardi, A., Vecchio, R., Borrello, M., Caracciolo, F., & Cembalo, L. (2019). Willingness to pay for insect-based food: The role of information and carrier. *Food Quality and Preference*, *72*, 177–187.

Lukiwati, D. R. (2010). Teak caterpillars and other edible insects in Java. *Forest Insects as Food: Humans Bite Back*, *99*.

MacEvilly, C. (2000). Bugs in the system. *Nutrition Bulletin*, *25*(4), 267–268.

Malm, M., & Liceaga, A. M. (2021). Physicochemical Properties of Chitosan from Two Commonly Reared Edible Cricket Species, and Its Application as a Hypolipidemic and Antimicrobial Agent. *Polysaccharides*, *2*(2), 339–353.

Melgar-Lalanne, G., Hernández-Álvarez, A. J., & Salinas-Castro, A. (2019). Edible insects processing: Traditional and innovative technologies. *Comprehensive Reviews in Food Science and Food Safety*, *18*(4), 1166–1191.

Mendoza-Salazar, A., Santiago-López, L., Torres-Llanez, M. J., Hernández-Mendoza, A., Vallejo-Cordoba, B., Liceaga, A. M., et al. (2021). In Vitro Antioxidant and Antihypertensive Activity of Edible Insects Flours (Mealworm and Grasshopper) Fermented with Lactococcus lactis Strains. *Fermentation*, 7(3), 153.

Meyer-Rochow, V. B., Gahukar, R. T., Ghosh, S., & Jung, C. (2021). Chemical composition, nutrient quality and acceptability of edible insects are affected by species, developmental stage, gender, diet, and processing method. *Food*, *10*(5), 1036.

Mills, E. C., & Mackie, A. R. (2008). The impact of processing on allergenicity of food. *Current Opinion in Allergy and Clinical Immunology*, *8*(3), 249–253.

Mishyna, M., Martinez, J.-J. I., Chen, J., & Benjamin, O. (2019). Extraction, characterization and functional properties of soluble proteins from edible grasshopper (Schistocerca gregaria) and honey bee (Apis mellifera). *Food Research International*, *116*, 697–706.

Mouritsen, O. G., Duelund, L., Calleja, G., & Frøst, M. B. (2017). Flavour of fermented fish, insect, game, and pea sauces: Garum revisited. *International Journal of Gastronomy and Food Science*, *9*, 16–28.

Mutungi, C., Irungu, F., Nduko, J., Mutua, F., Affognon, H., Nakimbugwe, D., et al. (2019). Postharvest processes of edible insects in Africa: A review of processing methods, and the implications for nutrition, safety and new products development. *Critical Reviews in Food Science and Nutrition*, *59*(2), 276–298.

Nakagaki, B. J., & Defoliart, G. R. (1991). Comparison of diets for mass-rearing Acheta domesticus (Orthoptera: Gryllidae) as a novelty food, and comparison of food conversion efficiency with values reported for livestock. *Journal of Economic Entomology*, *84*(3), 891–896.

Nino, M. C., Reddivari, L., Ferruzzi, M. G., & Liceaga, A. M. (2021). Targeted Phenolic Characterization and Antioxidant Bioactivity of Extracts from Edible Acheta domesticus. *Food*, *10*(10), 2295.

Novo, S. (1997). *Cocina mexicana o historia gastronómica de la ciudad de México*. Porrua Hnos.

Nyangena, D. N., Mutungi, C., Imathiu, S., Kinyuru, J., Affognon, H., Ekesi, S., et al. (2020). Effects of traditional processing techniques on the nutritional and microbiological quality of four edible insect species used for food and feed in East Africa. *Food*, *9*(5), 574.

Oonincx, D. G., van Itterbeeck, J., Heetkamp, M. J., van den Brand, H., van Loon, J. J., & van Huis, A. (2010). An exploration on greenhouse gas and ammonia production by insect species suitable for animal or human consumption. *PLoS One*, *5*(12), e14445.

Osimani, A., Milanović, V., Cardinali, F., Roncolini, A., Garofalo, C., Clementi, F., et al. (2018). Bread enriched with cricket powder (Acheta domesticus): A technological, microbiological and nutritional evaluation. *Innovative Food Science & Emerging Technologies, 48*, 150–163.

Otero, P., Gutierrez-Docio, A., Del Hierro, J. N., Reglero, G., & Martin, D. (2020). Extracts from the edible insects Acheta domesticus and Tenebrio molitor with improved fatty acid profile due to ultrasound assisted or pressurized liquid extraction. *Food Chemistry, 314*, 126200.

Pal, P., & Roy, S. (2014). Edible insects: future of human food–a review. *International Letters of Natural Sciences, 21*.

Pali-Schöll, I., Meinlschmidt, P., Larenas-Linnemann, D., Purschke, B., Hofstetter, G., Rodríguez-Monroy, F. A., et al. (2019). Edible insects: Cross-recognition of IgE from crustacean-and house dust mite allergic patients, and reduction of allergenicity by food processing. *WAO Journal, 12*.

Phiriyangkul, P., Srinroch, C., Srisomsap, C., Chokchaichamnankit, D., & Punyarit, P. (2015). Effect of food thermal processing on allergenicity proteins in Bombay locust (Patanga succincta). *International Journal of Food Engineering, 1*(1), 23–28.

Purschke, B., Brüggen, H., Scheibelberger, R., & Jäger, H. (2018). Effect of pre-treatment and drying method on physico-chemical properties and dry fractionation behaviour of mealworm larvae (Tenebrio molitor L.). *European Food Research and Technology, 244*(2), 269–280.

Ramos-Elorduy, J. (2009). Anthropo-entomophagy: Cultures, evolution and sustainability. *Entomological Research, 39*(5), 271–288.

Raubenheimer, D., & Rothman, J. M. (2013). Nutritional ecology of entomophagy in humans and other primates. *Annual Review of Entomology, 58*, 141–160.

Ribeiro, J. C., Cunha, L. M., Sousa-Pinto, B., & Fonseca, J. (2018). Allergic risks of consuming edible insects: A systematic review. *Molecular Nutrition & Food Research, 62*(1), 1700030.

Rumpold, B. A., & Schlüter, O. K. (2013). Nutritional composition and safety aspects of edible insects. *Molecular Nutrition & Food Research, 57*(5), 802–823.

Schösler, H., De Boer, J., & Boersema, J. J. (2012). Can we cut out the meat of the dish? Constructing consumer-oriented pathways towards meat substitution. *Appetite, 58*(1), 39–47.

Searchinger, T., Walte, R., Hanson, C., & Ranganathan, J. (2019). *World Resources Report: Creating a Sustainable Food Future. A Menu of Solutions to Feed Nearly 10 Billion People by 2050* (p. 564). World Resources Institute.

Severini, C., Azzollini, D., Albenzio, M., & Derossi, A. (2018). On printability, quality and nutritional properties of 3D printed cereal based snacks enriched with edible insects. *Food Research International, 106*, 666–676.

Shockley, M., Lesnik, J., Allen, R. N., & Muñoz, A. F. (2018). Edible insects and their uses in North America; past, present and future. In *Edible insects in sustainable food systems* (pp. 55–79). Springer.

Smetana, S., Leonhardt, L., Kauppi, S.-M., Pajic, A., & Heinz, V. (2020). Insect margarine: Processing, sustainability and design. *Journal of Cleaner Production, 264*, 121670.

Srivastava, S., Babu, N., & Pandey, H. (2009). Traditional insect bioprospecting-As human food and medicine. *Indian Journal of Traditional Knowledge, 8*(4), 485–494.

Ssepuuya, G., Aringo, R., Mukisa, I., & Nakimbugwe, D. (2016). Effect of processing, packaging and storage-temperature based hurdles on the shelf stability of sautéed ready-to-eat Ruspolia nitidula. *Journal of Insects as Food and Feed, 2*(4), 245–253.

Steinfeld, H., Gerber, P., Wassenaar, T., Castel, V., & de Haan, C. (2006). *Livestock's long shadow: environmental issues and options.* Food & Agriculture Org.

Tiencheu, B., Womeni, H. M., Linder, M., Mbiapo, F. T., Villeneuve, P., Fanni, J., et al. (2013). Changes of lipids in insect (Rhynchophorus phoenicis) during cooking and storage. *European Journal of Lipid Science and Technology*, *115*(2), 186–195.

Tranter, H. (2013). *Insects Creeping into English Diets: Introducing Entomophagy to School Children in a Provincial Town*. Norwich: University of East Anglia.

Tucker, C. A. (2014). The significance of sensory appeal for reduced meat consumption. *Appetite*, *81*, 168–179.

Tuorila, H., Lähteenmäki, L., Pohjalainen, L., & Lotti, L. (2001). Food neophobia among the Finns and related responses to familiar and unfamiliar foods. *Food Quality and Preference*, *12*(1), 29–37.

Van Broekhoven, S., Bastiaan-Net, S., de Jong, N. W., & Wichers, H. J. (2016). Influence of processing and in vitro digestion on the allergic cross-reactivity of three mealworm species. *Food Chemistry*, *196*, 1075–1083.

Van Huis, A. (2017). Did early humans consume insects. *Journal of Insects as Food and Feed*, *3*(3), 161–163.

van Huis, A., Dicke, M., & Van Loon, J. (2015). *Insects to feed the world*. Wageningen Academic Publishers.

Van Huis, A., Van Itterbeeck, J., Klunder, H., Mertens, E., Halloran, A., & Vantomme, P. (2013). *Edible Insects: Future Prospects for Food and Feed Security*. Food and Agriculture Organization of the United Nations. FAO Forestry Paper, FAO, Rome (187 pp).

Vandeweyer, D., Lenaerts, S., Callens, A., & Van Campenhout, L. (2017). Effect of blanching followed by refrigerated storage or industrial microwave drying on the microbial load of yellow mealworm larvae (Tenebrio molitor). *Food Control*, *71*, 311–314.

Vetter, R. (1995). A case of ingestant allergy from eating a grasshopper. *The Food Insects Newsletter*, *8*(5).

Volpicella, M., Leoni, C., Dileo, M. C., & Ceci, L. R. (2019). Progress in the Analysis of Food Allergens through Molecular Biology Approaches. *Cell*, *8*(9), 1073.

Wong, L., Huang, C. H., & Lee, B. W. (2016). Shellfish and house dust mite allergies: is the link tropomyosin? *Allergy, Asthma & Immunology Research*, *8*(2), 101–106.

Wynants, E., Crauwels, S., Verreth, C., Gianotten, N., Lievens, B., Claes, J., et al. (2018). Microbial dynamics during production of lesser mealworms (Alphitobius diaperinus) for human consumption at industrial scale. *Food Microbiology*, *70*, 181–191.

Yhoung-aree, J. (2008). Edible insects in Thailand: nutritional values and health concerns. In *Forest insects as food: humans bite back. Proceedings of a workshop on Asia-Pacific resources and their potential for development, Chiang Mai, Thailand, 19-21*.

Yi, L., Van Boekel, M. A., Boeren, S., & Lakemond, C. M. (2016). Protein identification and in vitro digestion of fractions from Tenebrio molitor. *European Food Research and Technology*, *242*(8), 1285–1297.

Zhao, X., Vázquez-Gutiérrez, J. L., Johansson, D. P., Landberg, R., & Langton, M. (2016). Yellow mealworm protein for food purposes-extraction and functional properties. *PLoS One*, *11*(2), e0147791.

Zielińska, E., Karaś, M., & Baraniak, B. (2018). Comparison of functional properties of edible insects and protein preparations thereof. *LWT*, *91*, 168–174.

Further reading

Finke, M. D. (2004). *Nutrient content of insects*. Springer.

CHAPTER FIVE

Fungal protein

Pedro F. Souza Filho[†,*]
Laboratory of Biochemical Engineering, Chemical Engineering Department, Federal University of Rio Grande do Norte (UFRN), Natal, RN, Brazil
*Corresponding author: e-mail address: pedro.souzafilho@ufpe.br

Contents

Abstract

This chapter reviews the uses of processed fungal cells as protein-rich foods in substitution to meat. Yeasts, mushrooms, and filamentous fungi have long been part of the human diet, improving the nutritional quality and taste of different foods. Recently, because of the populational growth and environmental impact caused by animal farming, the use of animal-derived protein has been raising concerns in public and scientific debates. Fungal biomass represents a nutritious, safe meat alternative that can help society in the reduction of greenhouse gases emissions and improve public health. Additionally, it provides all essential amino acids and has an interesting fatty acid profile. Current research is involved in finding new strains with improved efficiency, the investigation of new substrates (especially agro-industrial wastes) to reduce the environmental impact, and the development of new formulations to use mycoprotein in different dishes.

† Current address: Chemical Engineering Department, Federal University of Pernambuco (UFPE), Recife, PE, Brazil.

Advances in Food and Nutrition Research, Volume 101
ISSN 1043-4526
https://doi.org/10.1016/bs.afnr.2022.04.003

1. Introduction

In 2015, the United Nations launched an agenda toward sustainable development, raising attention to the social and climate impacts caused by human activity. The program established 17 goals with the message of ending poverty, promoting prosperity and people's well-being while protecting the environment to be achieved by 2030. Included in these goals are the end of hunger (goal 2) and good health and well-being (goal 3). Both goals can be achieved by providing not only food but high-quality meals to the population. However, as the 2020 UN report informs, from 2014 to 2019, the number of people facing food insecurity was on the rise, from 22.4% to 25.9% of the global population, which is equivalent to 2 billion people (United Nations, 2020). The term food insecurity can be defined as the lack of continuous access to food in adequate quantity and quality because of limited money or other resources (dos Anjos Coelho Kubo, da Costa, & Gubert, 2020; Gundersen & Ziliak, 2015). This represents a threat to the health of people, as several diseases are associated with this condition, e.g., birth defects, anemia, cognitive problems, behavioral problems (aggression, anxiety, depression, suicide ideation), poor sleep, higher risks of being hospitalized, asthma, and worse oral health (Gundersen & Ziliak, 2015). Moreover, it is expected that the COVID-19 pandemic will worsen this scenario, imposing a new threat to eradicate food insecurity (United Nations, 2020).

Together with food insecurity, malnutrition and obesity correspond to the most complex threats to the livelihoods of populations globally, making them the leading causes of poor health (Huizar, Arena, & Laddu, 2021). The most common malnutrition cases are caused by deficient intakes of protein, iron, iodine, and vitamin A (Stephenson, Latham, & Ottesen, 2000). Specifically, protein-energy malnutrition (PEM) will lead to muscle weakness, altered immune function, decreased functionality, and increased risk of infection. In the most severe cases, PEM can cause kwashiorkor disease, nutritional marasmus, and marasmic kwashiorkor (Stephenson et al., 2000). Recently, the cases of PEM have been increasing not only in the poor regions of the world but also among the elderly living in care homes in developed nations (British Dietetic Association, 2017; Mathewson, Azevedo, Gordon, Phillips, & Greig, 2021).

On the other hand, diets richer in proteins are associated with the beneficial reduction of weight and cardiometabolic risk factors, attenuation of the fat-free mass loss, reduction of waist circumference, and improved markers of blood glucose. One must notice, however, that the source of protein plays a role in the positive effects. Diets that prioritize animal protein intake, particularly meat, have been linked to increased mortality risk. This is due to the macronutrients and micronutrients consumed alongside the protein, which vary according to their origin (Sokolowski, Higgins, Vishwanathan, & Evans, 2020). Other disadvantages of animal proteins are their growing costs, limited supply, and the negative impacts on climate change, freshwater depletion, and biodiversity loss (Sá, Moreno, & Carciofi, 2020).

Therefore, alternative protein sources have been proposed to substitute animal protein in the human diet. In this context, microorganisms, such as bacteria, algae, and fungi, represent a reliable solution as they can efficiently convert substrates into proteins. According to the literature, the Single Cell Protein (SCP) protein yield per kg of input feed is tenfold higher than cattle. Additionally, the product contains a well-balanced amino acid profile and its production is independent of the environmental condition (Bekatorou, Psarianos, & Koutinas, 2006). Some restrictions apply to the choice of the microorganism to produce SCP, e.g., non-pathogenic, non-toxic, rapid growth, little nutritional requirements, easy processing system, and low levels of nucleic acids with good biological values (Gervasi et al., 2018).

2. Fungi

The Fungi Kingdom comprises heterotrophic eukaryotes whose forms vary from microscopic, single-celled yeasts to macroscopic multicellular mushrooms (Jayasiri et al., 2015). Some common characteristics of the Kingdom include the cell wall formed of β-glucan and chitin, the presence of the aminoadipidic pathway for the biosynthesis of lysine, and the presence of flattened mitochondrial cristae (Naranjo-Ortiz & Gabaldón, 2019). In nature, fungi play an essential role in nutrient recycling and support plant life in the form of endophytes and mycorrhizae, providing nitrogen, phosphorus, and other elements in exchange for photosynthesis-derived metabolites. On the other hand, they are also responsible for a plethora of diseases in animals and plants. In biotechnology, fungi can be used in the production of enzymes, alcoholic beverages, medicines, biocontrol agents, natural fertilizers, natural pigments, and cosmetics (Souza Filho, 2018).

According to their life cycle, fungi are divided into three major groups, namely, unicellular, macro filamentous, and multicellular filamentous fungi (Wakai, Arazoe, Ogino, & Kondo, 2017). The unicellular fungi are represented by the yeasts; they can convert glucose into alcohol, lactic acid, and carbon dioxide, and produce unique flavor substances (Lu et al., 2021). Thus, they have long been used to produce fermented food and beverages like beer, wine, fermented dairy products, and bread (Bekatorou et al., 2006). In general, microbial fermentation reduces the content of antinutritional and toxic compounds in the food, improves nutritional accessibility, prolongs the shelf life, alters the sensory characteristics of the products (Wikandari et al., 2021), reduces the volume, and shortens the cooking time (Nout & Aidoo, 2011). Moreover, food-grade yeasts can be used as nutritional supplements, food additives, conditioners, flavoring agents, and livestock feeds because of their high nutritional value proteins, enzymes, and vitamins (Bekatorou et al., 2006).

A classic example of yeast being used directly as food comes from World War II when a massive campaign of production and supply of the yeast *Saccharomyces cerevisiae* was organized, first to feed the army, and then the whole population (Matassa, Boon, Pikaar, & Verstraete, 2016). The use of yeast to enrich the soldiers' ration, by providing a rich source of vitamin B complex, was responsible for transforming breweries into essential industries for the American government (Jacobson, 2009). *Hansenula*, *Pitchia*, *Torulopsis*, and *Saccharomyces* are the most common yeast genera used for the production of SCP (Suman, Nupur, Anuradha, & Pradeep, 2015). A genetically-engineered yeast is used for producing soy leghemoglobin, a plant heme protein. The heme proteins are present in beef and may be partially responsible for the distinctive meat-like flavor (van Loo, Caputo, & Lusk, 2020).

The macro filamentous fungi are those that produce fruiting bodies known as mushrooms (Wakai et al., 2017). Although they have long been part of the human diet for their pleasant taste, they are not an important source of protein (Boland et al., 2013). The most cultivated edible mushrooms are shiitake (*Lentinula*), oyster mushroom (*Pleurotus*), wood ear mushroom (*Auricularia*), and button mushroom (*Agaricus bisporus*). In a nutritional aspect, mushrooms are poor in fat and digestible carbohydrates, are rich in fiber and protein (ca. 2% fresh weight), have a better amino acid profile than vegetables (e.g., potatoes and carrots), have a well-balanced composition of minerals and vitamins, and are low in calories (27–30 kcal/100 g) (Grimm & Wösten, 2018). Recent studies have investigated the replacement of animal

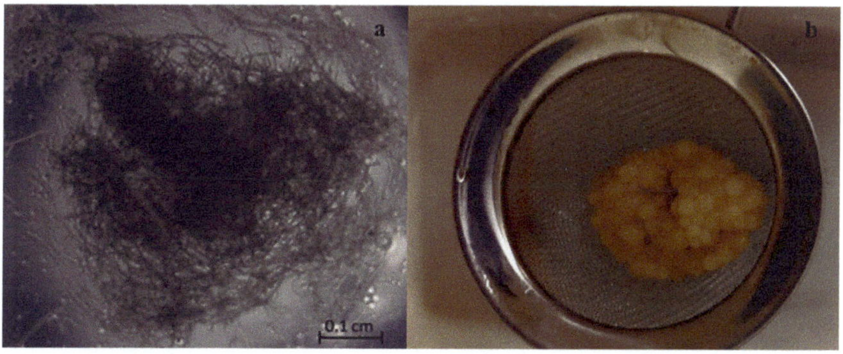

Fig. 1 (A) *Neurospora intermedia* mycelium observed in the microscope (Souza Filho, 2018); (B) *Rhizopus oryzae* pellets grown in submerged fermentation.

meat by *Pleurotus sapidus* and *Lentinula edodes* in the formulation of sausages (Ahlborn et al., 2019; Stephan, Ahlborn, Zajul, & Zorn, 2018; Wang et al., 2019).

Lastly, the multicellular filamentous fungi are characterized by their growth as septate hyphae, i.e., branching filaments (Fig. 1) that are divided into distinct sections, which get entangled to form the mycelium (Troiano, Orsat, & Dumont, 2020). Some species of filamentous fungi have been used in cuisine for centuries for the preparation of, e.g., tempeh (*Rhizopus* sp.), hamanatto, miso, shoyu (*Aspergillus oryzae*), oncom (*Neurospora intermedia*), and cheese (*Penicillium roqueforti* and *Penicillium camemberti*). Moreover, the RNA-depleted mycelium of the fungus *Fusarium venenatum* has been marketed since the 1980s as the meat substitute mycoprotein (Souza Filho, Andersson, Ferreira, & Taherzadeh, 2019; Whittaker, Johnson, Finnigan, Avery, & Dyer, 2020).

3. Production process

The technology used in the production of protein depends on the fungal morphology. While yeasts are preferably grown using the submerged fermentation technology, molds and mushrooms can be grown in solid-state fermentation systems, as the metabolism of filamentous fungi is less affected by the moisture content of the medium for a broader range of values (Sadh, Duhan, & Duhan, 2018).

The industrial production of yeast cells includes a propagation stage, followed by unitary operations of harvesting, concentration, and/or drying (Bekatorou et al., 2006). The scheme presented in Fig. 2 represents the most

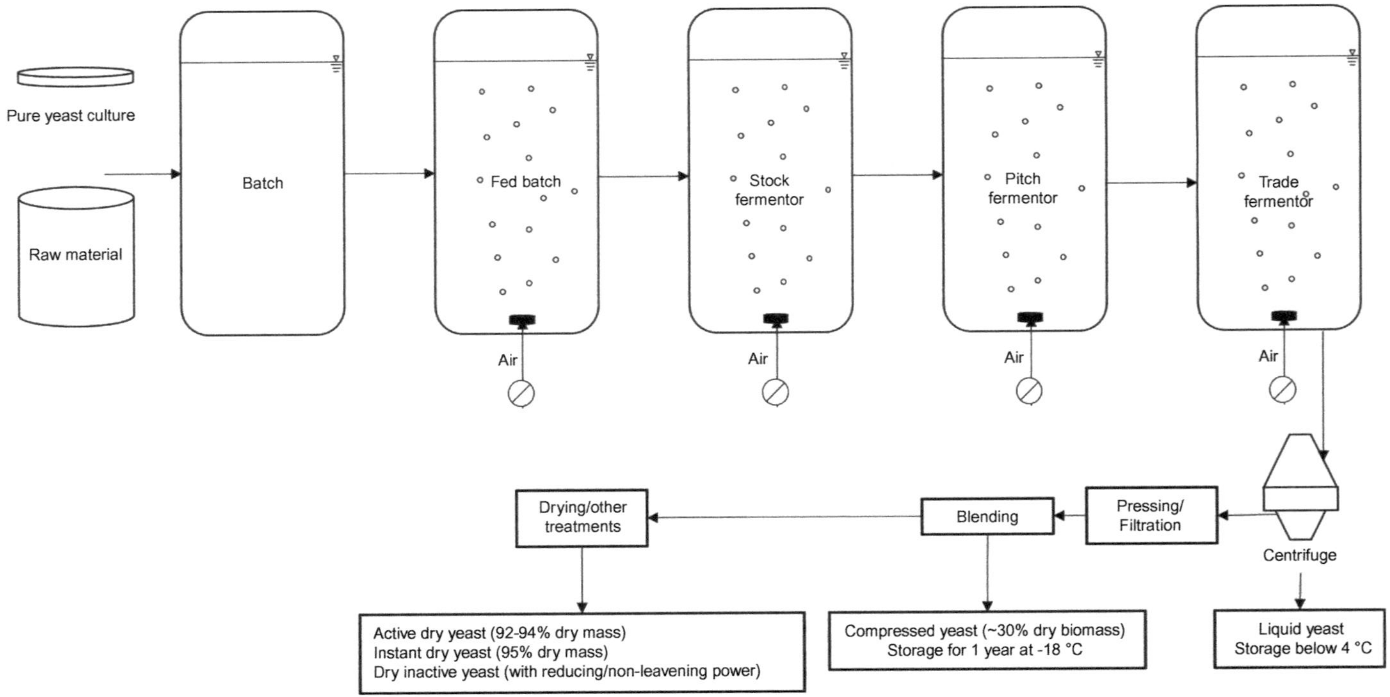

Fig. 2 Most important unitary operations in the production of baker's yeast. *Adapted from Bekatorou, A., Psarianos, C., & Koutinas, A. A. (2006). Production of Food grade yeast.* Food Technology and Biotechnology, *44(3), 407–415.*

important operations. Molasses, a byproduct of sugar manufacturing, are the most common media for yeast growth. Furthermore, other food-industry, agricultural or forestry byproducts can also be used, such as starch, distiller's wash, whey, fruit and vegetable wastes, wood, and straw (Bekatorou et al., 2006).

Production of yeast SCP from food waste has also been investigated (Gervasi et al., 2018). Currently, the spent brewer's *S. cerevisiae* are marketed as extracts e.g. Marmite® (Unilever and Sanitarium Health Food), Vegemite® (Bega Cheese Ltd.), Cenovis® (Gustav Gerig AG), and Vitam-R® (VITAM Hefe-Produkt GmbH) (Ritala, Häkkinen, Toivari, & Wiebe, 2017). *Candida utilis* inactive cells are commercialized using the name Torula. They contain more than 50% of proteins and can be used directly as a meat substitute or as an additive in processed foods, baby food, meat products, doughs, etc. (Bekatorou et al., 2006).

The production of mushrooms is carried out in solid-state. The substrate used depends on the type of mushroom: wood-inhabiting mushrooms (such as *Lentinula* and *Auricularia*) are cultivated in sawdust or other lignocellulosic materials, and soil-inhabiting mushrooms (e.g., *Agaricus* and *Macrolepiota*) are grown using composts. In the composting process, bacteria and fungi break down the raw materials, leaving available nutrients for the mushroom to use. Some of the main raw materials used include wheat straw, hay, corncobs, oat, barley, rice straw, and spent mushroom, i.e., the residue left after the cultivation of mushroom. Moreover, the material can be supplemented with manure, rice bran, urea, ammonium phosphate, and calcium sulfate. After composting is finished, the material is pasteurized and inoculated with the fungal spores. Wood-inhabiting mushrooms are capable of consuming the nutrients directly from the lignocellulosic materials. Therefore, the composting process is not necessary. Generally, the material is disposed of in polypropylene bags, inside which the fungi grow until the fruiting bodies start appearing (Hyde et al., 2019).

The filamentous fungus *Fusarium venenatum* has been investigated since the 1960s to develop a meat analog from fungal proteins (Stephan et al., 2018). In 1985, Marlow Foods (UK) launched the Quorn™ brand with meat substitute products (Ritala et al., 2017). Recently, several other companies have started to invest in this area, including Mycorena (using the fungus *A. oryzae*), Sustainable Bioproducts LLC (*Fusarium oxysporum*), MycoTechnology (*Lentinula edodes*) (Meyer et al., 2020), and 3FBIO (*F. venenatum*).

In the Quorn process, *F. venenatum* is grown in a defined medium containing glucose, salts of ammonium, potassium, magnesium, phosphate, and trace elements (e.g., biotin). The cultivation is carried out in external loop airlift bioreactors with a total volume of 150 m^3, at continuous mode, with the temperature kept at 28–30 °C, and the pH at 6. The control of the chemostat is performed by online monitoring the CO_2 evolution rate in the exit gas. At these conditions, the specific growth rate of the microorganism is 0.17–0.20 h^{-1}. The biomass concentration in the reactor is approximately 10 g/L, yielding a production of 300–350 kg biomass per hour. Every 6 h, tests for detection of mycotoxins are performed to guarantee harmful products are not present in the final mycoprotein (Denny, Aisbitt, & Lunn, 2008; Wiebe, 2002).

The harvested biomass is then submitted to a heat treatment to reduce its RNA content from 10% (*w*/w dry weight) to less than 2%. The fungal biomass (still in the broth) is rapidly heated to temperatures above 68 °C, kept for 30–45 min, further heated to 90 °C, and centrifuged. In this process, endogenous RNAses are activated and degrade the RNA molecules into monomers. Additionally, it stops the fungal growth, induces diffusion of a fraction of the cell components through the cell wall, including RNA nucleotides and proteinaceous material (Lonchamp, Clegg, & Euston, 2019). An excessive amount of RNA in the diet can increase the levels of uric acid in the body, representing a risk factor for gout (Denny et al., 2008; Jonas et al., 2001). According to the literature, the consumption of up to 2 g of RNA per day can be considered safe. DNA, on the other hand, is not considered so harmful (to cause a similar effect on the uric acid content, the intake of DNA has to be twice the consumption of RNA) and, therefore, its consumption is not controlled. Comparatively, beef liver contains 2% of RNA, beef heart, 0.6%, and muscles even less (Jonas et al., 2001).

After centrifugation, the solid material, containing 20% of solids or more and looking like pastry dough, is cooled to 4 °C. The fibers of the material are mechanically aligned into a network and albumen from free-range eggs is added to stabilize it (Wiebe, 2002). Other ingredients are also added in this step, such as firming agents (e.g., calcium chloride and calcium acetate), stabilizers (e.g., carrageenan and sodium alginate), gelling agents (e.g., pectin), natural colorants, and flavorings (Denny et al., 2008). Recently, vegetable proteins have been found to replace the egg albumen in the mycoprotein formulation, which can allow the production of vegan products (Derbyshire & Delange, 2021).

The material is finally cooked and shaped like sausages, burgers, fillets, steaks, etc. Lastly, the product is frozen to be commercialized (Wiebe, 2002). The freezing is an important operation because the ice crystals formed push the fibers together, creating bundles that give the final product a more similar meat texture (Quorn, 2022).

4. Regulatory oversight

The laws and regulations that control production, marketing, and other activities related to food play an important role in protecting human health and life (Ismail, Hwang, & Joo, 2020). Therefore, alternative protein sources must be approved by local regulatory agencies before entering the market.

In 1985, the Ministry of Agriculture Fisheries and Food of the British Government finished the review of the Quorn safety data, describing from in vitro to human chronic studies, and approved the unrestricted sale of the Quorn products in the United Kingdom (Finnigan, 2011). During the 1990s, Quorn started distributing its products in other countries of the European Union. Currently, *F. venenatum* and *S. cerevisiae* are the only fungi with EU approval to be marketed as food. *Yarrowia lipolytica* and *Clostridium butyricum* can be used as a food supplement (Lähteenmäki-Uutela, Rahikainen, Lonkila, & Yang, 2021).

From 2002, the Food and Drug Administration (FDA) granted mycoprotein the Generally Regarded as Safe (GRAS) status, which allowed the introduction of the product in the American market in the same year (Jacobson & DePorter, 2018). The Food Standards Australia New Zealand (a statutory authority in the Australian Government) has recognized no safety concerns regarding *F. venenatum* mycoproteins. They, however, highlighted the occurrence of adverse events (gastrointestinal disturbances and allergies) at a very rare frequency (Food Standards Australia New Zealand, 2020).

5. The fungal protein market

According to the Transparency Market Research, the global market value of meat substitutes in 2021 was estimated to be USD 168 billion. Based on the increasing popularity of vegetarian and vegan diets, as well as the diets that avoid animal products, this market is expected to grow at a compound annual growth rate (CAGR) of 10% in the next 10 years, crossing the USD 434 billion by 2031. Particularly, the mycoprotein segment is projected to expand at an 8.9% CAGR (Transparency Market Research, n.d.). Richter (2019) has heard analysts who have anticipated a CAGR above

10% for the meat alternative sales up to 2029 USD, reaching a turnover of USD 140 billion in 2029. Among the meat substitutes, single-cell protein represents 15.8% of the market, a fraction that is predicted to be maintained by 2025 (Joseph, Searing, Watson, & McKeague, 2020).

The reduction of the cost involved in the production is one of the challenges to encourage the consumption of mycoprotein. The total cost of mycoproteins depends on the substrates, i.e., the components used by the fungus to grow (glucose and salts of ammonium, potassium, magnesium, phosphate). This cost can be reduced by the use of agro-industrial byproducts, such as molasses, starches, and fruit and vegetable wastes (Hashempour-Baltork, Khosravi-Darani, Hosseini, Farshi, & Reihani, 2020). For example, the investigation of a biorefinery using food waste for the production of SCP of *Kluyveromyces marxianus* has pointed to the substrate as responsible for 53% of the estimated total cost (Aggelopoulos et al., 2014).

6. Public acceptance

Bonny, Gardner, Pethick, and Hocquette (2015) have discussed how consumers are keener to purchase a new product when it is similar to another product they are already familiarized with. Thus, it is favorable for an alternative protein product to resemble meat in appearance nutrition, convenience, and meal solutions. A study involving citizens from the United Kingdom and the Netherlands has indicated that sensorial factors (texture, color, taste, smell, and appearance) have a strong effect on the acceptance of meat substitutes by people who are not regular consumers (Hoek et al., 2011). Other studies, summarized by Joseph et al. (2020) have identified that health concerns have driven around half of the consumers who substituted meat with protein alternatives in their diets. More than one third (36%) of those interviewees also cited the meat substitutes as good sources of protein, and approximately a quarter of the respondents have changed their habits due to environmental (27%) and animal welfare concerns (24%).

Furthermore, a study involving older adults (65 years old or above) from five EU countries has identified SCP as a more acceptable protein alternative than insects and meat grown in vitro (20% of acceptance vs 9% and 6%, respectively). However, the values are much lower compared to the acceptance of plant-based protein products (58%). Yet, 11% of the respondents had no opinion about their SCP acceptance, this being the highest value among the presented alternatives together with in vitro meat, evidencing unfamiliarity of this population with the product (Grasso, Hung, Olthof, Verbeke, & Brouwer, 2019).

7. Nutritional aspects

Mycoprotein is naturally rich in dietary fibers and proteins, containing all nine essential amino acids (Cherta-Murillo et al., 2020). According to the European Commission, the following claims can be made to mycoprotein: "high in protein" (because at least 20% of the energy value of the food is provided by protein), "low in fat" (it contains <3 g of fat per 100 g of solids), "low in saturated fat" (it contains <1.5 g of saturated fatty acids per 100 g of solids), and "high in fiber" (it contains at least 6 g of fiber per 100 g of solids) (Derbyshire & Ayoob, 2019) (Table 1).

Regarding the United States and Canada regulations, to claim that a food is "high" in some nutrient, a determined amount of this food (the Reference Amounts Customarily Consumed—RACC) should provide at least 20% of the Daily Value (DV) for that nutrient (Derbyshire & Ayoob, 2019). As for 2018, the RACC for meat substitutes corresponds to 85 g of the cooked material or 110 g of the raw material (U.S. Food & Drug Administration, 2018). Therefore, mycoproteins can be considered a food high in protein and fibers (see Table 2).

Besides its low content of fat, the fatty acid profile of mycoproteins is considered to be favorable, with a low content of saturates (Table 3). Moreover, the fungal protein contains all essential amino acids, is low in sodium (4 mg/100 g), and is a good source of selenium (20 μg/100 g) and zinc (7.6–9.0 mg/100 g) (Denny et al., 2008). The fiber profile is formed by 1/3 of chitin and 2/3 of β-glucans (Cherta-Murillo et al., 2020). This structure creates a "fibrous chitin–glucan matrix" not commonly present in the human diet and with low water solubility (88% insoluble) (Bottin et al., 2016). On the other hand, the fungal biomass is poor in vitamin B12; the iron content is low and is less bioavailable than red meat as it is non-heme (Denny et al., 2008). Nevertheless, the addition of iron or other substances on the mycoprotein formulation to increase its content or bioavailability has been suggested (Souza Filho et al., 2019).

Choline is another micronutrient present in low amounts (180 μg/100 g) compared to animal proteins (3.57 mg/100 g of beef, 5.67 mg/100 g of chicken, and 8.62 mg/100 g of salmon) and other meat alternatives (2.16 mg/100 g of peas, 9.72 mg/100 g of tofu, and 47.27 mg/100 g of soybeans) (Zeisel, Mar, Howe, & Holden, 2003).

Table 1 Nutritional content of fungal protein sources.

	Mycoprotein (raw, wet)			Shitake (cooked)	Quorn Mince (88% mycoprotein)
	(Cherta-Murillo et al., 2020; Ruxton & McMillan, 2010)	(Denny et al., 2008; Finnigan, 2011)	(Derbyshire & Delange, 2021)	(Derbyshire & Delange, 2021)	(Ruxton & McMillan, 2010)
Energy (kJ)	356	360	356	230	393
Carbohydrates (g)	9	1.7	3	12.3	4.5
Sugars (g)	*0*	*0.8*			*0.6*
Fat (g)	3	2.9	2.9	0.2	2
Saturates (g)	*0.7*	*0.6*	*0.7*	*0.1*	*0.5*
Monounsaturates (g)		*0.5*			
Polyunsaturates (g)		*1.8*			
Protein (g)	11	11.5	11	1.6	14.5
Fibers (g)	6	6	6		5.3
Vitamin B6 (mg)			0.1		0.12
Vitamin B9 (μg)			114		
Vitamin B12 (μg)		0	0.72	0	

Vitamin D (μg)	1.6				1.4
Sodium (mg)		4			
Calcium (mg)			48	3	
Phosphorus (mg)			290	29	
Iron (mg)		0.5	0.39	0.4	
Magnesium (mg)			49	14	
Zinc (mg)		9.0	7.6		0.78
Potassium (mg)			71	120	
Choline (μg)			180		
Selenium (μg)	20	20			17.4

Table 2 Quorn product nutritional profile compared to the Daily Value intake recommended by the FDA.

	Quorn mince (110g) (Ruxton & McMillan, 2010)	Daily value (U.S. Food & Drug Administration, n.d.)	%DV
Carbohydrates (g)	4.95	275	1.8
Total Fat (g)	2.2	78	2.8
Saturated Fat(g)	0.55	20	2.75
Protein (g)	15.95	50	31.9
Fibers (g)	5.83	28	20.8

Table 3 Amino acid and fatty acid profiles of mycoprotein (Finnigan, Needham, & Abbott, 2017).

Amino acid	g/100g Protein	Fatty acid	g/100g mycoprotein (dry)
Lysine	**8.3**	Palmitic (16:0)	1.3
Methionine	**2.1**	Stearic (18:0)	0.2
Threonine	**5.5**	Oleic (18:1)	1.4
Tryptophan	**1.6**	Linoleic (18:2)	4.3
Valine	**6.2**	Linolenic (18:3)	0.9
Leucine	**8.6**		
Isoleucine	**5.2**		
Phenylalanine	**4.9**		
Histidine	**3.5**		
Cysteine	0.8		
Arginine	7.3		
Tyrosine	4.0		
Aspartic Acid	10.3		
Serine	5.1		
Glutamic Acid	12.5		
Proline	4.5		
Glycine	4.5		
Alanine	6.0		

8. Health impact

The interest in the effects of mycoprotein consumption on human health has been the subject of several investigations. The main studied responses include satiety and energy intake, insulin release and glycemic profile, blood levels of cholesterol and lipids, and muscle mass.

The satiating properties of mycoproteins were first investigated by Turnbull, Walton, and Leeds (1993), who observed the decrease in the desire to eat 3 h after the ingestion of a mycoprotein meal, compared to a chicken meal. The effect was attributed to the fiber content of mycoprotein. The ingestion of fibers has been claimed to affect both the satiation (i.e., the satisfaction of appetite during feeding that marks the end of eating) and the satiety (inhibition of hunger as a result of having eaten). The satiation is caused by gastric distention and the work and time required for chewing. The effect on satiety is associated with the viscous nature of the soluble fibers, which prolong the intestinal phase of nutrient digestion and absorption (Slavin & Green, 2007). Cherta-Murillo et al. (2020) have further argued that the ingestion of fiber induces increased gastric distension, decreased gastric emptying, and hormone release, resulting in satiety.

Conversely, Burley, Paul, and Blundell (1993) have observed no difference in food consumption after a mycoprotein or chicken meal, although the energy intake after the mycoprotein consumption was decreased by 18%. Another study (Williamson et al., 2006) has also compared the effect of mycoprotein and chicken consumption on satiety of female individuals aged between 18 and 50 years and with body mass index ranging from 25 to 29.9 kg/m^3. Williamson et al. (2006) have observed that the consumption of mycoprotein led to a reduced food intake 20 min after (at lunchtime) that was not compensated by consuming more food at dinner. Bottin et al. (2016) have not found significant differences in hunger, desire to eat, prospective food intake, and gastric emptying between mycoprotein and chicken, although they have measured a 10% decrease in the energy intake after the mycoprotein meal.

Recently, researchers have compared mycoprotein with an animal protein (source not informed) on the appetite of overweight and obese individuals who have finished an 8-week weight loss process. Again, mycoprotein intake neither has shown any effect on acute or sustained reductions in appetite nor has improved weight loss maintenance (Andersen et al., 2020).

The consumption of mycoprotein has been investigated regarding its effects on glucose and insulin levels among non-diabetic individuals. A comparison between mycoprotein and soya flour (prepared in such a way that the only difference on the macronutrient level was the dietary fiber content) has indicated that, after 1 h, the glycemia and insulinemia were 13% and 36% lower, respectively (Turnbull & Ward, 1995). Compared with the same amount of protein intake, mycoprotein and milk have resulted in the same insulin response, both with a peak 15 min after the ingestion. However, the mycoprotein response lasted longer (45 min) than milk (30 min) (Dunlop et al., 2017). A similar result was observed by Monteyne et al. (2020), with the insulin peak being observed 15 and 30 min after the ingestion of milk or mycoprotein, respectively, and the return to the fasting level after 45 and 60 min, respectively. On the other hand, the consumption of chicken or mycoprotein yielded no difference in the glucose profile during the 3 h after the meal, while the insulin peak was significantly lower when the fungal protein was the choice of food (Bottin et al., 2016).

Udall, Lo, Young, and Scrimshaw (1984) have reported the decrease in the cholesterol level of subjects consuming mycoproteins for 30 days. In a study with 17 voluntaries, it was observed that a diet containing mycoproteins reduced the total cholesterol, low-density lipoprotein (LDL), and triglycerides by 13%, 9%, and 53%, respectively, and increased the high-density lipoprotein (HDL) by 11% after 3 weeks, opposed to the meat-containing diet, which increased LDL and triglycerides (12% and 11%, respectively), reduced the HDL (11%) and maintained the total cholesterol level (Turnbull, Leeds, & Edwards, 1990). The same authors also compared the effect of mycoprotein and soy in the lipid profile for a longer time (8 weeks) and observed a higher decrease in the total cholesterol and LDL levels of the individuals consuming mycoprotein, while no differences were observed for the HDL and triglyceride (Turnbull, Leeds, & Edwards, 1992). Ruxton and McMillan (2010) noticed the mycoprotein effect on the cholesterol levels is only significant in a group with a high initial total cholesterol level. A more complete analysis, considering 45 lipid concentrations on blood, was performed by Coelho et al. (2021). All the lipid fractions were significantly reduced in individuals consuming a mycoprotein-containing diet when compared to a diet including animal protein (meat and fish).

The study in vivo using rats also proved the efficiency of mycoprotein in reducing the lipid levels. The rats had a hyperlipidemia condition induced and then had mycoprotein added to their diets. Feeding 200 mg/kg/day, the rats had a reduction of 58%, 75%, 24%, and 75% on their levels of total cholesterol, triglycerides, LDL, and very-low-density lipids (VLDL), respectively. Comparably, 20 mg/kg/day of simvastatin decreased the total cholesterol and triglycerides level by 68% and 84%, respectively (Thomas et al., 2017).

The European Food Safety Authority (EFSA) has analyzed the health claims associated with the consumption of mycoprotein and has concluded that the reduction in LDL levels in the blood is caused only by the presence of β-glucans. The data relating the consumption of mycoprotein with an increase in satiety and, consequently, a reduction in the energy intake were considered inconclusive (European Food Safety Authority, 2011).

Moreover, some studies suggest that mycoprotein stimulates muscle protein synthesis. Comparing milk and mycoprotein by the same amount of the amino acid leucine, the fungal product stimulates resting and postexercise muscle protein synthesis at higher rates (Monteyne et al., 2020). Ingestion of mycoprotein has also been demonstrated to increase the amino acid levels in the blood, with a peak between 45 and 120 min, and maintaining the levels above those of fasting for up to 240 min. Comparatively, the consumption of milk yielded an amino acid level peak 30 min after ingestion and the levels returned to the fasting ones after 45 min (Dunlop et al., 2017). Additionally, a dose of 60–80 g of mycoprotein has given the optimal anabolic response in this study. It was also observed the excretion of guanidinoacetic acid (GAA) in urine was superior when consuming mycoprotein compared to chicken. GAA is a precursor for creatine and, therefore, is an essential substrate for muscle energy metabolism, besides being possibly involved in appetite regulation (Bottin et al., 2016).

The published data about the health benefits of a diet including mycoprotein points toward its power to fight hunger, reduce energy intake, and facilitate weight loss. However, long-term studies, with large sample sizes, are yet to be performed before one can fully elucidate how and to what extent mycoprotein can contribute to improving human health (Souza Filho et al., 2019).

Another benefit the consumption of meat alternatives can bring to public health is the reduction of the pathogens and emerging diseases associated with livestock farming, such as the avian and swine flu. The control of these diseases involves the costly vaccination of all animals and the slaughtering of those exposed to the pathogens. Alternatively, protein sources that reduce the human-animal interaction (e.g., fungi) have the potential to reduce the risk of epidemic zoonosis and emerging diseases (Bonny et al., 2015). Lastly, according to the Centers for Disease Control and Prevention (CDC, USA), the ingestion of animal-derived food (meat, poultry, egg, dairy, fish, and shellfish) has caused 48.1% of the foodborne acquired illness and 50.4% of the deaths between 1998 and 2008 (Centers for Disease Control and Prevention, 2018).

9. Safety concerns

The main problems related to fungal SCP are twofold: the slow growth compared to bacteria, which increase the risk of contamination; and the production of mycotoxins. These toxins, which are produced by certain fungal species, such as *Aspergillus parasiticus* and *Aspergillus flavus*, may cause many allergic reactions, diseases, and liver cancer. Therefore, a rigorous screening before the selection of a strain is of utmost importance, besides maintaining the control of the process conditions and the product formulation (Adedayo, Ajiboye, Akintunde, & Odaibo, 2011; Anupama, & Ravindra, P., 2000; Hashempour-Baltork, Khosravi-Darani, et al., 2020; Ritala et al., 2017).

For example, although some strains of *F. venenatum* are known to produce some mycotoxins, the strain used for the industrial production of mycoprotein, at the conditions of excess nutrients in the medium, do not produce mycotoxins (Whittaker et al., 2020). The yeast *Y. lipolytica* has also had its safety thoroughly studied, and its use in food applications has been demonstrated to be safe (Ritala et al., 2017).

10. Ethics

Public criticism of animal farming is on the rise, including the environmental, animal welfare, and public health aspects (van Loo et al., 2020). In this regard, mycoproteins contribute less to the environmental impact than animal sources and can be included in restricted diets, such as vegetarian, vegan (provided egg albumen is not used in formulation), and halal.

Another point of interest involves the use of genetically modified organisms (GMO) in SCP production. The European market does not accept well the use of GMOs, although in other markets there is more acceptance. It is possible that the scarcity of traditional protein sources coupled with the accumulation of GMO consumption data may increase their acceptance (Ritala et al., 2017).

11. Social aspects

The social impacts caused by the expansion of the consumption of mycoprotein have not been extensively discussed yet. However, it is expected that the whole meat alternative market will alter more profoundly the rural communities. A set of interviews with rural producers in the United States has identified some concerns in terms of possible barriers to transitioning into the new market model, such as (i) crop producer farmers who presently sell their products to feed the animals and would need to adapt to a new production system; or (ii) animal farmers who would need to have access to a line of credit to adapt their business. On the other hand, the rural communities could enjoy the creation of new job opportunities in the alternative protein production facilities, the reduction of the problems related to farming pollution (e.g., pollution of air and water by manure), and the destination of part of the land to environmental preservation such as carbon sequestration or biodiversity conservation (Newton & Blaustein-Rejto, 2021).

12. Environmental aspects

The contribution of the food sector to the emission of anthropogenic greenhouse gases is estimated to be between 19% and 29%. Therefore, solutions to mitigate this would include changes in the way we produce and consume food (Clune, 2019). In this context, besides considering the economical, nutritional, and safety aspects, meat-alternative single-cell proteins must cause a lower environmental impact than animal protein. The evaluation of the environmental impact caused by mycoprotein has been analyzed through life cycle assessment.

The production of 1 kg of mycoprotein has a similar environmental impact to 1 kg of chicken meat. Individually, 45% of the overall impact comes from the mycoprotein processing, 25% is the result of the preparation at the final consumer (cooking), and 21% is connected to the raw materials used in its production (Smetana, Mathys, Knoch, & Heinz, 2015). This study has identified 1 kg of mycoprotein and 1 kg of chicken with a similar impact in endpoint damage assessment categories of human health, resources availability, and ecosystem quality. When the comparison takes into account the calorific energy value or the content of digestible proteins, the results for mycoprotein are all worse than chicken and dairy-, insect-, gluten- and soy-based options. Mycoprotein can be considered an efficient meat alternative in the categories of land (<2 m^2.year/kg) and water use (~500 L/kg) and performs similarly to the dairy product in energy consumption (15–20 kWh/kg) (Smetana, Aganovic, Irmscher, & Heinz, 2018).

Some fungal species have the ecological role of decomposing dead organic matter, being important recyclers of organic material (Souza Filho, 2018). The production of mushrooms has taken advantage of this ability by using agricultural residues (e.g., lignocellulosic material and cow's manure) as substrate (Fig. 3). Similarly, several agro-industrial wastes have been studied in the production of mycoprotein, including rice straw (Upcraft et al., 2021), potato processing wastewater (Spiller et al., 2020), date waste (Hashempour-Baltork, Hosseini, Assarehzadegan, Khosravi-Darani, & Hosseini, 2020), brewer-spent grain (Stoffel et al., 2019), bread waste (Gmoser et al., 2020; Hellwig, Gmoser, Lundin, Taherzadeh, & Rousta, 2020; Svensson et al., 2021), and pea-processing byproduct (Souza Filho, Nair, Andersson, Lennartsson, &

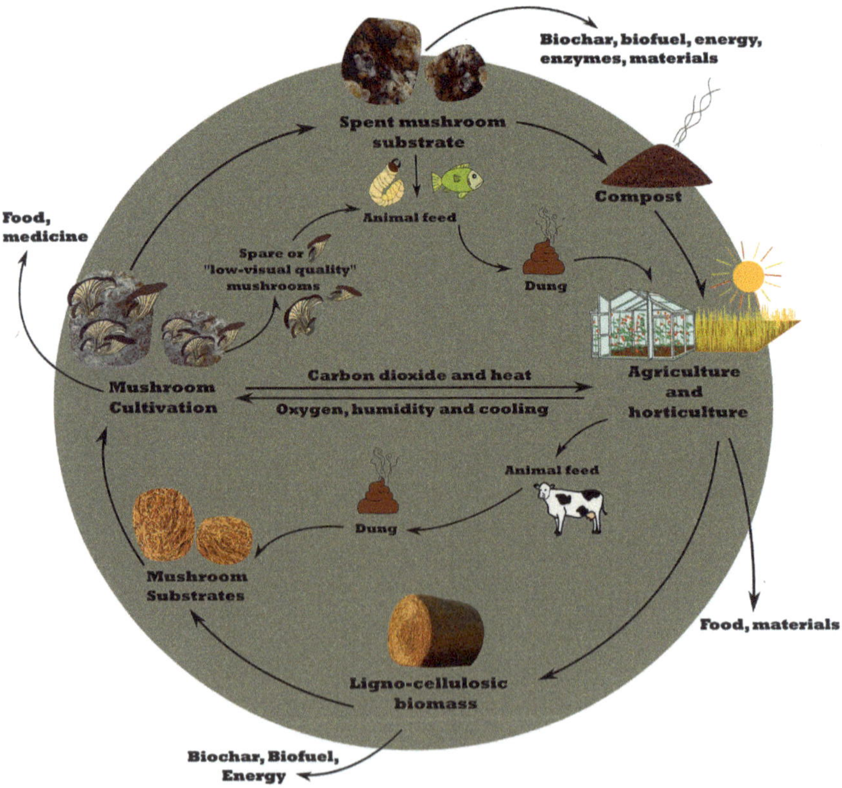

Fig. 3 Use of residues in the production of mushrooms (Grimm & Wösten, 2018).

Taherzadeh, 2018). The use of these residues can potentially reduce the environmental impact of mycoprotein production.

The use of rice straw as the substrate for mycoprotein production was evaluated using the LCA methodology and indicated significant advantages in the impact categories of global warming, terrestrial acidification, freshwater eutrophication, and arable land use in comparison to beef. Conversely, the water consumption category for mycoprotein was higher because of the high water demand of the rice culture (Upcraft et al., 2021). Lastly, the potential of GHG emission was determined to be dependent on the price of mycoprotein and the public acceptance, i.e., if the public acceptance increases and the price decreases, the GHG footprint savings would be even more pronounced (Ritchie, Reay, & Higgins, 2018).

13. Future trends

Current research on the use of fungal protein in the human diet includes the search for new species or strains that can efficiently convert the substrates into proteins (Whittaker et al., 2020). In this context, the use of GMOs may also be an alternative. Genetic modifications can increase the nutraceutical value of the biomass, e.g., by optimizing the amino acid composition or increasing the content of specific vitamins, fatty acids, etc. (Ritala et al., 2017).

As it has been discussed, new substrates (especially agro-industrial wastes) are also on the focus. They have the potential to reduce both the environmental impact caused by mycoprotein production and the costs of raw material. Lastly, new applications of mycoprotein on the formulations of varied dishes may improve its consumption, giving the consumer more options according to their needs.

References

Adedayo, M. R., Ajiboye, E. A., Akintunde, J. K., & Odaibo, A. (2011). Single cell proteins: As nutritional enhancer. *Advances in Applied Science Research*, *2*(5), 396–409.

Aggelopoulos, T., Katsieris, K., Bekatorou, A., Pandey, A., Banat, I. M., & Koutinas, A. A. (2014). Solid state fermentation of food waste mixtures for single cell protein, aroma volatiles and fat production. *Food Chemistry*, *145*, 710–716. https://doi.org/https://doi.org/10.1016/j.foodchem.2013.07.105.

Ahlborn, J., Stephan, A., Meckel, T., Maheshwari, G., Rühl, M., & Zorn, H. (2019). Upcycling of food industry side streams by basidiomycetes for production of a vegan protein source. *International Journal of Recycling of Organic Waste in Agriculture*, *8*(1), 447–455. https://doi.org/10.1007/s40093-019-00317-4.

Andersen, S. V., Sandby, K., Hjorth, M. F., Korndal, S. K., Ritz, C., Sjödin, A., et al. (2020). No effects on appetite or body weight in weight-reduced individuals of foods containing components previously shown to reduce appetite - results from the SATIN (satiety innovation) study. *Obesity Medicine, 17*, 100188. https://doi.org/10.1016/j.obmed.2020.100188.

Anupama, & Ravindra, P. (2000). Value-added food: Single cell protein. *Biotechnology Advances, 18*(6), 459–479. https://doi.org/10.1016/S0734-9750(00)00045-8.

Bekatorou, A., Psarianos, C., & Koutinas, A. A. (2006). Production of Food grade yeast. *Food Technology and Biotechnology, 44*(3), 407–415.

Boland, M. J., Rae, A. N., Vereijken, J. M., Meuwissen, M. P. M., Fischer, A. R. H., van Boekel, M. A. J. S., et al. (2013). The future supply of animal-derived protein for human consumption. *Trends in Food Science & Technology, 29*(1), 62–73. https://doi.org/https://doi.org/10.1016/j.tifs.2012.07.002.

Bonny, S. P. F., Gardner, G. E., Pethick, D. W., & Hocquette, J.-F. (2015). What is artificial meat and what does it mean for the future of the meat industry? *Journal of Integrative Agriculture, 14*(2), 255–263. https://doi.org/https://doi.org/10.1016/S2095-3119(14)60888-1.

Bottin, J. H., Swann, J. R., Cropp, E., Chambers, E. S., Ford, H. E., Ghatei, M. A., et al. (2016). Mycoprotein reduces energy intake and postprandial insulin release without altering glucagon-like peptide-1 and peptide tyrosine-tyrosine concentrations in healthy overweight and obese adults: A randomised-controlled trial. *British Journal of Nutrition, 116*(2), 360–374. https://doi.org/DOI: 10.1017/S0007114516001872.

British Dietetic Association. (2017). *The management of malnourished adults in all community and all health and care settings*. British Dietetic Association. (https://www.bda.uk.com/uploads/assets/fd662e80-a9a2-4882-857e73faa78c589c/managementmalnutritionadultspolicystatement2017.pdf).

Burley, V. J., Paul, A. W., & Blundell, J. E. (1993). Influence of a high-fibre food (myco-protein*) on appetite: Effects on satiation (within meals) and satiety (following meals). *European Journal of Clynical Nutrition, 47*, 409–418.

Centers for Disease Control and Prevention. (2018). *Attribution of foodborne illness: Findings*. (Https://Www.Cdc.Gov/Foodborneburden/Attribution/Attribution-1998-2008.Html).

Cherta-Murillo, A., Lett, A. M., Frampton, J., Chambers, E. S., Finnigan, T. J. A., & Frost, G. S. (2020). Effects of mycoprotein on glycaemic control and energy intake in humans: A systematic review. *British Journal of Nutrition, 123*(12), 1321–1332. https://doi.org/10.1017/S0007114520000756.

Clune, S. (2019). Calculating GHG impacts of meals and menus using streamlined LCA data. In *Environmental Nutrition* (pp. 157–178). Elsevier. https://doi.org/10.1016/B978-0-12-811660-9.00010-2.

Coelho, M. O. C., Monteyne, A. J., Dirks, M. L., Finnigan, T. J. A., Stephens, F. B., & Wall, B. T. (2021). Daily mycoprotein consumption for 1 week does not affect insulin sensitivity or glycaemic control but modulates the plasma lipidome in healthy adults: A randomised controlled trial. *British Journal of Nutrition, 125*(2), 147–160. https://doi.org/10.1017/S0007114520002524.

Denny, A., Aisbitt, B., & Lunn, J. (2008). Mycoprotein and health. *Nutrition Bulletin, 33*(4), 298–310. https://doi.org/10.1111/j.1467-3010.2008.00730.x.

Derbyshire, E., & Ayoob, K.-T. (2019). Mycoprotein: Nutritional and health properties. *Nutrition Today, 54*(1), 7–15. https://doi.org/10.1097/NT.0000000000000316.

Derbyshire, E. J., & Delange, J. (2021). Fungal protein – What is it and what is the health evidence? A systematic review focusing on mycoprotein. *Frontiers in Sustainable Food Systems, 5*. https://doi.org/10.3389/fsufs.2021.581682.

dos Anjos Coelho Kubo, S. E., da Costa, T. H. M., & Gubert, M. B. (2020). Intakes of energy, macronutrients and micronutrients of a population in severe food insecurity risk in Brazil. *Public Health Nutrition, 23*(4). https://doi.org/10.1017/S1368980019003057.

Dunlop, M. V., Kilroe, S. P., Bowtell, J. L., Finnigan, T. J. A., Salmon, D. L., & Wall, B. T. (2017). Mycoprotein represents a bioavailable and insulinotropic non-animal-derived dietary protein source: A dose–response study. *British Journal of Nutrition, 118*(9), 673–685. https://doi.org/10.1017/S0007114517002409.

European Food Safety Authority. (2011). Scientific opinion on the substantiation of health claims related to mycoprotein and maintenance of normal blood LDL-cholesterol concentrations (ID 1619) and increase in satiety leading to a reduction in energy intake (ID 1620) pursuant to article 13(1) of regulation (EC) no 1924/2006. *EFSA Journal, 9*. https://efsa.onlinelibrary.wiley.com/doi/epdf/10.2903/j.efsa.2011.2042.

Finnigan, T. J. A. (2011). Mycoprotein: Origins, production and properties. In *Handbook of Food proteins* (pp. 335–352). Elsevier. https://doi.org/10.1533/9780857093639.335.

Finnigan, T., Needham, L., & Abbott, C. (2017). Mycoprotein: A healthy new protein with a low environmental impact. In *Sustainable protein sources* (pp. 305–325). Elsevier. https://doi.org/10.1016/B978-0-12-802778-3.00019-6.

Food Standards Australia New Zealand. (2020). *Novel food - record of views formed in response to inquiries.* (https://www.foodstandards.gov.au/industry/novel/novelrecs/Documents/record-of-views-september-2020.pdf).

Gervasi, T., Pellizzeri, V., Calabrese, G., di Bella, G., Cicero, N., & Dugo, G. (2018). Production of single cell protein (SCP) from food and agricultural waste by using Saccharomyces cerevisiae. *Natural Product Research, 32*(6), 648–653. https://doi.org/10.1080/14786419.2017.1332617.

Gmoser, R., Fristedt, R., Larsson, K., Undeland, I., Taherzadeh, M. J., & Lennartsson, P. R. (2020). From stale bread and brewers spent grain to a new food source using edible filamentous fungi. *Bioengineered, 11*(1), 582–598. https://doi.org/10.1080/21655979.2020.1768694.

Grasso, A. C., Hung, Y., Olthof, M. R., Verbeke, W., & Brouwer, I. A. (2019). Older Consumers' readiness to accept alternative, more sustainable protein sources in the European Union. *Nutrients, 11*(8), 1904. https://doi.org/10.3390/nu11081904.

Grimm, D., & Wösten, H. A. B. (2018). Mushroom cultivation in the circular economy. *Applied Microbiology and Biotechnology, 102*(18), 7795–7803. https://doi.org/10.1007/s00253-018-9226-8.

Gundersen, C., & Ziliak, J. P. (2015). Food insecurity and health outcomes. *Health Affairs, 34*(11). https://doi.org/10.1377/hlthaff.2015.0645.

Hashempour-Baltork, F., Hosseini, S. M., Assarehzadegan, M., Khosravi-Darani, K., & Hosseini, H. (2020). Safety assays and nutritional values of mycoprotein produced by fusarium venenatum IR372C from date waste as substrate. *Journal of the Science of Food and Agriculture, 100*(12), 4433–4441. https://doi.org/10.1002/jsfa.10483.

Hashempour-Baltork, F., Khosravi-Darani, K., Hosseini, H., Farshi, P., & Reihani, S. F. S. (2020). Mycoproteins as safe meat substitutes. *Journal of Cleaner Production, 253*, Elsevier Ltd. https://doi.org/10.1016/j.jclepro.2020.119958.

Hellwig, C., Gmoser, R., Lundin, M., Taherzadeh, M. J., & Rousta, K. (2020). Fungi burger from stale bread? A case study on perceptions of a novel protein-rich Food product made from an edible fungus. *Food, 9*(8), 1112. https://doi.org/10.3390/foods9081112.

Hoek, A. C., Luning, P. A., Weijzen, P., Engels, W., Kok, F. J., & de Graaf, C. (2011). Replacement of meat by meat substitutes. A survey on person- and product-related factors in consumer acceptance. *Appetite, 56*(3), 662–673. https://doi.org/10.1016/j.appet.2011.02.001.

Huizar, M. I., Arena, R., & Laddu, D. R. (2021). The global food syndemic: The impact of food insecurity, malnutrition and obesity on the healthspan amid the COVID-19 pandemic. *Progress in Cardiovascular Diseases, 64*. https://doi.org/10.1016/j.pcad.2020.07.002.

Hyde, K. D., Xu, J., Rapior, S., Jeewon, R., Lumyong, S., Niego, A. G. T., et al. (2019). The amazing potential of fungi: 50 ways we can exploit fungi industrially. *Fungal Diversity*, *97*(1), 1–136. https://doi.org/10.1007/s13225-019-00430-9.

Ismail, I., Hwang, Y.-H., & Joo, S.-T. (2020). Meat analog as future food: A review. *Journal of Animal Science and Technology*, *62*(2), 111–120. https://doi.org/10.5187/jast.2020.62.2.111.

Jacobson, L. (2009). Beer goes to war. *Food, Culture & Society*, *12*(3), 275–312. https://doi.org/10.2752/175174409X431996.

Jacobson, M. F., & DePorter, J. (2018). Self-reported adverse reactions associated with mycoprotein (Quorn-brand) containing foods. *Annals of Allergy, Asthma & Immunology*, *120*(6), 626–630. https://doi.org/10.1016/j.anai.2018.03.020.

Jayasiri, S. C., Hyde, K. D., Ariyawansa, H. A., Bhat, J., Buyck, B., Cai, L., et al. (2015). The faces of fungi database: Fungal names linked with morphology, phylogeny and human impacts. *Fungal Diversity*, *74*(1), 3–18. https://doi.org/10.1007/s13225-015-0351-8.

Jonas, D. A., Elmadfa, I., Engel, K.-H., Heller, K. J., Kozianowski, G., König, A., et al. (2001). Safety considerations of DNA in Food. *Annals of Nutrition and Metabolism*, *45*(6), 235–254. https://doi.org/10.1159/000046734.

Joseph, P., Searing, A., Watson, C., & McKeague, J. (2020). Alternative proteins: Market research on consumer trends and emerging landscape. *Meat and Muscle Biology*, *4*(2). https://doi.org/10.22175/mmb.11225.

Lähteenmäki-Uutela, A., Rahikainen, M., Lonkila, A., & Yang, B. (2021). Alternative proteins and EU food law. *Food Control*, *130*, 108336. https://doi.org/10.1016/j.foodcont.2021.108336.

Lonchamp, J., Clegg, P. S., & Euston, S. R. (2019). Foaming, emulsifying and rheological properties of extracts from a co-product of the Quorn fermentation process. *European Food Research and Technology*, *245*(9), 1825–1839. https://doi.org/10.1007/s00217-019-03287-z.

Lu, S., Zheng, F., Wen, L., He, Y., Wang, D., Wu, M., et al. (2021). Yeast engineering technologies and their applications to the food industry. *Food Biotechnology*, *35*(3), 252–271. https://doi.org/10.1080/08905436.2021.1942037.

Matassa, S., Boon, N., Pikaar, I., & Verstraete, W. (2016). Microbial protein: Future sustainable food supply route with low environmental footprint. *Microbial Biotechnology*, *9*(5), 568–575. https://doi.org/https://doi.org/10.1111/1751-7915.12369.

Mathewson, S. L., Azevedo, P. S., Gordon, A. L., Phillips, B. E., & Greig, C. A. (2021). Overcoming protein-energy malnutrition in older adults in the residential care setting: A narrative review of causes and interventions. *Ageing Research Reviews*, *70*. https://doi.org/10.1016/j.arr.2021.101401.

Meyer, V., Basenko, E. Y., Benz, J. P., Braus, G. H., Caddick, M. X., Csukai, M., et al. (2020). Growing a circular economy with fungal biotechnology: A white paper. *Fungal Biology and Biotechnology*, 7(1), 5. https://doi.org/10.1186/s40694-020-00095-z.

Monteyne, A. J., Coelho, M. O. C., Porter, C., Abdelrahman, D. R., Jameson, T. S. O., Jackman, S. R., et al. (2020). Mycoprotein ingestion stimulates protein synthesis rates to a greater extent than milk protein in rested and exercised skeletal muscle of healthy young men: A randomized controlled trial. *The American Journal of Clinical Nutrition*, *112*(2), 318–333. https://doi.org/10.1093/ajcn/nqaa092.

Naranjo-Ortiz, M. A., & Gabaldón, T. (2019). Fungal evolution: Diversity, taxonomy and phylogeny of the fungi. *Biological Reviews*, *94*(6), 2101–2137. https://doi.org/https://doi.org/10.1111/brv.12550.

Newton, P., & Blaustein-Rejto, D. (2021). Social and economic opportunities and challenges of plant-based and cultured meat for rural producers in the US. *Frontiers in Sustainable Food Systems*, *5*. https://doi.org/10.3389/fsufs.2021.624270.

Nout, M. J. R., & Aidoo, K. E. (2011). Asian fungal fermented Food. In M. Hofrichter (Ed.), *Industrial applications* (pp. 29–58). Springer. https://doi.org/10.1007/978-3-642-11458-8_2.

Quorn. (2022). *Quorn – The production of alternative first-class protein source for a balanced, sustainable diet.* White Paper (Retrieved January 30, 2022, from https://assets.ctfassets.net/uexfe9h31g3m/42NIsYMOOAkioo6OwUoWCu/737474ac34ccc7188370336a1fa5078a/1_QUORN_-_GENERAL.pdf).

Richter, F. (2019). *Alternative Meat Market Poised for Growth.* Statista. (https://www.statista.com/chart/18394/meat-substitute-sales-in-selected-countries/).

Ritala, A., Häkkinen, S. T., Toivari, M., & Wiebe, M. G. (2017). Single cell protein—State-of-the-art, industrial landscape and patents 2001–2016. *Frontiers in Microbiology, 8*. https://doi.org/10.3389/fmicb.2017.02009.

Ritchie, H., Reay, D. S., & Higgins, P. (2018). Potential of meat substitutes for climate change mitigation and improved human health in high-income markets. *Frontiers in Sustainable Food Systems, 2*. https://doi.org/10.3389/fsufs.2018.00016.

Ruxton, C. H. S., & McMillan, B. (2010). The impact of mycoprotein on blood cholesterol levels: A pilot study. *British Food Journal, 112*(10), 1092–1101. https://doi.org/10.1108/00070701011080221.

Sá, A. G. A., Moreno, Y. M. F., & Carciofi, B. A. M. (2020). Plant proteins as high-quality nutritional source for human diet. *Trends in Food Science & Technology, 97*. https://doi.org/10.1016/j.tifs.2020.01.011.

Sadh, P. K., Duhan, S., & Duhan, J. S. (2018). Agro-industrial wastes and their utilization using solid state fermentation: A review. *Bioresources and Bioprocessing, 5*(1), 1. https://doi.org/10.1186/s40643-017-0187-z.

Slavin, J., & Green, H. (2007). Dietary fibre and satiety. *Nutrition Bulletin, 32*(s1), 32–42. https://doi.org/10.1111/j.1467-3010.2007.00603.x.

Smetana, S., Aganovic, K., Irmscher, S., & Heinz, V. (2018). Agri-Food waste streams utilization for development of more sustainable Food substitutes. In *Designing sustainable technologies, products and policies* (pp. 145–155). Springer International Publishing. https://doi.org/10.1007/978-3-319-66981-6_17.

Smetana, S., Mathys, A., Knoch, A., & Heinz, V. (2015). Meat alternatives: Life cycle assessment of most known meat substitutes. *The International Journal of Life Cycle Assessment, 20*(9), 1254–1267. https://doi.org/10.1007/s11367-015-0931-6.

Sokolowski, C. M., Higgins, S., Vishwanathan, M., & Evans, E. M. (2020). The relationship between animal and plant protein intake and overall diet quality in young adults. *Clinical Nutrition, 39*(8). https://doi.org/10.1016/j.clnu.2019.11.035.

Souza Filho, P. F. (2018). *Fungi-based biorefinery model for food industry waste - Progress toward a circular economy.* University of Boras Digital Academic Archive (Doctoral thesis, University of Boras] http://urn.kb.se/resolve?urn=urn:nbn:se:hb:diva-14888).

Souza Filho, P. F., Andersson, D., Ferreira, J. A., & Taherzadeh, M. J. (2019). Mycoprotein: Environmental impact and health aspects. *World Journal of Microbiology and Biotechnology, 35*(10), 147. https://doi.org/10.1007/s11274-019-2723-9.

Souza Filho, P. F., Nair, R. B., Andersson, D., Lennartsson, P. R., & Taherzadeh, M. J. (2018). Vegan-mycoprotein concentrate from pea-processing industry byproduct using edible filamentous fungi. *Fungal Biology and Biotechnology, 5*(1), 5. https://doi.org/10.1186/s40694-018-0050-9.

Spiller, M., Muys, M., Papini, G., Sakarika, M., Buyle, M., & Vlaeminck, S. E. (2020). Environmental impact of microbial protein from potato wastewater as feed ingredient: Comparative consequential life cycle assessment of three production systems and soybean meal. *Water Research, 171*, 115406. https://doi.org/10.1016/j.watres.2019.115406.

Stephan, A., Ahlborn, J., Zajul, M., & Zorn, H. (2018). Edible mushroom mycelia of Pleurotus sapidus as novel protein sources in a vegan boiled sausage analog system: Functionality and sensory tests in comparison to commercial proteins and meat sausages.

European Food Research and Technology, *244*(5), 913–924. https://doi.org/10.1007/s00217-017-3012-1.

Stephenson, L. S., Latham, M. C., & Ottesen, E. A. (2000). Global malnutrition. *Parasitology*, *121*(S1). https://doi.org/10.1017/S0031182000006478.

Stoffel, F., de OliveiraSantana, W., Gregolon, J. G. N., Kist, T. B. L., Fontana, R. C., & Camassola, M. (2019). Production of edible mycoprotein using agroindustrial wastes: Influence on nutritional, chemical and biological properties. *Innovative Food Science & Emerging Technologies*, *58*, 102227. https://doi.org/10.1016/j.ifset.2019.102227.

Suman, G., Nupur, M., Anuradha, S., & Pradeep, B. (2015). Single cell protein production: A review. *International Journal of Current Microbiology and Applied Sciences*, *4*(9), 251–262.

Svensson, S. E., Bucuricova, L., Ferreira, J. A., Souza Filho, P. F., Taherzadeh, M. J., & Zamani, A. (2021). Valorization of bread waste to a fiber- and protein-rich fungal biomass. *Fermentation*, 7(2), 91. https://doi.org/10.3390/fermentation7020091.

Thomas, A. B., Shetane, T. D., Singha, R. G., Nanda, R. K., Poddar, S. S., & Shirsat, A. (2017). Employing central composite Design for Evaluation of biomass production by fusarium venenatum: In vivo antioxidant and antihyperlipidemic properties. *Applied Biochemistry and Biotechnology*, *183*(1), 91–109. https://doi.org/10.1007/s12010-017-2432-5.

Transparency Market Research. (n.d.). Meat Alternatives Market Trends, 2021–2031. Https://Www.Transparencymarketresearch.Com/Meat-Alternatives-Market.Html.

Troiano, D., Orsat, V., & Dumont, M. J. (2020). Status of filamentous fungi in integrated biorefineries. *Renewable and Sustainable Energy Reviews*, *117*, 109472. https://doi.org/https://doi.org/10.1016/j.rser.2019.109472.

Turnbull, W. H., Leeds, A. R., & Edwards, G. D. (1990). Effect of mycoprotein on blood lipids. *The American Journal of Clinical Nutrition*, *52*(4), 646–650. https://doi.org/10.1093/ajcn/52.4.646.

Turnbull, W. H., Leeds, A. R., & Edwards, D. G. (1992). Mycoprotein reduces blood lipids in free-living subjects. *The American Journal of Clinical Nutrition*, *55*(2), 415–419. https://doi.org/10.1093/ajcn/55.2.415.

Turnbull, W. H., Walton, J., & Leeds, A. R. (1993). Acute effects of mycoprotein on subsequent energy intake and appetite variables. *The American Journal of Clinical Nutrition*, *58*(4), 507–512. https://doi.org/10.1093/ajcn/58.4.507.

Turnbull, W. H., & Ward, T. (1995). Mycoprotein reduces glycemia and insulinemia when taken with an oral-glucose-tolerance test. *The American Journal of Clinical Nutrition*, *61*(1), 135–140. https://doi.org/10.1093/ajcn/61.1.135.

U.S. Food & Drug Administration. (2018). Reference Amounts Customarily Consumed: List of Products for Each Product Category: Guidance for Industry. https://www.fda.gov/media/102587/download.

U.S. Food & Drug Administration. (n.d.). Daily Value on the New Nutrition and Supplement Facts Labels. https://Www.Fda.Gov/Food/New-Nutrition-Facts-Label/Daily-Value-New-Nutrition-and-Supplement-Facts-Labels#referenceguide.

Udall, J. N., Lo, C. W., Young, V. R., & Scrimshaw, N. S. (1984). The tolerance and nutritional value of two microfungal foods in human subjects. *The American Journal of Clinical Nutrition*, *40*(2), 285–292. https://doi.org/10.1093/ajcn/40.2.285.

United Nations. (2020). *The Sustainable Development Goals Report*, 2020. https://unstats.un.org/sdgs/report/2020/The-Sustainable-Development-Goals-Report-2020.pdf.

Upcraft, T., Tu, W.-C., Johnson, R., Finnigan, T., van Hung, N., Hallett, J., et al. (2021). Protein from renewable resources: Mycoprotein production from agricultural residues. *Green Chemistry*, *23*(14), 5150–5165. https://doi.org/10.1039/D1GC01021B.

van Loo, E. J., Caputo, V., & Lusk, J. L. (2020). Consumer preferences for farm-raised meat, lab-grown meat, and plant-based meat alternatives: Does information or brand matter? *Food Policy, 95*, 101931. https://doi.org/https://doi.org/10.1016/j.foodpol.2020.101931.

Wakai, S., Arazoe, T., Ogino, C., & Kondo, A. (2017). Future insights in fungal metabolic engineering. *Bioresource Technology, 245*, 1314–1326. https://doi.org/https://doi.org/10.1016/j.biortech.2017.04.095.

Wang, L., Guo, H., Liu, X., Jiang, G., Li, C., Li, X., et al. (2019). Roles of Lentinula edodes as the pork lean meat replacer in production of the sausage. *Meat Science, 156*, 44–51. https://doi.org/10.1016/j.meatsci.2019.05.016.

Whittaker, J. A., Johnson, R. I., Finnigan, T. J. A., Avery, S. V., & Dyer, P. S. (2020). The biotechnology of Quorn mycoprotein: Past, present and future challenges. In H. Nevalainen (Ed.), *Grand challenges in fungal biotechnology* (pp. 59–79). Springer International Publishing. https://doi.org/10.1007/978-3-030-29541-7_3.

Wiebe, M. (2002). Myco-protein from fusarium venenatum : A well-established product for human consumption. *Applied Microbiology and Biotechnology, 58*(4), 421–427. https://doi.org/10.1007/s00253-002-0931-x.

Wikandari, R., Kinanti, D. A., Permatasari, R. D., Rahmaningtyas, N. L., Chairunisa, N. R., Sardjono, H., et al. (2021). Correlations between the chemical, microbiological characteristics and sensory profile of fungal fermented Food. *Fermentation*, 7(4), 261. https://doi.org/10.3390/fermentation7040261.

Williamson, D. A., Geiselman, P. J., Lovejoy, J., Greenway, F., Volaufova, J., Martin, C. K., et al. (2006). Effects of consuming mycoprotein, tofu or chicken upon subsequent eating behaviour, hunger and safety. *Appetite, 46*(1), 41–48. https://doi.org/10.1016/j.appet.2005.10.007.

Zeisel, S. H., Mar, M.-H., Howe, J. C., & Holden, J. M. (2003). Concentrations of choline-containing compounds and betaine in common foods. *The Journal of Nutrition, 133*(5), 1302–1307. https://doi.org/10.1093/jn/133.5.1302.

CHAPTER SIX

Should and will "cultured meat" become a reality in our plates?

Marie-Pierre Ellies-Oury[a,b,*], Sghaier Chriki[c], and Jean-François Hocquette[b]
[a]Bordeaux Sciences Agro, Gradignan, France
[b]INRAE, University of Clermont-Ferrand, VetAgro Sup, Saint Genès Champanelle, France
[c]ISARA – Agro School for Life, Agroecology and Environment Unit, Lyon, France
*Corresponding author: e-mail address: marie-pierre.ellies@agro-bordeaux.fr

Contents

Abstract

Produced from proliferating cells in bioreactors with a controlled culture medium, "cultured meat" has been presented by its supporters, who are mainly private actors (start-ups), as a sustainable solution to meet the growing demand for animal proteins without weaknesses of animal husbandry in terms of environmental impact, animal welfare or even health. The aim of this chapter is to take stock of current knowledge on the potential benefits and pitfalls of this novel product. Since robust scientific arguments are lacking on these aspects, there is no consensus on the health and nutritional qualities of

Advances in Food and Nutrition Research, Volume 101
ISSN 1043-4526
https://doi.org/10.1016/bs.afnr.2022.04.005

"cultured meat" for human consumption and on its potential low environmental impact. In addition, many issues related to the market, legislation, ethics and consumer perception remain to be addressed. The way in which this new product is regarded appears to be influenced by many factors related mainly to its price, as well as to the perception of safety, sensory traits but also environmental and nutritional issues. Therefore, research by universities and public research institutes indicates that "cultured meat" production does not present any major advantages in economic, nutritional, sensory, environmental, ethical or social terms compared to conventional meat. Thus, a more balanced diet by diversifying our sources of plant and animal proteins, consuming other meat substitutes, and reducing food losses and waste appear to be more effective short-term solutions to the urgent need of producing enough food for the growing human population (while reducing environmental degradation and animal suffering).

Abbreviations

CH_4 methane
EFSA European Food Safety Authority
ESC embryonic stem cell
GFI Good Food Institute
GHG greenhouse gas emissions
GMO genetically modified organism
GWP global warming power
IPCC Intergovernmental Panel on Climate Change
iPSC induced pluripotent stem cell
USDA United States Department of Agriculture

1. Introduction: Meat production issues and the context of "cultured meat" development

Despite the significant economic weight of the livestock sector (Dumont et al., 2016), livestock farming and the consumption of animal products are experiencing an identity, societal and economic crisis. Controversies are being called into question by many citizens and consumers:

– *Farming methods and animal welfare*: in addition to this mistrust of animal products, there is a rejection of certain farming methods, in particular "intensive farming methods," based on the search for high labor productivity (made possible by the "rationalisation" of farming practices thanks to technical progress). These practices have been accompanied by the enlargement of agricultural structures and specialization, with a view to achieving economies of scale. Monogastric farms (pigs, poultry) are particularly singled out because of the disconnection between the land

and the farm and the living conditions of the animals (confinement, high concentration of animals, mutilations), which are deemed not to respect the well-being and natural behavior of the animals. These issues have been strongly voiced by abolitionist movements through a large-scale communication campaign against livestock farming.
- *Environmental impacts*: the low feed efficiency of animals for the production of animal proteins is presented as being a limitation to meeting the food challenge. Indeed, animals mobilize large areas (especially ruminants) and also use large quantities of raw materials (especially monogastric animals) in direct competition with human food. Thus, on a global scale, 33% of cereal production, 74% of maize production and 83% of soya production is used for animal feed (Rauw et al., 2020). In addition, animals release nitrogenous elements as well as potentially polluting mineral elements when they are emitted in quantities that exceed the recycling capacity of the soil (particularly in regions with high concentrations of livestock).
- *The quality of animal products*: although animal products have valuable nutritional characteristics (notably sources of iron and B vitamins) (Rémond, 2019), their consumption is being questioned in developed countries, particularly since the International Agency for Research on Cancer highlighted a "probable" link between red meat consumption and the risk of colorectal cancer (Duru, Le Bras, & Grillot, 2021). In this context, the European Commission and the European Food Safety Authority (EFSA), using a holistic approach to livestock farming, have shown the need to reduce the current over-consumption of animal products, particularly meat (Sijpestijn, Wezel, & Chriki, 2022). The questioning of meat quality is sometimes extended to sensory quality (lack of typicality and regular quality) or hygiene quality (presence of antibiotics, risks of epizootics and zoonoses), especially in the era of COVID.

It can also be noted that in some countries, a significant proportion of agricultural production is no longer used for human (or animal) food. According to United States Department of Agriculture (USDA) data, 37.6% of the corn produced in the United States in 2017–2018 was used for ethanol production, a proportion that has remained relatively stable over the last 10 years (Olivier, 2021).

The conventional meat industry faces major environmental sustainability challenges in terms of land use, freshwater consumption and pollution, greenhouse gas emissions (GHG) and biodiversity loss.

Over the last five decades, in order to improve animal production, breeding strategies have been widely implemented and animals have been pushed to their biological limits in order to produce more food, resulting in a loss of natural biodiversity, contrary to agroecological sustainability. Moreover, with increasing urbanisation and climate change, the land available for agriculture without further deforestation will continue to decrease. The high water footprint, together with water pollution and water scarcity, call for a more sustainable way to improve the efficiency of conventional meat production.

In addition to the negative effect of livestock farming on water pollution, the emission of greenhouse gases associated with global warming is also a big problem that attracts great societal concern. The livestock industry contributes 12–18% of total GHG emissions, including 37% of all methane emissions mainly associated with ruminants, making it a significant contributor to global climate change (Gomez-Zavaglia, Mejuto, & Simal-Gandara, 2020). A meat-rich diet releases more than twice the daily carbon dioxide equivalent of a vegetarian diet (Scarborough et al., 2014). A reduction in meat consumption and/or the adoption of alternatives to meat have therefore been suggested as ways to reduce GHG emissions and their impact on global warming and climate change. However, the environmental impact of conventional meat and other meat substitutes remains controversial. Other studies have found different estimates of GHGs associated with direct emissions from livestock, which are lower (e.g., <3% of anthropogenic GHGs) than previous estimates (Pitesky, Stackhouse, & Mitloehner, 2009).

Given the aforementioned challenges facing the conventional meat industry, a shift in the market share of conventional meat may occur in the future with the rapid development of new meat alternatives, mainly including plant-, mycoprotein-, insect- and cell-based products, which are being touted for their potential to mitigate the environmental, health and safety, and animal welfare issues associated with conventional meat production.

Of these alternatives, "cultured meat" is of interest to all sectors of society because of its potential benefits: (1) similarity to conventional meat; (2) improved animal welfare; (3) natural resource savings; (4) potential reduction in greenhouse gas emissions; (5) potential public health and safety; (6) provision of more protein to meet the growing population and food demand; (7) reforestation and wildlife protection. However, these potential benefits are controversial (Chriki & Hocquette, 2020). Cell-based meat is produced using animal cell culture technology, i.e., from animal cells rather than slaughtered animals. Using a series of revolutionary techniques and technologies, cell meat is, according to its proponents, the only one that can produce true muscle-based animal meat (Lee, Yong, Kim, Choi, & Jo, 2020; Siegrist & Hartmann, 2020). This is presented as a major advantage over other meat substitutes. Although "cultured meat" is closer to real meat than other meat substitutes, technically speaking, this new product is not exactly real animal meat, defined as "animal flesh (especially mammals) that can be consumed as food." Furthermore, in EU legislation (as indicated on http://foodhealthlegal.eu/), "cultured meat" is considered a new product, not real meat (Chriki, Ellies-Oury, & Hocquette, 2022).

2. Where do we stand with its development and marketing?

2.1 What is it? How is it made?

So-called "cultured meat" is produced from a small number of cells that can proliferate and differentiate into components of meat. These cells can be stem cells or myoblasts still in the proliferation stage, isolated from embryos or biopsied from live animals of any species, which means that, theoretically, any type of meat can be synthetized. Stem cell lines include embryonic stem cells (ESCs), adult stem cells and induced pluripotent stem cells (iPSCs) (Chen et al., 2022). These cells can also be cell lines immortalized due to some genetic modification, which means that new batches of cells do not have to be taken from animals in this case (Soice & Johnston, 2021) (Fig. 1). One scientific issue is the choice of this starting material: indeed, some companies like Aleph Farms prefer to work with animal cells to avoid any GMO products that are not well accepted by consumers, but others may prefer to avoid using animals altogether for their own protection. Some companies may seek to produce "meat" from rare or unusual species to produce for instance Wagyu beef (which is the most expensive beef) like Orbillion or deer meat or more exotic species.

In all instances, the cells proliferate a very large number of times at physiological temperature in bioreactors on scaffolds, thus producing large quantities of muscle cells from a very small number of cells. This is the basic principle of this technology which avoids slaughtering a large number of animals as is the case today. Proliferation is carried out in a culture medium rich in nutrients, but also containing hormones and growth factors, which are necessary for the proliferation of muscle cells and their differentiation. In traditional cell culture in research laboratories, these hormones and growth

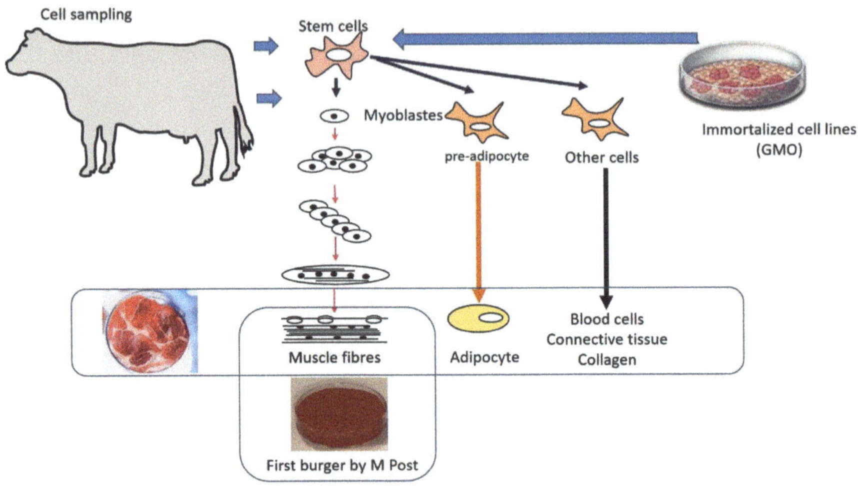

Fig. 1 Principles of the technology. *Adapted from Soice, E., & Johnston, J. (2021). Immortalizing cells for human consumption. International Journal of Molecular Sciences, 22(21), 11660. https:/doi.org/10.3390/ijms222111660.*

factors are combined with a small amount of foetal bovine serum obtained after slaughtering pregnant cows and added to the culture medium. This process was developed by Professor Mark Post who presented the first synthetic burger in a famous London restaurant in 2013. One very important scientific question relates to how to design a culture medium without foetal bovine serum, which according to recent results is likely (Messmer et al., 2022). However, growth factors are expensive, and a significant reduction in the cost of culture media is needed for economically viable production of "cultured meat" (Chen et al., 2022). More generally, the question is how to develop an animal component-free medium for cell culture and active research is being carried out in this area (Guan et al., 2021).

After proliferation, the stem cells are transferred to bioreactors to grow on scaffolds and differentiate into muscle tissue. A scaffold allows the cells to attach and then assemble and organise themselves into a tissue that is similar to a natural animal muscle. The scaffold material must of course be food-safe, edible, or biodegradable and derived from animal-free sources since the final product is intended for human consumption (Chen et al., 2022; Guan et al., 2021).

At this stage, it is important to point out the different technical strategies. The simplest one is to produce ground meat only from the myoblasts that produce muscle fibres. The first *in vitro* burger presented by Mark Post in

London in 2013 was produced in this way. Nowadays, the company Mosa Meat uses myosatellites from live cattle to manufacture beef patties. The most difficult strategy is to produce the muscle tissue in its complexity. This is indeed possible because stem cells can be converted not only into myoblasts to synthetize muscle fibres, but also into pre-adipocytes or other cell types such as fibroblasts (to produce collagen). Thus, the different cell types (myoblasts, adipocytes, fibroblasts) have to be cultured together to produce a piece of muscle which is still thin but more similar to a real muscle. However, this is not a simple matter of coculturing, as fibroblasts simply cocultured with myoblasts tend to overgrow and kill off the myoblasts. The challenge of growing connective tissue as a scaffold for organized muscle structure and blood vessels in coculture with muscle and fat cells is a complex and hugely underestimated issue. However, Aleph Farms has developed this advanced approach by coculturing multiple cell types consisting of myosatellite, endothelial cells and extracellular secretory cells within a three-dimensional porous scaffold to produce a fully formed steak (reviewed by Chen et al., 2022). Genetic engineering technologies can modify starting cell lines not only to optimize proliferation and differentiation, but also to achieve desirable qualities of "cultured meat" production, such as suspension growth, adaptation to serum-free medium or improved nutritional qualities (such as optimum amounts of iron in muscle fibres). Some authors have also suggested using insect cell lines suitable for suspension growth, tolerating nutrient starvation and better immortalisation for "cultured meat" production. Regardless of the technology used, there is concern about the development of genetic instability during cell proliferation because continued cell division may accumulate gene mutations, which may result in undesirable cells from genomic alterations similar to cancerous cells. As a result, research needs to design regular monitoring of genetic stability for quality assurance (Chen et al., 2022).

To summarize the current technology, while the basic principle looks very simple (a high rate of cell proliferation allows the production of a large quantity of muscle fibres, thus providing the possibility of supplying enough meat for the growing human population), many options are technically possible. This explains the different strategies developed by the various private companies working in this area. In addition, many technical issues are still uncertain and may impair the success of the technology. Finally, because the product does not look like muscle tissue in any way, and due to the fact that muscle ageing (which occurs after the animals are slaughter) is not considered, the final product resembles muscle fibres more than meat.

2.2 Who produces it? Who markets it? Who has plans to do so in the short-term?

The idea of "cultured meat" for human consumption was mentioned in 1932 by Winston Churchill. Later (1950–1999), Willem Van Eelen suggested the idea of using tissue culture to produce meat and this idea was granted a patent on the concept of stem cells. From 2000 onwards, different scientists started to culture muscle tissue in order to produce meat. In 2004, Jason Matheny founded New Harvest, an organization to promote research on "cultured meat." In 2008, moral arguments were reviewed and developed for and against "cultured meat." In 2013, the first "cultured meat"-based burger was cooked and tasted by a sensory panel in London. It was presented as a "proof of concept" by Prof. Mark Post (Zhang et al., 2021). It was made up of more than 10,000 muscle fibres and cost nearly $330,000 (€250,000) (Guan et al., 2021). From 2015, different private companies such as Memphis Meat (USA), Super Meat (Israel) and Mosa Meat (The Netherlands) were founded and claimed they would serve "cultured meat" to society in the next 5 years (Zhang et al., 2021). The Good Food Institute (GFI), a non-profit organisation working internationally to accelerate alternative protein innovation, is also promoting "cultured meat". It was created in 2016 by the non-profit organisation "Mercy For Animals." By the end of 2020, approximately 60 "cultured meat" companies have been set up worldwide. In the last 2 years, dozens of "cultured meat" start-ups have been founded and millions of dollars invested. Most of the companies are in North America, but also in Europe, Asia and Israel. Memphis Meat remains the leader and Mosa Meat in second position in terms of investments (Guan et al., 2021). The strategy of Memphis Meat, which includes science, food futures and politics, aims to render "cultured meat" productive, edible and acceptable. Indeed, the "cultured meat" producer is presented as an important new participant (like the farmer) in our agri-food system (Stephens, 2022).

At present, about 1 billion dollars has been invested in total (Fassler, 2021). This does seem like a lot, but little compared to the global market share of all meat substitutes ($36 billion) and the total global market share ($844 billion) based on data provided by Joseph, Searing, Watson, and McKeague (2020). However, more than 190 patents published in English related to "cultured meat" were collected from 1997 to 2020 (Ng, Singh, Yap, Tay, & Choudhury, 2021). In 2020, only 327 scientific publications were found in the scientific literature whereas 12,900 press articles on "cultured meat" were found throughout public databases. This could be interpreted as evidence that the media and the public are highly interested in this topic despite a weak scientific background. Another conclusion from this analysis is that "cultured meat" is mainly developing in the United States and the United Kingdom, with other countries, such as China, observing the trend for potential future applications. Furthermore, research has mainly focused on technical aspects or potential consumer acceptance (Chriki, Ellies-Oury, Fournier, Liu, & Hocquette, 2020). Indeed, other authors have confirmed that only a small amount of research has briefly touched upon the safety and regulatory aspects of "cultured meat" (Ketelings, Kremers, & de Boer, 2021).

Prepared independently for GFI by the research consultant organisation CE Delft, a report indicates that, after addressing a series of technical and economic barriers, the production price could be significantly reduced (by a factor of 4000), so that "cultured meat" could be fully affordable within the next 9 years. However, it has been observed that "cultivated meat companies have repeatedly missed product launch deadlines" (Fassler, 2021).

2.3 What role does research play?

Research on "cultured meat" is currently being developed in different areas. A great deal of the research is aimed at either improving the technology or improving the intrinsic quality traits of the product (safety, taste) while other research teams are more focused on the potential benefits and weaknesses of the product in terms of environmental issues and consumer acceptance, for example.

In terms of technology, reaching price parity with normal meat requires cost reductions (reviewed by Chen et al., 2022) and also a scale-up of the process (reviewed by Bellani et al., 2020). The culture medium is the main cost factor, especially hormones and growth factors. The use of protein engineering to produce these components is being explored to reduce costs and avoid using foetal bovine serum (Chen et al., 2022). Media recycling can further reduce costs and Future Meat is developing such a system. Perfusion in the culture medium is a key issue for a reproducible manufacturing process (Bellani et al., 2020). The choice of the starting material in terms of cell populations and media optimization for each cell line are also elements that optimize the process for efficiency. For instance, besides stem cells or primary cultured cells from animal samples, immortalized cell lines have the advantage of being infinitely divisible without animals. However, several questions remain unanswered about the extent to which these cells are safe, suitable for industrial-scale production, and acceptable to consumers (Soice & Johnston, 2021). New scientific disciplines such as advanced artificial intelligence, machine learning and computer modelling are being used to develop this kind of research, in particular monitoring for production scale-up.

There is also a great deal of research on scaffolds to support cell adherence and their compatibility with food consumption (reviewed by Chen et al., 2022). This implies working on materials (synthetic, natural or composite) that may be sources of scaffolds (e.g., polymers, plant- or microbial-derived materials). Commercial scaffolds with non-animal biomaterials are not available yet and the current scientific research on such scaffolds is scarce. For companies aiming to reproduce the muscle tissue, overcoming the thickness limit of the product remains a major challenge. Another issue is the large-scale production of "cultured meat" which involves the design of bioreactors that provide a controlled environment for different parameters such as temperature, pH, oxygen, carbon dioxide and culture medium components. These bioreactors should be optimized either for increases in volume to maintain cell proliferation at the exponential growth stage or for differentiation thereafter. These bioreactors should contain scaffolds, provide a homogeneous environment and have low energy and consumable requirements. The efficiency of bioreactors is often based on continuous perfusion, automation and real-time monitoring to support continuous nutrient delivery and waste removal but also to reduce operating costs and the risk of contamination. Bioreactors also need to be regularly agitated with cells on aggregates or microcarriers. This is because livestock, when administered with antibiotics and hormones, use physiological processes to flush out the residues with the help of the liver and kidneys, but in "cultured meat" when antibiotics are added, residue removal requires an artificial mechanism (Iyer & Iyer, 2020).

Another area of research concerns the characterization of the final product in terms of safety as well as nutritional and sensory quality traits. Indeed, Fraeye, Kratka, Vandenburgh, and Thorrez (2020) observed that "cultured meat" prototypes are currently not available for technological, sensory and nutritional assessment by independent authorities. However, based on the current state of knowledge, it can be assumed that "cultured meat" currently differs significantly from traditional meat (e.g., with a low iron content). A great deal of research is therefore needed before the composition of "cultured meat" could resemble that of traditional meat. This is likely to increase both the cost of the medium, but also the environmental footprint of the entire process. In brief, the absence of any transparent process makes it very difficult to assess "cultured meat" (Fraeye et al., 2020).

However, the main research priorities of the "cultured meat" industry are related to safety and several organisations/companies have defined research priorities linked to:

(i) standards for safe residue levels of relevant inputs,
(ii) concentration of inputs (e.g., growth factors, antibiotics, scaffold, novel inputs) in the product depending on the process,
(iii) recycling of the medium that concentrates al lot of hazardous inputs/residues,
(iv) the range of genetic modification approaches and outcomes that affect safety,
(v) safety assessment of the final product, etc. Most of the hazards are not expected to be novel; but some may be specific to "cultured meat," which is a novel process (Ong et al., 2021).

This research into the hazards and risks of "cultured meat," and more generally on risk assessments, is crucial to determine whether "cultured meat" is safe or not. This is a prerequisite, especially for EU legislation. Despite the importance of this aspect, little research on the safety of "cultured meat" has been published so far (Ketelings et al., 2021).

Verbeke, Hung, Baum, and De Steur (2021) confirmed that improved animal welfare and perceived unnaturalness were the main motives and barriers to "cultured meat" acceptance, respectively. However, perceived barriers were initially twice as powerful in shaping consumers' willingness to eat "cultured meat" compared to motives. Verbeke, Sans, and Van Loo (2015) indicated that "cultured meat" acceptance will be based on technological, economic and psychological criteria, consumer ownership of the topic (which in turn depends on media coverage), consumer attention and consumer confidence in science, politics and society.

To conclude this section, it is important to emphasise that surveys of consumer attitudes towards "cultured meat" are currently largely based on hypothetical constructs. It will be interesting to see how consumer attitudes change with their experience of real products and their costs.

3. What are the challenges of commercial "cultured meat" production?

3.1 Mass production at low cost

Scientific start-ups have successfully grown muscle cells on a small scale in flat-bottomed petri dishes. Indeed, the first artificial burger made by Mark

Post in 2013 was made by scraping the muscle cells that lined the bottom of these petri dishes and stacking a large number of the resulting cell layers. This process carries with it all the limitations of large-scale production, both in terms of the process itself and the availability of culture medium components as explained above.

First, regarding the process itself, it is necessary to line the bottom of the dishes with a biomaterial (containing collagen) which serves as a temporary support for the myoblastic cells to multiply, align and fuse into muscle fibres and also to contract as would a muscle naturally attached to the animal's skeleton. The composition of this biomaterial must be compatible with human food and of a reasonable cost for large-scale production (Post & Hocquette, 2017; Rorheim, Mannino, Baumann, & Caviola, 2016).

The production of large quantities of muscle fibres also requires the design of new types of large capacity bioreactors. Thus, the development of incubators with suspended microbeads (of the order of 100 μm in diameter) enabling cell attachment is currently being studied (Bodiou, Moutsatsou, & Post, 2020). Other three-dimensional culture techniques are being considered, such as incubators with cell aggregates that can be attached to branched supports and/or incubators with variable volumes depending on the stage of culture (Moritz, Verbruggen, & Post, 2015) or even large-scale air reactors (Li et al., 2020). The objective is to ensure that the cells are dispersed and occupy as much of the volume of the incubator as possible so that they are not restricted in their multiplication.

In all cases, the circulation or renewal of the culture medium, the distribution of gases (O_2, CO_2), the absence of contamination by microorganisms (bacteria and fungi), the determination and control of the optimum culture temperature will be, among others, essential parameters. The optimization of the cultivation conditions is a crucial point both from both a technical and an economic point of view. Technically, it is a question of reproducing the hormonal, biochemical and mechanical stimuli to promote cell multiplication and differentiation (Post & Hocquette, 2017). Economically, it is a question of minimising production costs, especially as this application of biotechnology is presented as sustainable (the economic dimension being a pillar of sustainability) (Young et al., 2013).

3.2 Environmental assessment

"Cultured meat" production is described by its developers as less impactful on the environment and climate and more sustainable than traditional meat, in that it produces less greenhouse gases, consumes less water and uses less land (Chriki & Hocquette, 2020; Lynch & Pierrehumbert, 2019; Post, 2012; Tuomisto & Teixeira de Mattos, 2011; Welin, 2013).

It is recognized that livestock contribute significantly to GHG emissions with 14.5% of total emissions (Poore & Nemecek, 2018). More specifically, 60% of these GHGs are related to ruminants, notably cattle and buffalo. Of these GHGs, enteric methane (CH_4), mainly due to ruminants, accounts for 40% of emissions alone (Gerber, Mottet, Opio, Falcucci, & Teillard, 2015). Among the various GHGs, the limitation of methane emissions is presented as the most important advantage of "cultured meat," since the warming power of methane is greater than that of CO_2. However, it can be suspected that "cultured meat" will not be emission-free, even if the quantities emitted seem to be more limited (Tuomisto, 2019; Tuomisto & Teixeira de Mattos, 2011) (however, there is a lack of consensus on this point (Mattick, Landis, Allenby, & Genovese, 2015)). Thus, the use of fossil fuels is not negligible for the production of "cultured meat," especially due to the use of incubators to allow cell multiplication at physiological temperature. However, while the environmental impact of CH_4 in the short-term is largely negative, it should not be overlooked that the CO_2 emitted by "laboratory meat" has a significantly longer half-life, and will therefore persist longer in the atmosphere, thus reducing the long-term advantage of this "synthetic meat" (Lynch & Pierrehumbert, 2019).[a]

Although it is commonly said that 15,000 L of fresh water are needed to produce 1 kg of beef, it is important to put these figures into perspective as 95% of this water corresponds to evapotranspiration (and therefore strongly linked to rainfall) by surfaces intended for livestock such as meadows and pastures. This water is often referred to as "virtual" water insofar as it would also be present in the absence of animals (Doreau, Corson, & Wiedemann, 2012).

A more accurate estimate of the quantities of water actually consumed to produce 1 kg of beef consists of taking into account only the water actually consumed by the animals, buildings and crop irrigation. This value generally varies between 550 and 700 L/kg, far from the 15,000 L/kg previously announced (Doreau et al., 2012). As "cultured meat" is a new

[a] It is important to remember that the effect of GHG emissions on the temperature of the atmosphere is expressed by their global warming power (GWP), which is the ratio between the energy returned to the ground in 100 years by 1 kg of gas and that which would be returned by 1 kg of CO_2 emitted at the same time. It depends on the concentrations and lifetimes of the gases. The GWP of 28 CH_4 and N_2O is 265 according to the IPCC (Intergovernmental Panel on Climate Change) report in 2013.

product, not yet developed on an industrial scale, the quantities of water required for its production are currently unknown and will need to be studied in detail. Nevertheless, "cultured meat" has been marketed as consuming less water (367–521 L per kg, according to Tuomisto and de Mattos (2011)). These same researchers seem to recognize that the use of the figure 15,000 L leads to a greatly exaggerated evaluation of the difference. However, this error is widely reported in the media (Tuomisto, 2019; Tuomisto, Ellis, & Hasstrup, 2014).

The figures relating to the surface area required for meat production must be modulated and correctly interpreted, not only in quantitative terms but also in terms of global impacts. Globally, the unfrozen land surface (13.2 billion hectares) is composed of 62% non-agricultural land (cities, deserts, forests) and 38% agricultural land. This 38% of agricultural land is divided into 25% of non-cultivable agricultural land used for livestock production in the form of pasture (grasslands, mountain areas, steppes, savannahs), 9% of cultivable agricultural land not needed for livestock production and 4% of cultivable agricultural land used indirectly by livestock for the production of animal feed. Thus, of the 2.5 billion hectares used for livestock production, more than half (1.3 billion) is non-arable land, which can only be used by animals (Mottet et al., 2017). While it is obvious that "cultured meat" production will require less land in absolute terms, it is important to specify that livestock farming provides many services, particularly environmental (such as landscape maintenance or the maintenance of plant and animal biodiversity in grasslands) or social services, such as maintaining a rural population in areas abandoned by today's predominantly urban consumers (Dumont et al., 2016, 2017; Duru, Donnars, Ryschawy, Therond, & Dumont, 2017). Thus, a reduction in livestock farming following the development of "cultured meat" production could have many unforeseeable consequences on the characteristics of the landscape and the maintenance of the human population in the countryside (Suthar, Riddhi, Theertha, Anupama, & Deepti, 2019; Wilks & Phillips, 2017).

This inventory of the assumed environmental impact of "cultured meat" therefore highlights the need to modulate the less impactful and more sustainable character of this product put forward by its promoters.

3.3 Quality of this new product

Proponents of "cultured meat" claim that it will be safer than traditional meat, because it is produced in a fully controlled environment, and free from any other organisms, including pathogens. The strategic choice of naming the biotechnology product "clean meat" refers to this notion of a sterile product, free of any health hazard (Ong, Choudhury, & Naing, 2020).

Although traditional meat is generally protected from contamination by digestive and urinary contents as a result of strict breeding, slaughter and processing procedures, the risk of contamination, particularly bacterial contamination, is still present (even if it is minimal), as shown by certain health crises, particularly in the manufacture of minced steak. This is why drastic measures have been taken to remove all carcasses potentially contaminated by cases of perforated intestines. Similarly, following two major health scares concerning minced beef steak in 2005 and 2011 (presence of *E. Coli* O157:H7 bacteria), traceability and batch sequencing have been strengthened and supplemented by the development of tests on other highly pathogenic strains of *E. Coli*, in particular O26:H11, O103:H2, O111:H8, O145:H28) (Wasilenko et al., 2014). "Cultured meat" is unlikely to come into contact with intestinal pathogens such as *E. coli, Salmonella* or *Campylobacter* (Bhat & Fayaz, 2011; Shapiro, 2018). However, as with any new technology, there is no such thing as zero risk, as the consequences and risks are not yet fully predictable or even unknown. Any error is therefore likely to have deleterious consequences, especially for consumers who are most sensitive to the new health risks that may arise. In terms of health, there is also the question of the use of antibiotics, which is strongly criticised in conventional livestock farming because of the increased risk of resistance (Oliver, Murinda, & Jayarao, 2011). This is why most "cultured meat" manufacturers now claim to have abandoned the use of antibiotics, as is traditionally the case for cell cultures, in order to avoid bacterial contamination (Chriki & Hocquette, 2020). However, the proof of concept by Dr. Post in 2013 was indeed achieved using antibiotics.

Given the multiplication of cells, some unexpected biological mechanisms could occur during "cultured meat" production. In particular, the high frequency of cell multiplication is likely to induce dysregulation, as is frequently observed in the case of cancer cells. Although it is likely that these lines will be largely eliminated during the process, the risk still exists. This raises the question of whether the consumption of such cells could induce risks to the metabolism and health of the consumer of "cultured meat" (Chriki & Hocquette, 2020). It can be assumed that this is probably not the case due to the degradation of the ingested cells by gastric acidity. However, the acceptability of "cultured meat" to the consumer may be diminished if this biological mechanism of cell dysregulation is brought to their attention (Hocquette, 2016).

While meat has an advantageous nutritional composition, notably due to some of its well-known constituents such as protein, iron, zinc and vitamin B12, it is also frequently considered too fatty, but also too rich in saturated fatty acids (Young et al., 2013). According to the European Commission, it is difficult to ascertain how close the macro- and micronutrient content and protein composition of "cultured meat" is to that of traditional meat. On the other hand, in order to obtain meat with optimal nutritional properties (bioavailable proteins, trace elements, vitamin B12, etc.), different strategies have recently been developed by start-ups to improve the nutritional value of "cultured meat" (Bhat & Bhat, 2011; Bhat, Kumar, & Bhat, 2017; Kadim, Mahgoub, Baqir, Faye, & Purchas, 2015). In particular, adjusting the fatty acid composition in the production process of "cultured meat" has been considered, and more specifically incorporating a reduction of saturated fatty acids in favour of polyunsaturated fatty acids (particularly omega-3). The addition of fatty acids can, for example, be achieved by coculturing adipocytes derived from adipose stem cells. However, this requires the integration of exogenous antioxidants in order to limit the risks of oxidation and rancidity, as these antioxidants are naturally present in the ruminant diet (in grazed grass in particular). In addition to the fatty acid composition, an increase in the myoglobin content, but also in trace elements (Zinc and Selenium in particular), should also be considered in order to improve the nutritional properties of the meat *cultured,* but also to promote cell growth. Finally, as traditional meat is one of the main sources of vitamin B12 (Williams, 2007) in order to replace it, "cultured meat" must also contain this vitamin to satisfy the dietary needs of consumers (Obeid et al., 2019). Once absorbed, vitamin B12 must also be transported across the cell membrane. These constraints are likely to present a significant challenge to the manufacture of muscle tissue in culture.

Further studies are needed at this stage to better assess the possibilities of potentiating the beneficial effects of micronutrients in cultured cells. A reduction in the health benefits of these compounds cannot be excluded, as the properties of micronutrients vary according to the support matrix in which they evolve.

3.4 Cost of "cultured meat" and start-ups in the sector

In France (Ellies-Oury, Lee, Jacob, & Hocquette, 2019) as in many European countries (CRÉDOC, 2018), the price of meat appears to be a major barrier to its consumption. This is why the price of "cultured meat" could be considered as a significant obstacle to its commercialization (Verbeke et al., 2015), even though since the first *in vitro* hamburger developed within the framework of the "Dutch Cultured meat Project" in August 2013 (and estimated to cost €250,000 for 142 g; Post, 2014), prices have decreased consistently.

In December 2020, the Singapore restaurant "1880" served lab-grown chicken for €19. The Israeli start-up Aleph Farm is planning to enter the "high quality market" with thin slices of "meat" (about 5 mm thick) in 2022 obtained from stem cells, which would cost around €46 per piece to produce. Similarly, the American start-up Modern Meadow is developing "cultured meat" chips using 3D printing at a cost of around €100 per slice. According to a Mosa Meat spokeswoman,[b] "*once we scale up production, we expect the cost of producing a hamburger to be about* €9." This technology "*could even, in the long run, become cheaper than a traditional burger.*" Similarly, the objective targeted in the European project Meat4All[c] coordinated by Biotech Foods Sl, is a selling price of around €6/kg from 2023. A number of companies have invested in the research and development of "cultured meat" in recent years. Although nowhere near mass market prices, "cultured meat" has the potential to make its entry into restaurants in the next few years. We can assume this cost would be marginally lower in the coming years, but it will remain more expensive than farmed meat.

3.5 Can we do without animals to produce "cultured meat"?

3.5.1 The need to slaughter animals

Theoretically, one of the main potential benefits of "cultured meat" is that it produces huge amounts of muscle from a small number of animal cells, either from farm animals, or from cell lines, which proliferate in giant incubators. Thus, this process is likely to use far fewer animals to produce a lot of meat.

In this way, some authors indicate that the production of "cultured meat" would be more "animal-friendly," as conventional meat could be considered "cruel" to animals by some consumers (Ong et al., 2020). Some animal advocates would thus be able to accept the concept of "cultured meat," or even describe this product as "victimless meat," "cruelty-free meat" or "slaughter-free meat" (Chriki et al., 2020). Communication through the "animal welfare" argument thus seems to be a favorable alternative for the development,

---

[b] Reuters, 2019. The $280,000 lab-grown burger could be a more palatable $10 in 2 years. Reuters. URL https://www.reuters.com/article/us-food-tech-labmeat-idUSKCN1U41W8 (accessed 10.4.21).

[c] https://cordis.europa.eu/project/id/958660/fr

appropriation and acceptability of "cultured meat" by consumers (Baum, Verbeke, & De Steur, 2022; Piochi, Micheloni, & Torri, 2022).

Despite the potential to dramatically reduce animal use, there are a number of animal ethics-based objections to the development of "cultured meat." The first objection is that, because the process is based on biopsies from living animals to extract stem cells, there will still be a need for some animals. The living conditions of these animals should be questioned since they will be regularly subjected to biopsies to provide stem cells on a continuous basis. Therefore, specific welfare issues will arise for these animals. Some people may have moral problems with this process of regular biopsies, as they did with genetic modifications of livestock. The selection of donor animals and the details of the biopsy processes must be optimised, as this is a key bottleneck in the "cultured meat" production process (Melzener, Verzijden, Buijs, Post, & Flack, 2021).

The development of pluripotent stem cell lines from livestock species has seen progress in recent years with a view to avoiding regular sampling of living animals and to work with immortal cell lines with infinite proliferative capacity. The starting cell lines can come from (1) a recently slaughtered animal from which muscle satellite cells would be isolated or (2) an embryo or umbilical cord from which embryonic stem cells and mesenchymal stem cells would be isolated respectively (Ketelings et al., 2021).

This implies genetic engineering techniques which will target the genes involved in proliferation. This also requires genetic modification for the cells to be able to differentiate into the desired tissue. From a theoretical point of view, this will avoid the use of animals. However, further evaluation is required to screen the (unexpected) effects of different cell reprogramming approaches on cell phenotype and biological functions. Many of these approaches have not been explored in livestock species (Reiss, Robertson, & Suzuki, 2021).

3.5.2 The need for foetal bovine serum

Regardless of the starting point, the cell culture process requires the use of a growth medium rich in suitable nutrients for cell proliferation and differentiation. One of the most commonly used components at present is foetal bovine serum (used by Post to produce the first cultured burger and more recently used by Eat Just to produce "cultured nuggets" marketed in Singapore). This is taken from the foetus of pregnant cows at slaughter. Thus, there appears to be a need to (i) maintain farm animals to develop the production of foetal bovine serum, and (ii) slaughter pregnant cows so that the serum can be collected.

Many companies have committed to eliminating the use of foetal bovine serum, a high-cost, biologically variable ingredient that presents technological, ethical and sustainability issues for "cultured meat" production (Chen et al., 2022), and some have reported to have already done so.

These foetal bovine serum-free media would be based on (1) microbial fermentations to synthesise recombinant growth proteins, or (2) non-animal extracts, e.g., developing plant-proteins and amino acids derived from plant hydrolysates.

Recently, an omics approach has identified key cellular signals that allow myocytes to develop in the absence of animal serum (Domigan, Feisst, & Ogilvie, 2022; Messmer et al., 2022).

3.5.3 The future of animals if we do not need them

To conclude this section, it is important to add that if livestock are removed and replaced by "cultured meat," a number of ecosystem services (i.e., the beneficial impacts of livestock) as well as the production of by-products from farm animals (hide, blood, viscera) would be lost (Dumont et al., 2020; Olivier, 2021). Indeed, livestock farming systems have numerous functions: besides supplying proteins for human nutrition, they provide income for rural populations and thus support a large part of the world's rural communities. Moreover, livestock produce not only meat, milk, and eggs, but also wool, fiber, and leather. They also provide socio-cultural services including tourist events such as those associated with transhumance and traditional production, and products with a local image and sense of *terroir* such as Protected Designation of Origin cheeses and other products (Dumont et al., 2020).

Furthermore, there are still many small family farms that depend on livestock for their survival, particularly in developing countries, and it is likely that the development of "cultured meat" by FoodTech companies would jeopardize the economic and food survival of these farmers (Chriki et al., 2022).

However, to nuance these potential consequences on rural life and farming, it is important to note that there are two possible paths for the development of "cultured meat":

- either "cultured meat" completely replaces the conventional meat,
- or, it is developed in parallel, as a product adding to the current offer of alternative proteins, or as a high-end product.

In the latter case, the consequences are not the same in terms of animal population associated loss of animal biodiversity.

3.6 Is "cultured meat" really meat? What are the legal issues?

One of the main issues that arise in the "cultured meat" sector is the nature of the product: is it meat? Indeed, the naming and labeling of this novel biotechnology may have significant implications and consequences for consumer acceptance of "cultured meat" as well as for future labeling policies (Asioli, Bazzani, & Nayga, 2021).

According to the definition of the American Meat Science Association (Boler & Woerner, 2017), but also for biological or technical reasons, or for semantic and commercial reasons (Ong et al., 2020), "cultured meat" is not really meat. Indeed, meat is defined as *"edible tissues from an animal consumed as food"* and *"to be considered meat, 'in vitro meat' must be originally sourced from an animal cell, be inspected and considered safe for consumption, and be comparable in composition and sensory characteristics to meat derived naturally from animals"* according to Boler and Woerner (2017). Consequently, according to the authors of this article, the use of the word "meat" has induced an ambiguity favorable to the advocates of "cultured meat," who seek to get rid of the negative aspects associated with meat (environmental degradation and animal suffering), while taking advantage of the positive values of meat for consumers (strength, vitality, healthiness, etc.). The start-ups have thus succeeded in imposing the name "meat" for these cultivated muscle fibers into the everyday language. Indeed, the main keywords used in press articles are "meat" and, to a lesser extent, "food" (Chriki et al., 2020).

In the European Union, meat is defined in Annex I of regulation No 853/2004 of the European Parliament and The Council as "*skeletal muscle of mammalian and avian species recognised as fit for human consumption, with naturally embedded or adhering tissues.*" According to the European Commission, meat is formed from different tissues, among them skeletal muscle, bone, connective tissues, blood vessels, adipocytes or even nerves. Thus, meat industry stakeholders are increasingly trying to ban the use of the term "meat" in the labeling of "cultured meat" products. In addition, it appears that the American legislation is moving towards the recognition of "cultured meat" as "meat," and thus eligible for this designation, however, this is not yet decided (Boler & Woerner, 2017).

As long as no genetic modification is involved, the approval of the European Commission, on the advice of the European Food Safety

Authority (EFSA) under the Novel Foods Regulation is required before "cultured meat" can be marketed in Europe. The main purpose of this regulation is to ensure that novel foods are safe to eat which is not fully demonstrated at this stage for "cultured meat." There is therefore a need for them to be correctly labeled (so as not to mislead consumers) and not to be nutritionally disadvantageous compared to the foods they seek to replace. Similar food regulations related to "cultured meat" in other parts of the world are under discussion (Chen et al., 2022). In addition, "cultured meat" products with a genetically modified component are currently unlikely to be approved in Europe due to the heavy restrictions on genetically modified foods already in place and the general negative public perception of this technology.

Finally, unlike America and Oceania, it is very important to recall that the use of hormones is prohibited for meat production in Europe (Directive 96/22/EU; 29 April 1996). Adding exogenous hormones to hormones naturally present in a growing animal is forbidden in Europe on the grounds of the precautionary principle, even if these man-made hormones are identical to natural hormones. Hence, it can hardly be expected that the European legislator would authorize the use of synthetic hormones for the production of "cultured meat." However, this could be considered different since the meat culture is not conducted on a living animal (Chriki et al., 2022).

3.7 What is the religious status of "cultured meat"?

Cultured "meat," like any other new technology, raises numerous questions not only ethical and philosophical but also religious (Chriki & Hocquette, 2020). Mainly because of its nebulous status, religious authorities are still debating whether cultured "meat" is *Kosher* (consumable under Jewish dietary laws), *Halal* (for Muslim consumers, lawful or permitted, i.e., compliant with Islamic laws), or what to do if there is no animal available for ritual practices (Hindu consumers).

This debate on the religious status of "cultured meat" seems to be less present or absent in *Christianity*, which allows its followers to eat any kind of meat because all meat is clean according to its religious practice (Jagadeesan & Bin Salem, 2020). Indeed, one of the principles of Christianity is: "*Not that which enters into the mouth defiles the man; but that which proceeds out of the mouth, that defiles the man*" (Matthew 15:11).

However, for religions that advocate vegetarian diets, such as *Hinduism, Buddhism and Jainism*, the question could be also debated according to the status given to this innovation, depending on whether or not it is recognized as coming from animal flesh (Jagadeesan & Bin Salem, 2020).

Concerning the *Jewish* religion, rabbinical opinions are divided, because the religious status of cultured "meat" is determining by three factors (Kenigsberg & Zivotofsky, 2020):

(1) the source of the original cells,

(2) the nature of the growth medium, and finally

(3) the exact nature of the process involved.

Thus, some experts think that "cultured meat" can only be considered *Kosher* if the original cells were taken from a slaughtered *Kosher* animal. Others assume that regardless of the source of the cells used to produce the "cultured meat," they will certainly lose their original identity. Therefore, the outcome cannot be defined as forbidden for consumption (Kenigsberg & Zivotofsky, 2020), even it is the cultured pork that is forbidden for consumption in its original meat (Purdy, 2022).

For *Islam*, the crucial question is whether the cultured "meat" is compliant with Islamic laws or not, most commonly referred to as "*Halal* or not". Since "meat" culturing is a recent invention, the traditional Islamic jurist to whom Muslims often refer has never discussed its *Halal* status. It is therefore contemporary Islamic jurists who are now in charge of this issue.

Since its introduction in 2013, there are three types of Muslim opinions on the consumption of cultured "meat" (Hamdan, Ramli, Zaman Huri, Abd Rahman, & Abdullah, 2021):

(1) The first view completely rejects the concept of cultured "meat" because it is contrary to the nature of meat God has created, which is out of the real context.

(2) The second view maintains a non-bias approach, neither agreeing nor disagreeing until cultured "meat" is understood and introduced into the market.

(3) The third view allows *Halal* cultured "meat" production by identifying the source of the cells and the growth medium used in the culture process.

Accordingly, cultured "meat" is only considered *Halal* if the stem cells are extracted from a *Halal* slaughtered animal, such as cow, camel, chicken and others (all types of livestock except pig), and provided no blood or serum is used in the process. Indeed, serum should be avoided unless one can prove that the meat will not be altered as a result of contact with the serum (being potentially unclean) (Hamdan, Post, Ramli, & Mustafa, 2018).

However, for this novel product, the stem cells are currently taken from living animals, so the cultured "meat" that is produced from these cells is unclean and forbidden for consumption because one of the principles of Islam is that "*Whatever is cut from a living animal is dead (and consequently, considered as unclean)*" (Hamdan et al., 2018). Nonetheless, marine life is not included in this ban as other *hadiths* (religious text) have stated that marine life is permitted and *Halal* despite the condition of their carcass. Indeed, even if the stem cells are taken when the marine life is still alive, any "cultured meat" originating from marine life is considered permissible (Hamdan et al., 2021).

In conclusion, this recent issue clearly requires further study from religious authorities, especially since muscle fiber culture techniques are evolving in order to respond to the issues detailed in this article.

4. What do consumers think?

4.1 Methodology of the surveys

Consumer acceptance is a complex issue that is currently the subject of investigation by social scientists. A main approach to addressing this issue is to conduct online surveys to examine how consumers will react to "cultured meat." Based on consumers' perceptions and attitudes toward "cultured meat," and their willingness to try it, eat it regularly or buy it, these surveys have been conducted in different countries among various consumer groups, mainly focused on Europe (Dupont & Fiebelkorn, 2020; Hocquette, 2015; Palmieri, Perito, & Lupi, 2020; Siegrist & Hartmann, 2020) and the United States (Wilks & Phillips, 2017), but not exclusively (Bogueva & Marinova, 2020; Chriki et al., 2021; Heidemann, Taconeli, Reis, Parisi, & Molento, 2020; Hwang, You, Moon, & Jeong, 2020; Liu, Hocquette, Ellies-Oury, Chriki, & Hocquette, 2021; Reis, Heidemann, Borini, & Maiolino Molento, 2020; Zhang, Li, & Bai, 2020), emphasising that there were multiple factors influencing acceptance as well as rejection of "cell-based meat."

4.2 Main results

In a survey of 1890 scientists and students from all over the world by French authors, Hocquette et al. (2015) reported that the majority of respondents considered the technique for the production of "cultured meat" as realistic and feasible. However, no majority emerged to think that this novel product would be a good solution from a sensory or nutritional point of view or that it would actually contribute to solving problems of environmental degradation or animal welfare. Only around 10% of the respondents would be willing to consume "cultured meat" regularly, with the vast majority preferring to reduce their meat consumption, although a third of them were hesitant.

Another survey conducted in the United States of over 670 people confirmed the high proportion of hesitancy and the fact that only a minority (around 10 to 20%) of respondents favoured regular consumption of "cultured meat" (Wilks & Phillips, 2017).

A more recent survey in Germany confirmed the discrepancy between willingness to "try the product" and willingness to "consume it regularly," with ultimately a very moderate acceptance of "cultured meat" (Weinrich, Strack, & Neugebauer, 2020).

Another survey in Ireland showed that urban consumers were more receptive to "cultured meat" and more concerned about the environmental impacts of "current meat production" practices (Shaw & Mac, 2019). Rural consumers were more concerned about the potentially negative effects that "cultured meat" production could have on agriculture and the farmers' lifestyles (Chriki et al., 2021; Heidemann et al., 2020; Shaw & Mac, 2019). This result is confirmed by another study showing, in Italy and the United Kingdom, that the potential consumer of "cultured meat" is young, highly educated, somewhat familiar with the concept, meat-eating and inclined to reduce consumption (Baum et al., 2022; Mancini & Antonioli, 2019; Piochi et al., 2022).

Furthermore, age is one of the major factors influencing people's food choices (Grasso, Hung, Olthof, Verbeke, & Brouwer, 2019). This is in line with some authors who have argued that young people, who are plausibly the real future target, may be more concerned about animal welfare issues leading to a greater willingness to try alternative proteins (Chriki et al., 2021; De Backer & Hudders, 2015; Heidemann et al., 2020). However, in an Australian survey conducted of Generation Z, (18–24 years of age), considered to be familiar with technologies and with a sense of environmental responsibility, Bogueva and Marinova (2020) reported that 72% of respondents from this younger generation are not ready to accept "cultured meat." The way in which the questions are worded or the difference in age or maturity of the respondents may explain this type of different observation.

Gender influences both concerns about the ethical and environmental problems caused by the meat industry, opinions on meat consumption and the overall perception of "cultured meat" (Verbeke et al., 2021; Wilks & Phillips, 2017). Women are the most concerned about the ethical and environmental issues related to livestock and are more willing to substitute conventional meat for "cultured meat" (Chriki et al., 2021; Heidemann et al., 2020; Piochi et al., 2022). However, in another recent survey (Baum et al., 2022), acceptance of "cultured meat" was higher among men (Wilks & Phillips, 2017). This type of contradiction can be explained by an interaction between age and gender (Hocquette et al., 2015).

In our labs, we carried out a survey to investigate the opinions of the Chinese (Liu et al., 2021), Brazilian (Chriki et al., 2021), and French (Hocquette, Liu, Ellies-Oury, Chriki, & Hocquette, 2022) consumers on "cultured meat," with more than 4600 respondents in each country. About 40% in Brazil, 47% in China and 79% in France were not willing to eat "cultured meat" regularly. The majority of respondents (68% in France, 71% in Brazil and 86% in China) were willing to pay less for it compared to "conventional meat,." Emotional resistance such as the perception of "absurdity and/or disgust" (for 15% and 18% of Chinese and Brazilian respondents respectively and for 59% of French respondents) would lead to emotional resistance and subsequently result in no willingness to eat "cultured meat" regularly. The main concern is unnaturalness with specific concerns regarding food safety in China and the negative consequences on traditional farming in France. Respondents would not only expect "cultured meat" to be safe, tasty and with a low environmental impact, but also to be cheap and of high nutritional quality, mainly in China. Approximately 51% of the respondents in Brazil considered that "cultured meat" should not be called "meat" for marketing purposes, while data is not available in China and France.

To conclude, consumer acceptance of "cultured meat" is on average low and will depend on a combination of many factors mainly related to price, safety, sensory traits and also environmental and nutritional issues depending on the country.

5. Conclusion and outlook

In this unfavorable context against animal husbandry, the meat substitutes currently commercialized are increasingly popular with consumers. The same could be said of so-called "cultured meat," which regularly generates media interest, especially in Anglo-Saxon countries including the United States, although other European countries, Israel and certain Asian countries are also starting to show an interest in it.

Made from stem cells or proliferating muscle cells, "cultured meat" has been presented by its supporters, who are mainly private stakeholders (start-ups), as a food similar to conventional meat but more sustainable, even if strong scientific arguments are lacking to support this assessment. In addition, many difficulties remain to be overcome. First, only muscle fibres (similar to minced meat) are produced. Consequently, we should not call it "meat". Second, there are very few published scientific studies related to its environmental impact, with no unanimity on this point. In addition, there is a lack of knowledge of the real environmental impact of livestock with a lot of misunderstanding in the public media. Therefore, claims that "cultured meat" will be more environmentally friendly than conventional meat are not robust enough. The third point is that, because the technology is still in its infancy, it is impossible to prove that it will generate safe, healthy and tasty products. Indeed, no nutritional data are available. Safety is subject to controversy based on various assumptions, both positive and negative. Taste has not been studied using rigorous approaches and is highly dependent on the ingredients added at the end of the process. The product may be also perceived as not natural. Furthermore, while it is advertised as a solution to animal ethical issues, it brings with it new ethical issues for both animals and humans (Table 1). Positive claims about the potential benefits of "cultured meat" are driven by private companies motivated by financial gain or by activists motivated by animal welfare or environmental issues.

To conclude, research by universities and public research institutes indicates that "cultured meat" production does not present any major advantages in economic, nutritional, sensory, environmental, ethical or social terms compared to conventional meat. Consuming other meat substitutes, a better-balanced diet by diversifying our sources of plant and animal proteins and reducing food losses and waste appear to be more effective solutions on a short-term basis to the urgent problems listed above (i.e., providing enough food for the growing human population while reducing environmental degradation and animal suffering). It should also be remembered that livestock farming is not limited to the production of meat and that "cultured meat," unlike today's livestock, would not produce the other environmental and cultural services provided by the diversity of the current livestock farming systems.

Table 1 The remaining questions associated with "cultured meat."

Questions related to the technology
Is there any potential ethical drift in this new technology? Are there any unpredictable genetic modifications in these meat cells, especially when using cell lines?
Ethical questions
Is it possible and/or acceptable to produce "meat" from cells from pets or even human beings? What will be the life and the future of the remaining farm animals? These animals will be biopsied regularly and this raises new welfare questions. Alternatively, if cell lines are used, farm animals will no longer be needed. At least for a transitional period, the number of farm animals should be decreased by slaughtering them and/or limiting their reproduction. Animal biodiversity will be reduced with fewer farm animals.
Questions related to pastures and farmers
What will be the future of pastures, agriculture, landscapes and the countryside without any or fewer farm animals? This will be associated with a decline in plant biodiversity, rural life and so on. What will be the future of farmers, especially in underdeveloped countries without any or fewer farm animals?
Agro-food system issues
How and by whom will our food production be managed and controlled? Do we want FoodTech companies to manage our food production and quality instead of local farmers? What will be the potential benefits and weaknesses of this potential major change?

References

Asioli, D., Bazzani, C., & Nayga, R. M., Jr. (2021). Are consumers willing to pay for in-vitro meat? An investigation of naming effects. *Journal of Agricultural Economics*, *00*, 1–20. https://doi.org/10.1111/1477-9552.12467.

Baum, C. M., Verbeke, W., & De Steur, H. (2022). Turning your weakness into my strength: How counter-messaging on conventional meat influences acceptance of cultured meat. *Food Quality and Preference*, *97*, 104485. https://doi.org/10.1016/j.foodqual.2021.104485.

Bellani, C. F., Ajeian, J., Duffy, L., Miotto, M., Groenewegen, L., & Connon, C. J. (2020). Scale-up Technologies for the Manufacture of adherent cells. *Frontiers in Nutrition*, 7. https://www.frontiersin.org/article/10.3389/fnut.2020.575146.

Bhat, Z. F., & Bhat, H. (2011). Animal-free meat biofabrication. *American Journal of Food Technology*, *6*(6), 441–459.

Bhat, Z. F., & Fayaz, H. (2011). Prospectus of cultured meat-advancing meat alternatives. *Journal of Food Science and Technology-Mysore, 48*(2), 125–140. https://doi.org/10.1007/s13197-010-0198-7.

Bhat, Z. F., Kumar, S., & Bhat, H. F. (2017). In vitro meat: A future animal-free harvest. *Critical Reviews in Food Science and Nutrition, 57*(4), 782–789. https://doi.org/10.1080/10408398.2014.924899.

Bodiou, V., Moutsatsou, P., & Post, M. J. (2020). Microcarriers for upscaling cultured meat production. *Frontiers in Nutrition,* 7. https://doi.org/10.3389/fnut.2020.00010.

Bogueva, D., & Marinova, D. (2020). Cultured meat and Australia's generation Z. *Frontiers in Nutrition,* 7, 148. https://doi.org/10.3389/fnut.2020.00148.

Boler, D. D., & Woerner, D. R. (2017). What is meat? A perspective from the American meat science association. *Animal Frontiers,* 7(4), 8–11. https://doi.org/10.2527/af.2017.0436.

Chen, L., Guttieres, D., Koenigsberg, A., Barone, P. W., Sinskey, A. J., & Springs, S. L. (2022). Large-scale cultured meat production: Trends, challenges and promising biomanufacturing technologies. *Biomaterials, 121274.* https://doi.org/10.1016/j.biomaterials.2021.121274.

Chriki, S., Ellies-Oury, M.-P., Fournier, D., Liu, J., & Hocquette, J.-F. (2020). Analysis of scientific and press articles related to cultured meat for a better understanding of its perception. *Frontiers in Psychology Eating Behavior, 11,* 1845. https://doi.org/10.3389/fpsyg.2020.01845.

Chriki, S., Ellies-Oury, M.-P., & Hocquette, J.-F. (2022). Is "cultured meat" a viable alternative to slaughtering animals and a good comprise between animal welfare and human expectations? *Animal Frontiers, 12*(1), 35–42. https://doi.org/10.1093/af/vfac002.

Chriki, S., & Hocquette, J.-F. (2020). The myth of cultured meat: A review. *Frontiers in Nutrition,* 7, 7.

Chriki, S., Payet, V., Pflanzer, S. B., Ellies-Oury, M.-P., Liu, J., Hocquette, É., et al. (2021). Brazilian Consumers' attitudes towards so-called "cell-based meat". *Food, 10*(11), 2588. https://doi.org/10.3390/foods10112588.

CRÉDOC. (2018). *Combien de végétariens en Europe ?* (p. 36) [Synthèse de l'étude « Panorama de la consommation végétarienne en Europe »].

De Backer, C. J. S., & Hudders, L. (2015). Meat morals: Relationship between meat consumption consumer attitudes towards human and animal welfare and moral behavior. *Meat Science, 99,* 68–74. https://doi.org/10.1016/j.meatsci.2014.08.011.

Domigan, L. J., Feisst, V., & Ogilvie, O. J. (2022). Recipes for cultured meat. *Nature Food, 3*(1), 9–10. https://doi.org/10.1038/s43016-021-00437-z.

Doreau, M., Corson, M. S., & Wiedemann, S. G. (2012). Water use by livestock: A global perspective for a regional issue? *Animal Frontiers, 2*(2), 9–16.

Dumont, B., Dupraz, P., Aubin, J., Benoit, M., Chatellier, V., Bouamra-Mechemache, Z., et al. (2016). *Rôles, impacts et services issus des élevages en Europe.* (Synthèse de l'expertise scientifique collective).

Dumont, B., Modernel, P., Benoit, M., Ruggia, A., Soca, P., Dernat, S., et al. (2020). Mobilizing ecological processes for herbivore production: Farmers and researchers learning together. *Frontiers in Sustainable Food Systems, 4*(544828). https://doi.org/10.3389/fsufs.2020.544828.

Dumont, B., Ryschawy, J., Duru, M., Benoit, M., Delaby, M., & Dourmad, J. Y. (2017). Les bouquets de services, un concept clé pour raisonner l'avenir des territoires d'élevage. *INRA Productions Animales, 30*(4), 407–422.

Dupont, J., & Fiebelkorn, F. (2020). Attitudes and acceptance of young people toward the consumption of insects and cultured meat in Germany. *Food Quality and Preference, 85,* 103983. https://doi.org/10.1016/j.foodqual.2020.103983.

Duru, M., Donnars, C., Ryschawy, J., Therond, O., & Dumont, B. (2017). La "grange": Un cadre conceptuel pour appréhender les bouquets de services rendus par l'élevage dnas les territoires. *INRA Productions Animales, 30*(4), 273–284.

Duru, M., Le Bras, C., & Grillot, M. (2021). Une approche holistique de l'élevage, au cœur des enjeux de santé animale, humaine et environnementale. *Cahiers Agricultures, 30*, 26.

Ellies-Oury, M.-P., Lee, A., Jacob, H., & Hocquette, J.-F. (2019). Meat consumption – What French consumers feel about the quality of beef? *Italian Journal of Animal Science, 18*(1), 646–656. https://doi.org/10.1080/1828051X.2018.1551072.

Fassler, J. (2021). *Lab-grown meat is supposed to be inevitable. The science tells a different story*. The Counter. https://thecounter.org/lab-grown-cultivated-meat-cost-at-scale/.

Fraeye, I., Kratka, M., Vandenburgh, H., & Thorrez, L. (2020). Sensorial and nutritional aspects of cultured meat in comparison to traditional meat: Much to be inferred. *Frontiers in Nutrition*, 7. https://doi.org/10.3389/fnut.2020.00035.

Gerber, P. J., Mottet, A., Opio, C. I., Falcucci, A., & Teillard, F. (2015). Environmental impacts of beef production: Review of challenges and perspectives for durability. *Meat Science, 109*, 2–12.

Gomez-Zavaglia, A., Mejuto, J. C., & Simal-Gandara, J. (2020). Mitigation of emerging implications of climate change on food production systems. *Food Research International, 134*, 109256.

Grasso, A. C., Hung, Y., Olthof, M. R., Verbeke, W., & Brouwer, I. A. (2019). Older Consumers' readiness to accept alternative, more sustainable protein sources in the European Union. *Nutrients, 11*(8), 1904. https://doi.org/10.3390/nu11081904.

Guan, X., Lei, Q., Yan, Q., Li, X., Zhou, J., Du, G., et al. (2021). Trends and ideas in technology, regulation and public acceptance of cultured meat. *Future Foods, 3*, 100032. https://doi.org/10.1016/j.fufo.2021.100032.

Hamdan, M. N., Post, M. J., Ramli, M. A., & Mustafa, A. R. (2018). Cultured meat in Islamic perspective. *Journal of Religion and Health, 57*(6), 2193–2206. https://doi.org/10.1007/s10943-017-0403-3.

Hamdan, M. N., Ramli, M. A., Zaman Huri, N. M. F., Abd Rahman, N. N. H., & Abdullah, A. (2021). Will Muslim consumers replace livestock slaughter with cultured meat in the market? *Trends in Food Science and Technology*. https://doi.org/10.1016/j.tifs.2021.01.034.

Heidemann, M. S., Taconeli, C. A., Reis, G. G., Parisi, G., & Molento, C. F. M. (2020). Critical perspective of animal production specialists on cell-based meat in Brazil: From bottleneck to best scenarios. *Animals, 10*(9), 1678. https://doi.org/10.3390/ani10091678.

Hocquette, J.-F. (2015). Is it possible to save the environment and satisify consumers with artificial meat? *Journal of Integrative Agriculture, 14*(2), 206–207. https://doi.org/10.1016/S2095-3119(14)60961-8.

Hocquette, J.-F. (2016). Is in vitro meat the solution for the future? *Meat Science, 120*, 167–176.

Hocquette, A., Lambert, C., Sinquin, C., Peterolff, L., Wagner, Z., Bonny, S. P. F., et al. (2015). Educated consumers don't believe artificial meat is the solution to the problems with the meat industry. *Journal of Integrative Agriculture, 14*(2), 273–284. https://doi.org/10.1016/S2095-3119(14)60886-8.

Hocquette, É., Liu, J., Ellies-Oury, M.-P., Chriki, S., & Hocquette, J.-F. (2022). Does the future of meat in France depend on cultured muscle cells? Answers from different consumer segments. *Meat Science, 188*, 108776. https://doi.org/10.1016/j.meatsci.2022.108776.

Hwang, J., You, J., Moon, J., & Jeong, J. (2020). Factors affecting Consumers' alternative meats buying intentions: Plant-based meat alternative and cultured meat. *Sustainability, 12*(14), 5662. https://doi.org/10.3390/su12145662.

Iyer, R. D., & Iyer, G. K. (2020). Is cultured meat a viable alternative to conventional meat? *Journal of Management and Public Policy, 11*(2), 19–27.

Jagadeesan, P., & Bin Salem, S. (2020). Religious and regulatory concerns of animal free meat and milk. *ScienceOpen Preprints.* https://doi.org/10.14293/S2199-1006.1.SOR-.PP7B21S.v1.

Joseph, P., Searing, A., Watson, C., & McKeague, J. (2020). Alternative proteins: Market research on consumer trends and emerging landscape. *Meat and Muscle Biology, 4*(2). Article 2 https://doi.org/10.22175/mmb.11225.

Kadim, I. T., Mahgoub, O., Baqir, S., Faye, B., & Purchas, R. (2015). Cultured meat from muscle stem cells : A review of challenges and prospects. *Journal of Integrative Agriculture, 14*(2), 222–233.

Kenigsberg, J. A., & Zivotofsky, A. Z. (2020). A Jewish religious perspective on cellular agriculture. *Frontiers in Sustainable Food Systems, 3.* https://doi.org/10.3389/fsufs.2019.00128.

Ketelings, L., Kremers, S., & de Boer, A. (2021). The barriers and drivers of a safe market introduction of cultured meat: A qualitative study. *Food Control, 130*, 108299. https://doi.org/10.1016/j.foodcont.2021.108299.

Lee, H. J., Yong, H. I., Kim, M., Choi, Y. S., & Jo, C. (2020). Status of meat alternatives and their potential role in the future meat market—A review. *Asian-Australasian Journal of Animal Sciences, 33*(10), 1533–1543. Asian-Australasian Association of Animal Production Societies (AAAP). https://doi.org/10.5713/ajas.20.0419.

Li, X., Zhang, G., Zhao, X., Zhou, J., Du, G., & Chen, J. (2020). A conceptual air-lift reactor design for large scale animal cell cultivation in the context of in vitro meat production. *Chemical Engineering Science, 211*, 115269. https://doi.org/10.1016/j.ces.2019.115269.

Liu, J., Hocquette, É., Ellies-Oury, M.-P., Chriki, S., & Hocquette, J.-F. (2021). Chinese Consumers' attitudes and potential acceptance toward artificial meat. *Food, 10*(2), 353. https://doi.org/10.3390/foods10020353.

Lynch, J., & Pierrehumbert, R. (2019). Climate impacts of cultured meat and beef cattle. *Frontiers in Sustainable Food Systems, 3*, 5.

Mancini, M. C., & Antonioli, F. (2019). Exploring consumers' attitude towards cultured meat in Italy. *Meat Science, 150*, 101–110. https://doi.org/10.1016/j.meatsci.2018.12.014.

Mattick, C. S., Landis, A. E., Allenby, B. R., & Genovese, N. J. (2015). Anticipatory life cycle analysis of in vitro biomass cultivation for cultured meat production in the United States. *Environmental Science & Technology, 49*(19), 11941–11949.

Melzener, L., Verzijden, K. E., Buijs, A. J., Post, M. J., & Flack, J. E. (2021). Cultured beef: From small biopsy to substantial quantity. *Journal of the Science of Food and Agriculture, 101*(1), 7–14. https://doi.org/10.1002/jsfa.10663.

Messmer, T., Klevernic, I., Furquim, C., Ovchinnikova, E., Dogan, A., Cruz, H., et al. (2022). A serum-free media formulation for cultured meat production supports bovine satellite cell differentiation in the absence of serum starvation. *Nature Food, 3*(1), 74–85. https://doi.org/10.1038/s43016-021-00419-1.

Moritz, M. S. M., Verbruggen, S. E. L., & Post, M. J. (2015). Alternatives for large-scale production of cultured beef: A review. *Journal of Integrative Agriculture, 14*(2), 208–216. https://doi.org/10.1016/S2095-3119(14)60889-3.

Mottet, A., de Haan, C., Falcucci, A., Tempio, G., Opio, C., & Gerber, P. (2017). Livestock : On our plates or eating at our table? A new analysis of the feed/food debate. *Global Food Security, 14*, 1–8.

Ng, E. T., Singh, S., Yap, W. S., Tay, S. H., & Choudhury, D. (2021). Cultured meat—A patentometric analysis. *Critical Reviews in Food Science and Nutrition*, 1–11. https://doi.org/10.1080/10408398.2021.1980760.

Obeid, R., Heil, S. G., Verhoeven, M. M., Van Den Heuvel, E. G., De Groot, L. C., & Eussen, S. J. (2019). Vitamin B12 intake from animal foods, biomarkers, and health aspects. *Frontiers in Nutrition, 6.*

Oliver, S. P., Murinda, S. E., & Jayarao, B. M. (2011). Impact of antibiotic use in adult dairy cows on antimicrobial resistance of veterinary and human pathogens: A comprehensive review. *Foodborne Pathogens and Disease*, *8*(3), 337–355.

Olivier, A. (2021). La révolution agroécologique (Ecosociété). http://ecosociete.org/livres/la-revolution-agroecologique.

Ong, S., Choudhury, D., & Naing, M. W. (2020). Cell-based meat: Current ambiguities with nomenclature. *Trends in Food Science & Technology*, *102*, 223–231. https://doi.org/10.1016/j.tifs.2020.02.010.

Ong, K. J., Johnston, J., Datar, I., Sewalt, V., Holmes, D., & Shatkin, J. A. (2021). Food safety considerations and research priorities for the cultured meat and seafood industry. *Comprehensive Reviews in Food Science and Food Safety*, *20*(6), 5421–5448. https://doi.org/10.1111/1541-4337.12853.

Palmieri, N., Perito, M. A., & Lupi, C. (2020). Consumer acceptance of cultured meat: Some hints from Italy. *British Food Journal*. https://doi.org/10.1108/BFJ-02-2020-0092.

Piochi, M., Micheloni, M., & Torri, L. (2022). Effect of informative claims on the attitude of Italian consumers towards cultured meat and relationship among variables used in an explicit approach. *Food Research International*, *151*, 110881. https://doi.org/10.1016/j.foodres.2021.110881.

Pitesky, M. E., Stackhouse, K. R., & Mitloehner, F. M. (2009). Clearing the air: Livestock's contribution to climate change. *Advances in Agronomy*, *103*, 1–40.

Poore, J., & Nemecek, T. (2018). Reducing food's environmental impacts through producers and consumers. *Science*, *360*(6392), 987–992.

Post, M. J. (2012). Cultured meat from stem cells: Challenges and prospects. *Meat Science*, *92*(3), 297–301.

Post, M. J. (2014). Cultured beef: medical technology to produce food. *Journal of the Science of Food and Agriculture*, *94*, 1039–1041. https://doi.org/10.1002/jsfa.6474.

Post, M. J., & Hocquette, J.-F. (2017). Chapter 16 - new sources of animal proteins: Cultured meat. In P. P. Purslow (Ed.), *New aspects of meat quality* (pp. 425–441). Woodhead Publishing. https://doi.org/10.1016/B978-0-08-100593-4.00017-5.

Purdy, C. (2022). *Is cell-cultured pork kosher?*. Quartz: Jewish scholars think it's possible. https://qz.com/1238812/a-loophole-may-unite-jewish-scholars-split-on-clean-meat-pork-being-kosher/.

Rauw, W. M., Rydhmer, L., Kyriazakis, I., Øverland, M., Gilbert, H., Dekkers, J. C., et al. (2020). Prospects for sustainability of pig production in relation to climate change and novel feed resources. *Journal of the Science of Food and Agriculture*, *100*(9), 3575–3586.

Reis, G. G., Heidemann, M. S., Borini, F. M., & Maiolino Molento, C. F. (2020). Livestock value chain in transition: Cultivated (cell-based) meat and the need for breakthrough capabilities. *Technology in Society*, *62*, 101286. https://doi.org/10.1016/j.techsoc.2020.101286.

Reiss, J., Robertson, S., & Suzuki, M. (2021). Cell sources for cultivated meat: Applications and considerations throughout the production workflow. *International Journal of Molecular Sciences*, *22*(14), 7513. https://doi.org/10.3390/ijms22147513.

Rémond, D. (2019). Les protéines de la viande, intérêt pour la nutrition des seniors. *Médecine des Maladies Métaboliques*, *13*(3), 252–256. https://doi.org/10.1016/S1957-2557(19)30070-7.

Rorheim, A., Mannino, A., Baumann, T., & Caviola, L. (2016). Cultured meat: An ethical alternative to industrial animal farming. *Policy Paper by Sentience Politics*, *1*, 1–14.

Scarborough, P., Appleby, P. N., Mizdrak, A., Briggs, A. D., Travis, R. C., Bradbury, K. E., et al. (2014). Dietary greenhouse gas emissions of meat-eaters, fish-eaters, vegetarians and vegans in the UK. *Climatic Change*, *125*(2), 179–192.

Shapiro, P. (2018). *Clean meat: How growing meat without animals will revolutionize dinner and the world*. Simon and Schuster.

Shaw, E., & Mac Con Iomaire, M. (2019). A comparative analysis of the attitudes of rural and urban consumers towards cultured meat. *British Food Journal*, *121*(8), 1782–1800. https://doi.org/10.1108/BFJ-07-2018-0433.

Siegrist, M., & Hartmann, C. (2020). Perceived naturalness, disgust, trust and food neophobia as predictors of cultured meat acceptance in ten countries. *Appetite*, *155*, 104814. https://doi.org/10.1016/j.appet.2020.104814.

Sijpestijn, G. F., Wezel, A., & Chriki, S. (2022). Can agroecology help in meeting our 2050 protein requirements? *Livestock Science*, *256*, 104822. https://doi.org/10.1016/j.livsci.2022.104822.

Soice, E., & Johnston, J. (2021). Immortalizing cells for human consumption. *International Journal of Molecular Sciences*, *22*(21), 11660. https://doi.org/10.3390/ijms222111660.

Stephens, N. (2022). Join our team, change the world: edibility, productibility and fodd futures in cultured meat company recruitments videos. *Food, Culture & Society*, *25*(1), 1–17. https://doi.org/10.1080/15528014.2021.1884787.

Suthar, T. R., Riddhi, V. S., Theertha, K. N., Anupama, N. D., & Deepti, N. C. (2019). Review and future prospects for lab-grown meat. *RESEARCH REVIEW International Journal of Multidisciplinary*, *4*(04), 1885–1892.

Tuomisto, H. L. (2019). The eco-friendly burger. *EMBO Reports*, *20*(1).

Tuomisto, H. L., & de Mattos, M. J. T. (2011). Environmental impacts of cultured meat production. *Environmental Science & Technology*, *45*(14), 6117–6123. https://doi.org/10.1021/es200130u.

Tuomisto, H. L., Ellis, M. J., & Hasstrup, P. (2014). Environmental impacts of cultured meat: Alternative production scenarios. In *Proceedings of the 9th International Conference on Life Cycle Assessment in the Agri-Food Sector* (pp. 1360–1366).

Verbeke, W., Hung, Y., Baum, C. M., & De Steur, H. (2021). The power of initial perceived barriers versus motives shaping consumers' willingness to eat cultured meat as a substitute for conventional meat. *Livestock Science*, *253*, 104705. https://doi.org/10.1016/j.livsci.2021.104705.

Verbeke, W., Sans, P., & Van Loo, E. J. (2015). Challenges and prospects for consumer acceptance of cultured meat. *Journal of Integrative Agriculture*, *14*(2), 285–294. https://doi.org/10.1016/S2095-3119(14)60884-4.

Wasilenko, J. L., Fratamico, P. M., Sommers, C., DeMarco, D. R., Varkey, S., Rhoden, K., et al. (2014). Detection of Shiga toxin-producing Escherichia coli (STEC) O157:H7, O26, O45, O103, O111, O121, and O145, and salmonella in retail raw ground beef using the DuPont™ BAX® system. *Frontiers in Cellular and Infection Microbiology*, *4*. https://doi.org/10.3389/fcimb.2014.00081.

Weinrich, R., Strack, M., & Neugebauer, F. (2020). Consumer acceptance of cultured meat in Germany. *Meat Science*, *162*, 107924. https://doi.org/10.1016/j.meatsci.2019.107924.

Welin, S. (2013). Introducing the new meat. Problems and prospects. *Etikk I Praksis*, 7(1), 24–37.

Wilks, M., & Phillips, C. J. C. (2017). Attitudes to in vitro meat: A survey of potential consumers in the United States. *PLoS One*, *12*(2), e0171904. https://doi.org/10.1371/journal.pone.0171904.

Williams, P. (2007). Nutritional composition of red meat. *Nutrition and Dietetics*, *64*, S113–S119.

Young, J. F., Therkildsen, M., Ekstrand, B., Che, B. N., Larsen, M. K., Oksbjerg, N., et al. (2013). Novel aspects of health promoting compounds in meat. *Meat Science*, *95*(4), 904–911. https://doi.org/10.1016/j.meatsci.2013.04.036.

Zhang, L., Hu, Y., Badar, I. H., Xia, X., Kong, B., & Chen, Q. (2021). Prospects of artificial meat: Opportunities and challenges around consumer acceptance. *Trends in Food Science & Technology*, *116*, 434–444. https://doi.org/10.1016/j.tifs.2021.07.010.

Zhang, M., Li, L., & Bai, J. (2020). Consumer acceptance of cultured meat in urban areas of three cities in China. *Food Control*, *118*, 107390. https://doi.org/10.1016/j.foodcont.2020.107390.

CHAPTER SEVEN

Meat alternatives: A proofed commodity?

Fatma Boukid[a,*] and Mohammed Gagaoua[b]

[a]Food Safety and Functionality Programme, Institute of Agriculture and Food Research and Technology (IRTA), Monells, Spain
[b]Food Quality and Sensory Science Department, Teagasc Food Research Centre, Ashtown, Dublin, Ireland
*Corresponding author: e-mail address: fatma.boukid@irta.cat

Contents

Abstract

The demand for meat alternatives has been growing in the recent years and it is expected to keep expanding. Motives driving this growth are diverse and are mostly related to the health and ethical concerns over animal welfare and the possible impact of meat production on the environment. Meat alternatives are not anymore a niche market (i.e., vegetarians and vegans) but going mainstream (also consumed by meat eaters and flexitarians). This can be attributed to advances in formulations and ingredients and innovative technologies resulting in improved quality of products. This chapter summarizes the current knowledge about meat alternatives (from plant, fungi or/and algae). Plant-based alternatives are already a proofed commodity with a wide range of products available in the market. The search for more sustainable sources had put the spotlight on other sources such as fungi and algae due to their limited need to land, energy and water for biomass production. Fungi-based alternatives are steps ahead of those algae-based owing to fungi neutral flavor, comparable nutritional profile to meat

Advances in Food and Nutrition Research, Volume 101
ISSN 1043-4526
https://doi.org/10.1016/bs.afnr.2022.02.003

and no need to masking agents. Consumer acceptance is still a key challenge boosting research and development efforts to ensure a better quality, affordability, and sustainability.

1. Introduction

Proteins are fundamental elements for human health due to their content on essential amino acids. World Health Organization (WHO) set the daily protein intake at 0.8 g protein/kg body weight per day per for adults (WHO, 2002). The nutritional quality of proteins is not only a matter of content but also of amino acid composition and digestibility. Animal proteins are of high nutritional value having complete amino acid composition and high digestibility. They derive mainly from milk, eggs, cheese, and meat. However, high consumption animal-based products seemed to be related, although controversies exist, to health risks such as cardiovascular disease, cancer and an increased mortality rate (Hemler & Hu, 2019; Song et al., 2016; Zhong, Van Horn, & Greenland, 2020). Animal proteins raise in some countries mainly the developed ones ethical concerns over animal welfare and possible negative impact on the environment (Akaichi, Glenk, & Revoredo-Giha, 2019; Boukid, Rosell, Rosene, Bover-Cid, & Castellari, 2021; Hartmann & Siegrist, 2020). As such, the market of alternative proteins is witnessing a great expansion for multiple reasons including growing population expected to be 9 billion by 2050, health consciousness, niches following vegetarian, vegan and flexitarians diets, ethical concerns and sustainability (Batista et al., 2017; Boukid, Rosell, & Castellari, 2021; Souza Filho, Andersson, Ferreira, & Taherzadeh, 2019).

Meat alternatives are also called meat substitutes, meat analogs, fake meat, faux meat, and mock meat (Boukid, 2021). Meat alternatives are meatless products where the animal ingredients in processed meat products were replaced by non-animal substitutes with the intention to deliver a meat-like experience. The popularity of meat alternatives has increased rapidly in developed countries in recent years due to innovative technologies and the exploration of new ingredients simulating the taste, texture, and functionality of meat products. The marketed products have several forms such as burgers, patties, nuggets, and crumbles (Boukid & Castellari, 2021a, 2021b). Therefore, meat alternatives are not anymore destined only to niches "vegans and vegetarians" but also to meat reducers "flexitarians" and meat eaters (Boukid, 2021) and their global market expansion is following an exponential growth curve (Fig. 1). For example, from 1996 to 2020,

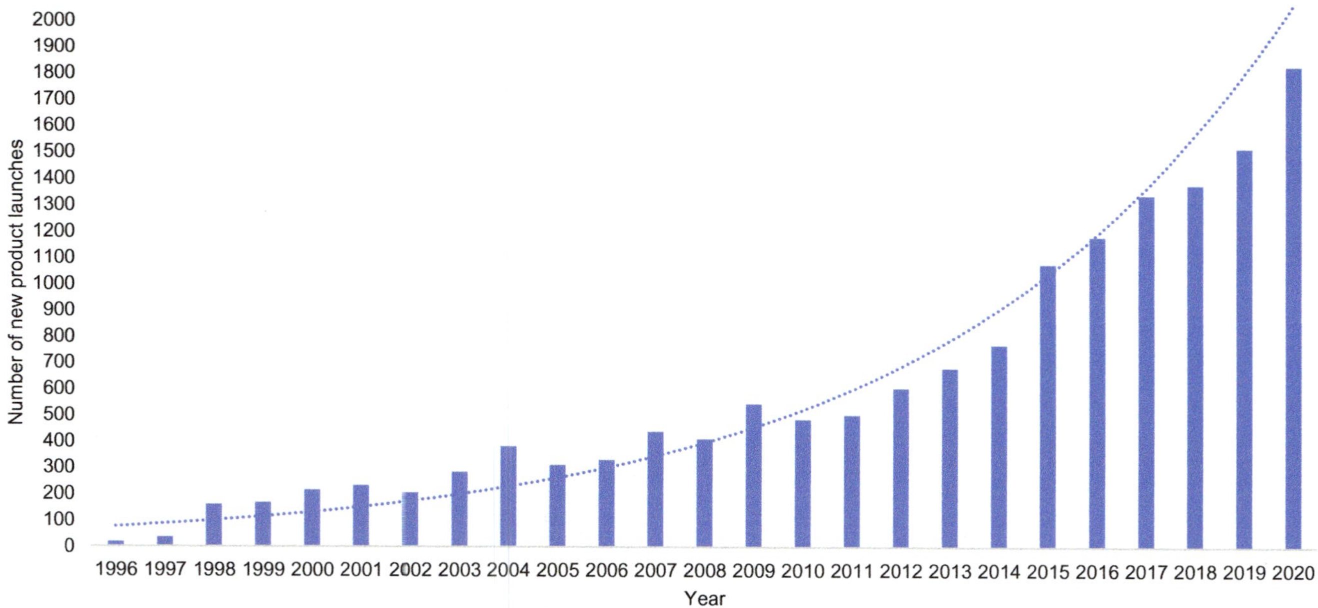

Fig. 1 Launches of alternative meat products in the global market. The figure is based on Mintel's GNPD database accessed on October 1st 2021 (Mintel, 2021), using the following keywords: food and drink categories; meat substitute subcategory; launched from 1996 to 2020. The search resulted in a sample of 15061 food (meat alternatives) products.

around 15,061 new products were launched in the market among which the third is claimed vegan (5008 product). The global market is estimated to account for USD 5.37 billion in 2021 and projected to reach USD 10.28 billion by 2028 at a compound annual growth rate (CAGR) of 10.48% (Fortune Business Insights, 2021).

As alternative to animal proteins, a large variety of sources such oil seeds, legumes and cereals and other alternative sources (i.e., fungi, bacteria, and algae) can be used to prepare new food products. As consumer demand for meatless alternatives grows, meat alternative market keeps expanding based on these sources by proposing plant-based, fungi-based, and algae-based products (Alcorta, Porta, Tárrega, Alvarez, & Vaquero, 2021). Among them, plant-based meat alternatives are a valid commodity with high popularity and easily available in supermarkets. Recently, these products are become available in fast foods franchises such as McDonald's UK&I that recently added plant-based burger to the menu (Mintel, 2020). Fungi- and algae-based meat are other new alternatives gaining interest within consumers. Advances in biotechnological tools deviated the innovation to cultured meat in the frame of cellular agriculture (Jairath, Mal, Gopinath, & Singh, 2021). However, this latter is still controversial for several reasons including safety, regulation and ethical concerns and consumer acceptance.

The rapid increase in the consumption of meat alternatives has raised concerns over their healthfulness and their efficiency in delivering a balanced nutrition. Beside nutritional quality, ultra-processed nature of these products is also a hot topic. Changing dietary behaviors to elicit a transition towards alternative protein sources is challenging in part owing to strongly held taste preferences, culinary traditions and social and cultural norms (Farsi, Uthumange, Munoz, & Commane, 2021). If these foods are developed to reduce meat consumption, ensuring an acceptable level of the nutritional and organoleptic qualities is a must. Based on the above, this chapter gives an overview about the main alternative proteins and discusses the current challenges and limitations in research and development of meat alternatives.

2. Main non-animal alternative protein sources used in meat alternatives

Table 1 summarizes the advantages and limitations of the main non-animal protein sources used to prepare meat alternatives. In the following subsections, we briefly describe those mainly used and the reasons of their use.

Table 1 Overview on non-animal protein sources as ingredients in meat alternatives.

Sources		Advantages	Limitations	References
Plant	Soybean	• High protein content • Balanced amino acid profile • High protein digestibility	• Allergies • Genetically modified organisms	Hendrawati, Audini, Ismiyati, Ramadhan, and Gustia (2021); Taylor, Remington, Panda, Goodman, and Baumert (2015); Xu et al. (2017)
	Gluten	• Rich in sulfur amino acids • Unique rheological properties enabling the creation of strong fibrous structures	• Lack of lysine, threonine, and tryptophan • Immune reactions such as celiac disease	Boukid, Mejri, Pellegrini, Sforza, and Prandi (2017); Boukid (2021); Schreuders et al. (2019); Shewry, Tatham, Forde, Kreis, and Miflin (1986)
	Peas	• Sustainable pulse crop • Hypoallergenic	• Contain anti-nutritional compounds • Deficient in sulfur amino acids such as methionine and cysteine	Boukid, Rosell, and Castellari (2021); Nosworthy et al. (2017)
Fungi		• Low need to land, energy and water • High nutritional value • Rapid growth	• High production cost • Extensive processing	Satari (2018); Lonchamp, Akintoye, Clegg, and Euston (2019); Lonchamp, Clegg, and Euston (2019)
Algae	Macroalgae	• Low need to land, energy and freshwater • High protein content • Grow in saline water	• Low protein digestibility • Off-flavors and color • Potential cross-allergies	Agregán et al. (2017); Bleakley and Hayes (2017)
	Microalgae	• Low need to land, energy and freshwater • Low Greenhouse gas emissions (GHG) • Low nutrient input • High protein content • Balanced amino acids composition	• Low protein digestibility • Off-flavors and color • Novel food regulation	Batista et al. (2019); Lafarga, Fernández-Sevilla, González-López, and Acién-Fernández (2020)
Insects		• Sustainable • Protein content comparable to meat	• Low acceptability • Unclear legislation • Allergies with certain compounds	de Gier and Verhoeckx (2018); Dion-Poulin, Laroche, Doyen, and Turgeon (2020)

2.1 Soybeans

Soybeans (*Glycine max*) are the main oilseed used for protein production and formulation of plant-based products owing to their high protein content (up to 40%). Canola (*Brassica napus* L.) and sunflower (*Helianthus annuus* L.) are also oilseeds that are gaining interest as alternative protein sources (18–26% proteins) as an approach for circular economy and reducing wastes (Alexandrino et al., 2017; Anwar, Ali, & Nasr, 2015; Kotecka-Majchrzak, Sumara, Fornal, & Montowska, 2020). After oil extraction, the by-products of oilseeds (*e.g.*, soybean, sunflower and canola) processing are rich in proteins (up to 65% protein) depending on the raw material and processing conditions (Bernard, 2016). Proteins can be recovered by hydraulic and screw pressing, or solvent extraction (Ji et al., 2019). The majority of proteins in oilseeds consist of storage proteins, mainly globulins and albumins (Boukid, Rosell, Rosene, et al., 2021). Soy proteins provide a balanced amino acid profile (18 amino acids) and all the essential amino acids (Gorissen et al., 2018). Nevertheless, there are raising concerns over the allergenicity of soy proteins, Glycinin and β-conglycinin (Robal et al., 2017; Taylor et al., 2015). Soy allergies cause a life-threatening allergic reaction such as anaphylaxis (Li, Zhu, Zhou, Peng, & Guo, 2016; Shriver & Yang, 2011). In terms of functional properties, soy proteins have excellent gelling, emulsifying ability and water- and oil-holding capacity (Barac, Pesic, Stanojevic, Kostic, & Bivolarevic, 2015; Guo, He, Wu, Zeng, & Chen, 2020). When added to alternative meat, it increases the firmness and juiciness of the final product owing to its capacity to retain water and the formation of a fibrous structure (Dekkers, Nikiforidis, & van der Goot, 2016; Kern, Scharfe, & Hinrichs, 2020).

2.2 Gluten

Wheat consists of 11–25% protein depending on species/varieties and cultivation conditions. Vital gluten is the commercial form of wheat proteins and it is recovered through wet extraction by removing starch, fiber, and fat (Bardini et al., 2018). Gluten is rich in sulfur amino acids but lacks lysine (Lys), threonine (Thr), and tryptophan (Trp) (Shewry et al., 1986). Although gluten can trigger immune reactions such as celiac disease in the case of genetically predisposed, gluten is commonly used in alternative meat formulation (Boukid, 2021; Boukid et al., 2017). Gluten has unique rheological properties enabling the creation of strong fibrous structures resulting in desirable alternative product textures (Schreuders et al., 2019).

2.3 Peas

Peas *(Pisum sativum* L.) belong to the family of pulses known as heathy, affordable, available and sustainable sources of proteins (Boukid, Zannini, Carini, & Vittadini, 2019). They are rich in proteins (20–25%) (Lan, Ohm, Chen, & Rao, 2019) composed mainly by globulins (legumin and vicilin), with minor amounts of convicilin proteins (Boukid, Rosell, & Castellari, 2021). The ratio between legumin/vicilin is an important factor related to the functionality of pea proteins (Liang & Tang, 2013; Mession, Roustel, & Saurel, 2017). Pea protein exhibits comparable oil- and water-binding capacity, emulsification and foaming properties of soy protein (Barac et al., 2015; Silva et al., 2019; Stone, Karalash, Tyler, Warkentin, & Nickerson, 2015). This is consolidating the potential use of pea proteins as a hypoallergenic substitute of soy proteins, for instance, in meat substitute formulations.

2.4 Fungi

Fungi belong to a separate kingdom to plants and animals, which includes the fruiting bodies of edible mushrooms. Utilization of fungi as an alternative protein source in human diets present several advantages including low land requirements and high consumers acceptability (Finnigan, Needham, & Abbott, 2016). Production methods of mycoprotein products are mainly based on fermentation of fungi in liquid culture medium (Stoffel et al., 2019). Mycoprotein is an interesting source of good-quality proteins. For instance, edible fungi (*Fusarium venenatum*) have an amino acid composition similar to that of milk and contain all the essential amino acids (Lonchamp, Akintoye, et al., 2019; Lonchamp, Clegg, & Euston, 2019). However, mycoprotein was reported to trigger serious allergic reactions (Jacobson & DePorter, 2018). Owing to their mild taste and low off-flavor, mycoprotein is used in commercial meat alternatives (Pojić, Mišan, & Tiwari, 2018).

2.5 Algae

Algae are classified into macroalgae and microalgae. Macroalgae are mostly known as a source of polysaccharides such as alginate, carrageenan, and agar, which are commonly used in meat alternative formulations as stabilizers, thickeners, and emulsifiers (Boukid & Castellari, 2021a, 2021b). Carrageenan is the most used macroalgal ingredient, but its safety is still a controversial subject (Bixler, 2017; Younes et al., 2018). In European Union (EU), carrageenan (E407) and processed Eucheuma seaweed (E407a) are authorized as food additives based on the Regulation (EU) No. 1333/2008 (EU., 2008). Macroalgae are also rich in proteins (up to 30%) but at lesser extent compared to microalgae (up to 67%) (Aiello et al., 2019; Biris-Dorhoi et al., 2020; Grossmann, Hinrichs, & Weiss, 2019). Microalgae, *Arthrospira platensis* (Spirulina) and *Chlorella* spp., are gaining interest as a protein source due to their low use of water, land, and nutrients for biomass production. Spirulina has a balanced amino acid profile and rich composition in essential amino acids (Lupatini et al., 2017). Concerns over the allergenicity of microalgae are scarcely studied. Nevertheless, it was reported that Spirulina might trigger some allergic reactions (Le, Knulst, & Röckmann, 2014). Spirulina present foaming, emulsifying, gelling and water/oil holding capacities comparable to that of egg white protein (Benelhadj, Gharsallaoui, Degraeve, Attia, & Ghorbel, 2016; Menegotto et al., 2019; Teuling, Schrama, Gruppen, & Wierenga, 2019; Yücetepe, Saroğlu, & Özçelik, 2019).

2.6 Edible insects

Edible insects are gaining attention as alternative food sources for human nutrition although their consumption is very ancient (Baiano, 2020). The global market of edible insects is expected to grow at a CAGR of 26.5% from 2020 to 2027 to reach around 4.63 USD billion (Globe News Wire, 2021). This market is fueled by the high nutritional value of insects, low environmental impact, and low risk of transmitting zoonotic diseases (Arppe, Niva, & Jallinoja, 2020; Borel et al., 2021; van Huis & Oonincx, 2017). The most consumed insects worldwide are beetles, files, grasshoppers, crickets, weaver ants, giant water bugs and bamboo caterpillar (Cadinu, Barra, Torre, Delogu, & Madau, 2020). These insects provide protein, fat, vitamins, and minerals comparable to those of meat. In addition to the reasons evoked above, insects are emerging alternative protein sources due to their high protein content (35%–60%) (Kim, Yong, Kim, Kim, & Choi, 2019). Compared to plant proteins, insect proteins have a better essential amino acids profile (high Lys, Trp, and Thr), and greater protein digestibility corrected for amino acid score (PDCAAS = 93%) (Oibiokpa, Akanya, Jigam, Saidu, & Egwim, 2018). Although edible insects are traditional foods in Asia, and in some countries in Africa and Latin America, their acceptance as food ingredients is still low in Europe and USA. In 2015, the novel food Regulation (EU) 2015/2283 clarified the legal status of edible insects as a food (European Parliament Council of the European Union, 2015). In USA, the status of insects is still unclear to be legally marketed as an approved food additive or generally recognized as safe (GRAS)

(Ferri, Di Federico, Damato, Proscia, & Grabowski, 2019; Lotta, 2019). On the other hand, safety issues of edible insects, such as anti-nutrients, microbial risk, and allergens are not fully deciphered (Palmer et al., 2020). This suggests further investigation especially of cross reactivity and effects of processing prior to include insects as ingredients in meat alternative formulations (De Marchi, Wangorsch, & Zoccatelli, 2021; Kim, Setyabrata, Lee, Jones, & Kim, 2016).

3. Plant-based meat alternatives

Plant-based meat alternatives are widely marketed, and they are a proofed commodity estimated to be around USD 4.3 billion in 2020 and is projected to reach USD 8.3 billion by 2025 (Markets and Markets, 2020). This market is driven by flexitarian consumers awareness about the health benefits associated with plant-based products and negative impact of over meat consumption (e.g., obesity, type-2 diabetes, heart disease, and certain cancers). Ethical and environmental concerns also are of great relevance more specifically for niche markets such as vegan and vegetarians. The main manufactures of plant-based alternatives in the global market are Beyond Meat (US), Impossible Foods Inc. (US), Maple Leaf Foods (Canada), The Vegetarian Butcher (Netherlands), Gardein Protein International (Canada), Morningstar Farms (US), Quorn Foods (US), Amy's Kitchen (US), Tofurky (US), Gold&Green Foods Ltd. (Finland), Sunfed (New Zealand), VBites Food Limited (UK), Alpha Foods (US), Planterra Foods (US), Before the Hooray Foods (US), and Yves Veggie Cuisine (Canada) 2025 (Markets and Markets, 2020). These alternatives come in different forms including burgers, sausages, and nuggets (Boukid & Castellari, 2021a, 2021b).

The vegan products are made from plant proteins (20–50%), starch (2–30%) and vegetal fat (0–5%) as well as flavoring and coloring ingredients (Boukid, 2021). Plant proteins can be incorporated into different forms namely flour (20–40%), or protein concentrates (65–70%) and isolates (85–90%) (Bedin, Torricelli, Gigliano, De Leo, & Pulvirenti, 2018; Preece, Hooshyar, Krijgsman, Fryer, & Zuidam, 2017). Texturization is a key step in meat alternatives manufacturing to transform native proteins with a globular shape into textured protein having a linear shape (Dekkers et al., 2016; Pietsch, Bühler, Karbstein, & Emin, 2019; Pietsch, Schöffel, Rädle, Karbstein, & Emin, 2019). In meat alternatives, soy proteins are the one mostly included for the reasons presented above to enable the formation of stable fibrous structures (Pietsch, Bühler, et al., 2019;

Schreuders et al., 2019). Wheat gluten allows the creation of a fibrous proteinaceous structure similar to that of meat thanks to their oil/fat binding and film-forming properties (Sun, Ge, He, Gan, & Fang, 2021). Pea proteins resulted in soft structure thanks to its gelling capacities, hence allowing to create a fibrous structure (Schreuders et al., 2019). It is worthy to note that a mixture (blend) of pea protein, gluten and other proteins can be used in the formulation of meat alternatives (Sha & Xiong, 2020). Impossible™ Foods Inc. and Beyond Meat® Inc., are the main players in plant-based meat alternatives (Mintel, 2020). The products of Impossible™ Foods Inc. use soy as the primary source of proteins, while Beyond Meat® uses a blend of pea protein and rice protein.

Hybrid meat products include both meat and plant-based ingredients (Grasso & Jaworska, 2020). These products preserve the animal flavor while reducing the amount of meat amounts in the formulation. Such products offer convenience and taste to flexitarian consumers (Alcorta et al., 2021; Hicks, Knowles, & Farouk, 2018). The purpose of hybrid meat products is to substitute a part of meat with a more sustainable source, while plant-based ingredients in meat products are added for their functionality such as extenders, fillers, and binders (Boukid & Castellari, 2021a, 2021b).

4. Fungi-based meat alternatives

The global market of fungi-based meat alternatives was valued at 891.34 Million USD in 2020 and estimated to grow with a CAGR of 8.28% from 2020 to 2027 (KTVN, 2021). This market is fueled by concerns over environmental impact of meat production, rapid fungal growth, and high protein content, as well as health benefits such as reducing blood cholesterol concentration and glycemic response (Smetana, Mathys, Knoch, & Heinz, 2015; Souza Filho, Nair, Andersson, Lennartsson, & Taherzadeh, 2018). The main industry players are Quorn/Monde Nissin Corporation, 3fbio Ltd and Mycorena (KTVN, 2021). Presently, mycoprotein is marketed in different forms such as sausages, patties, and burgers.

Fungi-based or mycoprotein meat is made from filamentous fungal biomass mimicking meat taste and texture (Table 2). In 2002, mycoprotein has been designated as GRAS by the Food and Drug Administration (FDA) (Finnigan et al., 2019). Currently, mycoprotein produced by an edible fungus (*Fusarium venenatum*) is marketed as Quorn™ meat substitutes (Derbyshire & Ayoob, 2019). *F. venenatum* is grown in a defined medium and treated to reduce its RNA content. The meat alternative is made by

Table 2 Summary of alternative meat products.

Product		Pros	Cons	References
Plant-based	Vegan products	• More sustainable than animal products • Suitable for vegan and vegetarian • Low cholesterol • Low saturated fatty acids	• Incomplete essential amino acids • Unpleasant taste • Highly processed	Boukid (2021); Dekkers, Boom, and van der Goot (2018); Mohamed (2017)
	Hybrid products	• More familiar to consumers • Taste like meat product depending on the substitution level • Suitable for meat eaters and meat lovers	• Not suitable for vegan and vegetarians • No clear regulation and labeling	Alcorta et al. (2021); Grasso and Jaworska (2020); Hicks et al. (2018)
Fungi-based		• High nutritional value • Minimally processed • Fibrous texture • Neutral flavor • No need to masking agents	• High production cost • Extensive processing	Ritchie, Laird, and Ritchie (2017); Hellwig, Gmoser, Lundin, Taherzadeh, and Rousta (2020)
Algae-based		• Nutritional value • Sustainable	• Off-flavors • Unfamiliar color	Caporgno et al. (2020); Grahl et al. (2018)
Insect-based		• Protein content comparable to meat • Sustainable	• Low acceptability • Unclear legislation	Fasel, Mene-Saffrane, Ruczynski, Komar, and Christe (2017); Sogari, Bogueva, and Marinova (2019)
Cultured meat		• Equivalent nutritive composition to traditional meat • Low GHG emissions, and low agricultural land, and water utilization • Efficient use of energy • Animal welfare • Less allergenic • Similar composition to meat	• Non "natural" product • Low consumer acceptance • Long term effect on health is still unknown • Use of hormone growth • Upscaling • Animal welfare	Jairath et al. (2021); Lynch and Pierrehumbert (2019); Stephens et al. (2018); van der Weele, Feindt, Jan van der Goot, van Mierlo, and van Boekel (2019)

mycoprotein, fiber, starch, and egg albumen (or potato protein in the vegan product) as well as coloring, flavoring, and binding agents to mimic the texture and aspect of meat. Compared to plant-based alternatives, fungi-based meat products have several advantages including the fibrous texture, neutral flavor and low use of land, energy and water as well as its minimally processed nature. Nevertheless, the costs of producing mycoprotein are very high, and therefore having lower prices might urge consumers to consume these products (Ritchie et al., 2017). The use of agro-industrial products to grow fungi can help in reducing production cost and environmental impact mainly by the introduction of potentially cost-effective processes (Hellwig et al., 2020).

5. Algae-based meat alternatives

Algae-based meat alternatives are still in infancy stages, and no data are currently available on market size (Gullón et al., 2020). Nevertheless, algae are already used in commercial meat substitute formulations such as Beef Jerky (USA), The Moving Mountains® Banger (UK) and NovaMeat (Spain). Algae-based meat alternatives are formulated using seaweeds or microalgae powders. A patented Chinese formulation of sausages is made of seaweed, corn starch, pork and other spice (Yuan, 2007). The incorporation of microalgae (chlorella and *Auxenochlorella protothecoides*) was reported to increase vitamins B and E in meatless extrudates (Caporgno et al., 2020). Blends of Spirulina (at 10, 30 or 50%) with textured soy resulted in meat alternatives with dark color and intense flavor (Grahl et al., 2018).

6. Insect-based meat alternatives

Beside proteins, edible insects are rich in unsaturated fatty acids, vitamins (riboflavin, pantothenic acid, and biotin), and minerals (Fe, Zn, Na, Ca, P, Mg, Mn, Cu). The profile of unsaturated fatty acids in edible insects was found comparable to that of poultry and fish (Orkusz, 2021). Several studies showed that the proteins of edible insects had interesting functional properties such as emulsion capacity and gelling ability to be used to partially replace meat in hybrid products (Azzollini, Wibisaphira, Lakemond, & Fogliano, 2019; Kim, Yong, Jang, Kim, & Choi, 2020; Mishyna, Keppler, & Chen, 2021). Partial replacement of pork meat with yellow mealworms resulted in increasing protein and reducing fat contents. It also increased hardness, gumminess, chewiness, and apparent viscosity of frankfurters (Choi et al., 2017). Cooked frankfurters made with 10% substitution maintained the sensorial quality at a level similar to that of the regular product (Choi et al., 2017). Hybrid sausages fortified with insect powders (5 and 10%) showed low cooking losses indicating a good water and fat stabilization. However, cooked hybrid cooked sausages had increased fluid release (both water and fat) when subjected to an external force showing a fragile structure (Scholliers, Steen, & Fraeye, 2020).

7. Cell-based meat

Cell-based meat is also called cultured meat or lab-grown referring to muscle cells produced by *in vitro* myogenesis involving currently animal muscle stem cell culture (Chen et al., 2021; Choi et al., 2021). Producing cultured meat is not recent and it was funded by the National Aeronautics and Space Administration (NASA) to produce *in vitro* cultured protein system or cultured meat for astronauts in 2002 (Benjaminson, Gilchriest, & Lorenz, 2002). Advances in cell culture process boosted the standardization of the production of cultured meat (Ben-Arye & Levenberg, 2019; Chen et al., 2021). Design approaches of cell-based alternatives ranged from commonly studied scaffold-based and self-organizing techniques to those more futurist such as three-dimensional (3D) printing and nanotechnology (Bhat, Kumar, & Fayaz, 2015). The scaffold-based technique involves the isolation of embryonic myoblasts or adult skeletal muscle cells taken from farm animals, primary cell culture inside a bioreactor, and then differentiation into muscle fibers onto a scaffold or carrier (e.g., collagen meshwork or microcarrier beads or gelatin or microstructured film) immersed in plant-based growth medium (Bhat et al., 2015; Orellana et al., 2020; Seah, Singh, Tan, & Choudhury, 2021). The stem cells fused into myotubes (small primitive muscle fibers) differentiate into myofibers, and then harvested and processed to develop the desired flavor (Mohammadian & Madadlou, 2018; Srutee, Sowmya, Uday, & A., 2021). This technique enables the production of a soft texture unlike the structured architecture of a steak, while using self-organizing technique, hence rending possible the production of a more rigid 3D structure (Bonny, Gardner, Pethick, & Hocquette, 2015; Sharma, Thind, & Kaur, 2015). In self-organizing technique, biopsy or explant taken from the donor animal is cultured in a nutrient medium under ambient conditions.

The resulting tissue have similar proportion of muscle, adipose and other cells to those of conventional meat with an enhanced organoleptic attributes (Ramani et al., 2021). However, biopsies need repeated collection of tissues which raises concerns over animal welfare and bioethics (Chen et al., 2021). Cell proliferation requires also hormones growth promotors which are prohibited in EU (Chriki & Hocquette, 2020). Furthermore, the lack of blood circulation in these tissues limits growth, and thus cultured cells go through necrosis if separated from the source of nutrients for long periods by more than 0.5 mm (Kang et al., 2021). Noteworthy, 3D printing is having a momentum in the food sector including alternative meat products. This technique may ensure a better regulation of nutritional composition whilst providing a realistic texture of cell-based meat (Handral, Tay, Chan, & Choudhury, 2020).

So far, cell-based meat is not commercialized because industrial production is still not economically viable due to the high production costs as well as lack of regulatory framework. In 2013, the first patty (approximately 85 g made by Dr. Post) costed USD 330,000 in 2013, while a beef meatball (approximately 1 kg made by Memphis Meat) made in 2017 costed USD 40,000 (Lee, Yong, Kim, Choi, & Jo, 2020; Stephens et al., 2018). Major companies and startups such as Memphis Meats (US), MosaMeat (Netherlands), Just, Inc (US), Integriculture (Japan), Finless Foods Inc. (US), Avant Meats Company Limited (China), Balletic Foods (US), Appleton Meats (Canada), Higher Steaks (UK), and others are planning to release their first prototypes in 2022 (Markets and Markets, 2021). According to a recent market report, the global cell-based meat market is expected to reach USD 214 million in 2025 and USD 593 million by 2032 (Markets and Markets, 2021). Finding equally efficient animal-free media to that animal-based, innovation in scaffolds (better cells proliferation, differentiation and tissue development), and improving process efficiency of bioreactors are the main challenges to overcome through basic research to fuel product innovation (Seah et al., 2021). In addition, avoiding the use of animal cells will ensure a better reputation of cell-based meats (Tso, Lim, & Forde, 2021). Cell-based meats have a nutritional composition and organoleptic properties comparable to those of conventional meat (Bhat et al., 2015; Bodiou, Moutsatsou, & Post, 2020). Since they are lab-cultured products, they don't require land use like conventional crops, and they use less water and produce less GHG emissions compared to traditional meat production (Jairath et al., 2021). These products can be suitable for people having adverse reactions to soy, peas, and gluten (Thavamani, Sferra, & Sankararaman, 2020).

8. Conclusion

The search for sustainable alternatives to animal proteins is gaining momentum. Plant proteins are leading ingredients in developing alternative meat products. The main challenges for plant-based proteins are overcoming the nutritional, organoleptic, and functional limitations to deliver similar experience to that of meat products. Several approaches are being developed to overcome these cons and to boost the position of plant-based meat products as a steady commodity. More alternatives are gradually rising presenting several advantages such as fungi- and algae-based alternatives. They are rich-protein sources and do not require high land or water/fertilizers input like traditional crops. Fungi have also a light color and no pronounced taste contrary to algae having strong color and umami flavor. This is giving privilege to fungal proteins in formulating meat alternatives, while algae can be more adequate for seafood alternatives. Innovative technologies to recover algal protein isolates will boost their applications in meat alternatives. Regardless of protein source, the application of alternative proteins needs to fulfill consumer requirements (nutritional, organoleptic and safety).

References

Agregán, R., Munekata, P. E., Domínguez, R., Carballo, J., Franco, D., & Lorenzo, J. M. (2017). Proximate composition, phenolic content and in vitro antioxidant activity of aqueous extracts of the seaweeds Ascophyllum nodosum, Bifurcaria bifurcata and Fucus vesiculosus. Effect of addition of the extracts on the oxidative stability of canola oil under accelerated storage conditions. *Food Research International*, *99*, 986–994. https://doi.org/10.1016/j.foodres.2016.11.009.

Aiello, G., Li, Y., Boschin, G., Bollati, C., Arnoldi, A., & Lammi, C. (2019). Chemical and biological characterization of Spirulina protein hydrolysates: Focus on ACE and DPP-IV activities modulation. *Journal of Functional Foods*. https://doi.org/10.1016/j.jff.2019.103592.

Akaichi, F., Glenk, K., & Revoredo-Giha, C. (2019). Could animal welfare claims and nutritional information boost the demand for organic meat? Evidence from non-hypothetical experimental auctions. *Journal of Cleaner Production*, *207*, 961–970. https://doi.org/10.1016/j.jclepro.2018.10.064.

Alcorta, A., Porta, A., Tárrega, A., Alvarez, M. D., & Vaquero, M. P. (2021). Foods for - plant-based diets: Challenges and inno;vations. *Foods*, *10*(2), 293. https://doi.org/10.3390/FOODS10020293.

Alexandrino, T. D., Ferrari, R. A., de Oliveira, L. M., de Cássia, S. C., Ormenese, R., & Pacheco, M. T. B. (2017). Fractioning of the sunflower flour components: Physical, chemical and nutritional evaluation of the fractions. *LWT - Food Science and Technology*, *84*, 426–432. https://doi.org/10.1016/j.lwt.2017.05.062.

Anwar, M. M., Ali, S. E., & Nasr, E. H. (2015). Improving the nutritional value of canola seed by gamma irradiation. *Journal of Radiation Research and Applied Sciences*, *8*(3), 328–333. https://doi.org/10.1016/j.jrras.2015.05.007.

Arppe, T., Niva, M., & Jallinoja, P. (2020). The emergence of the Finnish edible insect arena: The dynamics of an 'Active Obstacle'. *Geoforum, 108*, 227–236.

Azzollini, D., Wibisaphira, T., Lakemond, C. M. M., & Fogliano, V. (2019). Toward the design of insect-based meat analogue: The role of calcium and temperature in coagulation behavior of Alphitobius diaperinus proteins. *Lebensmittel-Wissenschaft und-Technologie, 100*, 75–82. https://doi.org/10.1016/j.lwt.2018.10.037.

Baiano, A. (2020). Edible insects: An overview on nutritional characteristics, safety, farming, production technologies, regulatory framework, and socio-economic and ethical implications. *Trends in Food Science & Technology, 100*, 35–50. https://doi.org/10.1016/J.TIFS.2020.03.040.

Barac, M. B., Pesic, M. B., Stanojevic, S. P., Kostic, A. Z., & Bivolarevic, V. (2015). Comparative study of the functional properties of three legume seed isolates: adzuki, pea and soy bean. *Journal of Food Science and Technology, 52*(5), 2779–2787. https://doi.org/10.1007/s13197-014-1298-6.

Bardini, G., Boukid, F., Carini, E., Curti, E., Pizzigalli, E., & Vittadini, E. (2018). Enhancing dough-making rheological performance of wheat flour by transglutaminase and vital gluten supplementation. *LWT - Food Science and Technology, 91*. https://doi.org/10.1016/j.lwt.2018.01.077.

Batista, A. P., Niccolai, A., Bursic, I., Sousa, I., Raymundo, A., Rodolfi, L., et al. (2019). Microalgae as functional ingredients in savory food products: Application to wheat crackers. *Foods (Basel, Switzerland), 8*(12). http://www.ncbi.nlm.nih.gov/pubmed/31771197.

Batista, A. P., Niccolai, A., Fradinho, P., Fragoso, S., Bursic, I., Rodolfi, L., et al. (2017). Microalgae biomass as an alternative ingredient in cookies: Sensory, physical and chemical properties, antioxidant activity and in vitro digestibility. *Algal Research, 26*, 161–171. https://doi.org/10.1016/J.ALGAL.2017.07.017.

Bedin, E., Torricelli, C., Gigliano, S., De Leo, R., & Pulvirenti, A. (2018). Vegan foods: Mimic meat products in the Italian market. *International Journal of Gastronomy and Food Science, 13*, 1–9. https://doi.org/10.1016/j.ijgfs.2018.04.003.

Ben-Arye, T., & Levenberg, S. (2019). Tissue engineering for clean meat production. *Frontiers in Sustainable Food Systems, 3*. https://doi.org/10.3389/FSUFS.2019.00046.

Benelhadj, S., Gharsallaoui, A., Degraeve, P., Attia, H., & Ghorbel, D. (2016). Effect of pH on the functional properties of Arthrospira (Spirulina) platensis protein isolate. *Food Chemistry, 194*, 1056–1063. https://doi.org/10.1016/j.foodchem.2015.08.133.

Benjaminson, M. A., Gilchriest, J. A., & Lorenz, M. (2002). In vitro edible muscle protein production system (MPPS): Stage 1, fish. *Acta Astronautica, 51*(12), 879–889. https://doi.org/10.1016/S0094-5765(02)00033-4.

Bernard, J. K. (2016). Oilseed and oilseed meals. *Reference Module in Food Science*. https://doi.org/10.1016/b978-0-08-100596-5.00756-3. Elsevier.

Bhat, Z. F., Kumar, S., & Fayaz, H. (2015). In vitro meat production: Challenges and benefits over conventional meat production. *Journal of Integrative Agriculture, 14*(2), 241–248.

Biris-Dorhoi, E. S., Michiu, D., Pop, C. R., Rotar, A. M., Tofana, M., Pop, O. L., et al. (2020). Macroalgae—A sustainable source of chemical compounds with biological activities. *Nutrients, 12*(10), 1–23.

Bixler, H. J. (2017). The carrageenan controversy. *Journal of Applied Phycology, 29*(5), 2201–2207. https://doi.org/10.1007/S10811-017-1132-4.

Bleakley, S., & Hayes, M. (2017). Algal proteins: Extraction, application, and challenges concerning production. *Foods, 6*(5), 33. https://doi.org/10.3390/foods6050033.

Bodiou, V., Moutsatsou, P., & Post, M. J. (2020). Microcarriers for upscaling cultured meat production. *Frontiers in Nutrition*, 7. https://doi.org/10.3389/FNUT.2020.00010.

Bonny, S. P. F., Gardner, G. E., Pethick, D. W., & Hocquette, J. F. (2015). What is artificial meat and what does it mean for the future of the meat industry? *Journal of Integrative*

Agriculture, *14*(2), 255–263. Chinese Academy of Agricultural Sciences https://doi.org/10.1016/S2095-3119(14)60888-1.

Borel, P., Hammaz, F., Morand-Laffargue, L., Creton, B., Halimi, C., Sabatier, D., et al. (2021). Using black soldier fly larvae reared on fruits and vegetables waste as a sustainable dietary source of provitamin a carotenoids. *Food Chemistry*, *359*.

Boukid, F. (2021). Plant-based meat analogues: From niche to mainstream. *European Food Research and Technology*, *247*(2), 297–308. https://doi.org/10.1007/S00217-020-03630-9.

Boukid, F., & Castellari, M. (2021a). Veggie burgers in the EU market: A nutritional challenge? *European Food Research and Technology*, *1*(10), 1–9. https://doi.org/10.1007/S00217-021-03808-9.

Boukid, F., & Castellari, M. (2021b). Food and beverages containing algae and derived ingredients launched in the market from 2015 to 2019: A front-of-pack labeling perspective with a special focus on Spain. *Foods*, *10*(1), 173. https://doi.org/10.3390/foods10010173.

Boukid, F., Mejri, M., Pellegrini, N., Sforza, S., & Prandi, B. (2017). How looking for celiac-safe wheat can influence its technological properties. *Comprehensive Reviews in Food Science and Food Safety*, *16*(5). https://doi.org/10.1111/1541-4337.12288.

Boukid, F., Rosell, C. M., & Castellari, M. (2021). Pea protein ingredients: A mainstream ingredient to (re)formulate innovative foods and beverages. *Trends in Food Science & Technology*, *110*, 729–742. https://doi.org/10.1016/j.tifs.2021.02.040.

Boukid, F., Rosell, C. M., Rosene, S., Bover-Cid, S., & Castellari, M. (2021). Non-animal proteins as cutting-edge ingredients to reformulate animal-free foodstuffs: Present status and future perspectives. *Critical Reviews in Food Science and Nutrition*, *137*, 1–31. https://doi.org/10.1080/10408398.2021.1901649.

Boukid, F., Zannini, E., Carini, E., & Vittadini, E. (2019). Pulses for bread fortification: A necessity or a choice? *Trends in Food Science & Technology*, *88*, 416–428. https://doi.org/10.1016/J.TIFS.2019.04.007.

Cadinu, L. A., Barra, P., Torre, F., Delogu, F., & Madau, F. A. (2020). Insect rearing: Potential, challenges, and circularity. *Sustainability (Switzerland)*, *12*(11). https://doi.org/10.3390/su12114567. MDPI AG.

Caporgno, M. P., Böcker, L., Müssner, C., Stirnemann, E., Haberkorn, I., Adelmann, H., et al. (2020). Extruded meat analogues based on yellow, heterotrophically cultivated Auxenochlorella protothecoides microalgae. *Innovative Food Science and Emerging Technologies*, *59*, 102275. https://doi.org/10.1016/j.ifset.2019.102275.

Chen, L., Guttieres, D., Koenigsberg, A., Barone, P. W., Sinskey, A. J., & Springs, S. L. (2021). Large-scale cultured meat production: Trends, challenges and promising biomanufacturing technologies. *Biomaterials*, *121274*. https://doi.org/10.1016/J.BIOMATERIALS.2021.121274.

Choi, Y.-S., Kim, T.-K., Choi, H.-D., Park, J.-D., Sung, J.-M., Jeon, K.-H., et al. (2017). Optimization of replacing pork meat with yellow worm (Tenebrio molitor L.) for Frankfurters. *Korean Journal for Food Science of Animal Resources*, *37*(5), 617. https://doi.org/10.5851/KOSFA.2017.37.5.617.

Choi, K.-H., Yoon, J. W., Kim, M., Lee, H. J., Jeong, J., Ryu, M., et al. (2021). Muscle stem cell isolation and in vitro culture for meat production: A methodological review. *Comprehensive Reviews in Food Science and Food Safety*, *20*(1), 429–457. https://doi.org/10.1111/1541-4337.12661.

Chriki, S., & Hocquette, J. F. (2020). The myth of cultured meat: A review. *Frontiers in Nutrition*, 7. https://doi.org/10.3389/FNUT.2020.00007.

de Gier, S., & Verhoeckx, K. (2018). Insect (food) allergy and allergens. *Molecular Immunology*, *100*, 82–106. https://doi.org/10.1016/J.MOLIMM.2018.03.015.

De Marchi, L., Wangorsch, A., & Zoccatelli, G. (2021). Allergens from edible insects: Cross-reactivity and effects of processing. *Current Allergy and Asthma Reports*, *21*(5). https://doi.org/10.1007/S11882-021-01012-Z/FIGURES/1.

Dekkers, B. L., Boom, R. M., & van der Goot, A. J. (2018). Structuring processes for meat analogues. *Trends in Food Science and Technology*, *81*, 25–36. Elsevier Ltd. https://doi.org/10.1016/j.tifs.2018.08.011.

Dekkers, B. L., Nikiforidis, C. V., & van der Goot, A. J. (2016). Shear-induced fibrous structure formation from a pectin/SPI blend. *Innovative Food Science and Emerging Technologies*, *36*, 193–200. https://doi.org/10.1016/j.ifset.2016.07.003.

Derbyshire, E., & Ayoob, K. T. (2019). Mycoprotein: Nutritional and health properties. *Nutrition Today*, *54*(1), 7–15. https://doi.org/10.1097/NT.0000000000000316.

Dion-Poulin, A., Laroche, M., Doyen, A., & Turgeon, S. L. (2020). Functionality of cricket and mealworm hydrolysates generated after pretreatment of meals with high hydrostatic pressures. *Molecules (Basel, Switzerland)*, *25*(22). https://doi.org/10.3390/MOLECULES 25225366.

EU. (2008). Regulation (EC) No 1333/2008 regulation (EC) No 1333/2008 of the European Parliament and of the Council on food additives. *Official Journal of the European Union*, *26*(3), 19–27.

European Parliament Council of the European Union. (2015). Novel foods. *Regulation (EU)*, 2015/2283. https://eur-lex.europa.eu/legal-content/EN/TXT/?uri=CELEX:32015R 2283.

Farsi, D. N., Uthumange, D., Munoz, J. M., & Commane, D. M. (2021). The nutritional impact of replacing dietary meat with meat alternatives in the UK: A modelling analysis using nationally representative data. *British Journal of Nutrition*, *1–11*. https://doi.org/10.1017/S0007114521002750.

Fasel, N. J., Mene-Saffrane, L., Ruczynski, I., Komar, E., & Christe, P. (2017). Diet induced modifications of fatty-acid composition in mealworm larvae (Tenebrio molitor). *Journal of Food Research*, *6*(5), 22. https://doi.org/10.5539/jfr.v6n5p22.

Ferri, M., Di Federico, F., Damato, S., Proscia, F., & Grabowski, N. T. (2019). Insects as feed and human food and the public health risk—A review. *Berliner und Munchener Tierarztliche Wochenschrift*, *132*(5–6), 191–218. Schluetersche Verlagsgesellschaft mbH and Co.KG https://doi.org/10.2376/0005-9366-18064.

Finnigan, T., Needham, L., & Abbott, C. (2016). Mycoprotein: A healthy new protein with a low environmental impact. *Sustainable Protein Sources*, 305–325. Elsevier Inc. https://doi.org/10.1016/B978-0-12-802778-3.00019-6.

Finnigan, T. J. A., Wall, B. T., Wilde, P. J., Stephens, F. B., Taylor, S. L., & Freedman, M. R. (2019). Mycoprotein: The future of nutritious nonmeat protein, a symposium review. *Current Developments in Nutrition*, *3*(6). https://doi.org/10.1093/cdn/nzz021.

Fortune Business Insights. (2021). *Meat Substitutes Market Size, Share | Global Report, 2021-2028*. https://www.fortunebusinessinsights.com/industry-reports/meat-substitutes-market-100239.

Globe News Wire. (2021). *Edible Insects Market by Product, Insect Type and Application—Global Forecast to 2027*. https://www.globenewswire.com/en/news-release/2021/01/08/2155707/28124/en/Edible-Insects-Market-by-Product-Insect-Type-and-Application-Global-Forecast-to-2027.html.

Gorissen, S. H. M., Crombag, J. J. R., Senden, J. M. G., Waterval, W. A. H., Bierau, J., Verdijk, L. B., et al. (2018). Protein content and amino acid composition of commercially available plant-based protein isolates. *Amino Acids*, *50*(12), 1685–1695. https://doi.org/10.1007/s00726-018-2640-5.

Grahl, S., Palanisamy, M., Strack, M., Meier-Dinkel, L., Toepfl, S., & Mörlein, D. (2018). Towards more sustainable meat alternatives: How technical parameters affect the sensory properties of extrusion products derived from soy and algae. *Journal of Cleaner Production*, *198*, 962–971. https://doi.org/10.1016/j.jclepro.2018.07.041.

Grasso, S., & Jaworska, S. (2020). Part meat and part plant: Are hybrid meat products fad or future? *Foods*, *9*, 1888. https://doi.org/10.3390/FOODS9121888.

Grossmann, L., Hinrichs, J., & Weiss, J. (2019). Solubility of extracted proteins from Chlorella sorokiniana, Phaeodactylum tricornutum, and Nannochloropsis oceanica: Impact of pH-value. *Lebensmittel-Wissenschaft und-Technologie*, *105*, 408–416. https://doi.org/10.1016/j.lwt.2019.01.040.

Gullón, B., Gagaoua, M., Barba, F. J., Gullón, P., Zhang, W., & Lorenzo, J. M. (2020). Seaweeds as promising resource of bioactive compounds: Overview of novel extraction strategies and design of tailored meat products. *Trends in Food Science and Technology*, *100*, 1–18. Elsevier Ltd. https://doi.org/10.1016/j.tifs.2020.03.039.

Guo, J., He, Z., Wu, S., Zeng, M., & Chen, J. (2020). Effects of concentration of flavor compounds on interaction between soy protein isolate and flavor compounds. *Food Hydrocolloids*, *100*. https://doi.org/10.1016/j.foodhyd.2019.105388.

Handral, H. K., Tay, S. H., Chan, W. W., & Choudhury, D. (2020). *3D printing of cultured meat products*. https://doi.org/10.1080/10408398.2020.1815172.

Hartmann, C., & Siegrist, M. (2020). Our daily meat: Justification, moral evaluation and willingness to substitute. *Food Quality and Preference*, *80*. https://doi.org/10.1016/j.foodqual.2019.103799.

Hellwig, C., Gmoser, R., Lundin, M., Taherzadeh, M. J., & Rousta, K. (2020). Fungi burger from stale bread? A case study on perceptions of a novel protein-rich food product made from an edible fungus. *Foods*, *9*(8), 1112. https://doi.org/10.3390/FOODS9081112.

Hemler, E., & Hu, F. (2019). Plant-based diets for cardiovascular disease prevention: All plant foods are not created equal. *Current Atherosclerosis Reports*, *21*(5), 18. https://doi.org/10.1007/s11883-019-0779-5.

Hendrawati, T. Y., Audini, K., Ismiyati, Ramadhan, A. I., & Gustia, H. (2021). Effects and characterization of different soybean varieties in yield and organoleptic properties of tofu. *Results in Engineering*, *11*.

Hicks, T. M., Knowles, S. O., & Farouk, M. M. (2018). Global provisioning of red meat for flexitarian diets. *Frontiers in Nutrition*, *5*. https://doi.org/10.3389/FNUT.2018.00050/TEXT.

Jacobson, M. F., & DePorter, J. (2018). Self-reported adverse reactions associated with mycoprotein (Quorn-brand) containing foods. *Annals of Allergy, Asthma and Immunology*, *120*(6), 626–630. https://doi.org/10.1016/j.anai.2018.03.020.

Jairath, G., Mal, G., Gopinath, D., & Singh, B. (2021). An holistic approach to access the viability of cultured meat: A review. *Trends in Food Science and Technology*, *110*, 700–710.

Ji, H., Han, F., Peng, S., Yu, J., Li, L., Liu, Y., et al. (2019). Behavioral solubilization of peanut protein isolate by atmospheric pressure cold plasma (ACP) treatment. *Food and Bioprocess Technology*, *12*(12), 2018–2027. https://doi.org/10.1007/S11947-019-02357-0.

Kang, D.-H., Louis, F., Liu, H., Shimoda, H., Nishiyama, Y., Nozawa, H., et al. (2021). Engineered whole cut meat-like tissue by the assembly of cell fibers using tendon-gel integrated bioprinting. *Nature Communications*, *12*(1), 1–12. https://doi.org/10.1038/s41467-021-25236-9.

Kern, C., Scharfe, M., & Hinrichs, J. (2020). Texturization of renneted casein-based gel particles by sheet die extrusion: Mechanical properties and numerical analysis of flow characteristics. *Journal of Food Engineering*, *278*. https://doi.org/10.1016/j.jfoodeng.2020.109938.

Kim, H. W., Setyabrata, D., Lee, Y. J., Jones, O. G., & Kim, Y. H. B. (2016). Pre-treated mealworm larvae and silkworm pupae as a novel protein ingredient in emulsion sausages. *Innovative Food Science and Emerging Technologies*, *38*, 116–123. https://doi.org/10.1016/J.IFSET.2016.09.023.

Kim, T. K., Yong, H. I., Jang, H. W., Kim, Y. B., & Choi, Y. S. (2020). Functional properties of extracted protein from edible insect larvae and their interaction with transglutaminase. *Foods*, *9*(5). https://doi.org/10.3390/FOODS9050591.

Kim, T. K., Yong, H. I., Kim, Y. B., Kim, H. W., & Choi, Y. S. (2019). Edible insects as a protein source: A review of public perception, processing technology, and research trends. *Food Science of Animal Resources*, *39*(4), 521–540. Korean Society for Food Science of Animal Resources https://doi.org/10.5851/kosfa.2019.e53.

Kotecka-Majchrzak, K., Sumara, A., Fornal, E., & Montowska, M. (2020). Oilseed proteins—Properties and application as a food ingredient. *Trends in Food Science and Technology*, *106*, 160–170. Elsevier Ltd. https://doi.org/10.1016/j.tifs.2020.10.004.

KTVN. (2021). *Global Mycoprotein Meat Substitute Market Size 2021 Industry Ana - KTVN Channel 2 - Reno Tahoe Sparks News, Weather, Video*. https://www.ktvn.com/story/44031580/global-mycoprotein-meat-substitute-market-size-2021-industry-analysis-by-top-manufacturers-development-revenue-and-regional-analysis-with-growth.

Lafarga, T., Fernández-Sevilla, J. M., González-López, C., & Acién-Fernández, F. G. (2020). Spirulina for the food and functional food industries. *Food Research International*, *137*, 109356. https://doi.org/10.1016/J.FOODRES.2020.109356.

Lan, Y., Ohm, J.-B., Chen, B., & Rao, J. (2019). Phase behavior, thermodynamic and microstructure of concentrated pea protein isolate-pectin mixture: Effect of pH, biopolymer ratio and pectin charge density. *Food Hydrocolloids*, *105556*. https://doi.org/10.1016/j.foodhyd.2019.105556.

Le, T. M., Knulst, A. C., & Röckmann, H. (2014). Anaphylaxis to Spirulina confirmed by skin prick test with ingredients of Spirulina tablets. *Food and Chemical Toxicology*, *74*, 309–310. https://doi.org/10.1016/j.fct.2014.10.024.

Lee, H. J., Yong, H. I., Kim, M., Choi, Y. S., & Jo, C. (2020). Status of meat alternatives and their potential role in the future meat market—A review. *Asian-Australasian Journal of Animal Sciences*, *33*(10), 1533–1543. https://doi.org/10.5713/AJAS.20.0419.

Li, H., Zhu, K., Zhou, H., Peng, W., & Guo, X. (2016). Comparative study of four physical approaches about allergenicity of soybean protein isolate for infant formula. *Food and Agricultural Immunology*, *27*(5), 604–623. https://doi.org/10.1080/09540105.2015.1129602.

Liang, H. N., & Tang, C. H. (2013). PH-dependent emulsifying properties of pea [Pisum sativum (L.)] proteins. *Food Hydrocolloids*, *33*(2), 309–319. https://doi.org/10.1016/j.foodhyd.2013.04.005.

Lonchamp, J., Akintoye, M., Clegg, P. S., & Euston, S. R. (2019). Functional fungal extracts from the Quorn fermentation co-product as novel partial egg white replacers. *European Food Research and Technology*. https://doi.org/10.1007/s00217-019-03390-1.

Lonchamp, J., Clegg, P. S., & Euston, S. R. (2019). Foaming, emulsifying and rheological properties of extracts from a co-product of the Quorn fermentation process. *European Food Research and Technology*, *245*(9), 1825–1839. https://doi.org/10.1007/s00217-019-03287-z.

Lotta, F. (2019). Insects as food: The legal framework. *Edible Insects in the Food Sector*, *105–118*. https://doi.org/10.1007/978-3-030-22522-3_8.

Lupatini, A. L., de Oliveira Bispo, L., Colla, L. M., Costa, J. A. V., Canan, C., & Colla, E. (2017). Protein and carbohydrate extraction from S. platensis biomass by ultrasound and mechanical agitation. *Food Research International*, *99*, 1028–1035. https://doi.org/10.1016/j.foodres.2016.11.036.

Lynch, J., & Pierrehumbert, R. (2019). Climate impacts of cultured meat and beef cattle. *Frontiers in Sustainable Food Systems*, *3*. https://doi.org/10.3389/FSUFS.2019.00005.

Markets and Markets. (2020). *Plant-based meat market size, share & growth trends | industry statistics and forecasts | markets and markets*. https://www.marketsandmarkets.com/Market-Reports/plant-based-meat-market-44922705.html.

Markets and Markets. (2021). *Cultured meat market | size | industry outlook | trends | forecast to 2032*. https://www.marketsandmarkets.com/Market-Reports/cultured-meat-market-204524444.html.

Menegotto, A. L., Souza, L. E. S.d., Colla, L. M., Costa, J. A. V., Sehn, E., Bittencourt, P. R. S., et al. (2019). Investigation of techno-functional and physicochemical properties of Spirulina platensis protein concentrate for food enrichment. *Lebensmittel-Wissenschaft und-Technologie, 114*. https://doi.org/10.1016/j.lwt.2019.108267.

Mession, J. L., Roustel, S., & Saurel, R. (2017). Interactions in casein micelle—Pea protein system (Part II): Mixture acid gelation with glucono-δ-lactone. *Food Hydrocolloids, 73*, 344–357. https://doi.org/10.1016/j.foodhyd.2017.06.029.

Mintel. (2020). *The future of animal proteins, meat alternatives: 2020 - Mintel*. https://clients.mintel.com/report/the-future-of-animal-proteins-meat-alternatives-2020?fromSearch=%3Ffreetext%3DPlant%2520meat%2520substitutes.

Mintel. (2021). *Mintel global new product database*. https://www.gnpd.com/sinatra/search_results/?search_id=3tMdwh5e47&page=0.

Mishyna, M., Keppler, J. K., & Chen, J. (2021). Techno-functional properties of edible insect proteins and effects of processing. *Current Opinion in Colloid & Interface Science, 101508*. https://doi.org/10.1016/J.COCIS.2021.101508.

Mohamed, R. T. S. Y. A. I. (2017). Opinions of non-vegetarian consumers among the Chinese community in Malaysia toward vegetarian food and diets. *Journal of Food Products Marketing, 23*(1), 80–98. https://doi.org/10.1080/10454446.2017.1244795.

Mohammadian, M., & Madadlou, A. (2018). Technological functionality and biological properties of food protein nanofibrils formed by heating at acidic condition. *Trends in Food Science and Technology, 75*, 115–128. Elsevier Ltd. https://doi.org/10.1016/j.tifs.2018.03.013.

Nosworthy, M. G., Franczyk, A. J., Medina, G., Neufeld, J., Appah, P., Utioh, A., et al. (2017). Effect of processing on the in vitro and in vivo protein quality of yellow and green split peas (Pisum sativum). *Journal of Agricultural and Food Chemistry, 65*(35), 7790–7796. https://doi.org/10.1021/acs.jafc.7b03597.

Oibiokpa, F. I., Akanya, H. O., Jigam, A. A., Saidu, A. N., & Egwim, E. C. (2018). Protein quality of four indigenous edible insect species in Nigeria. *Food Science and Human Wellness*, 7(2), 175–183. https://doi.org/10.1016/j.fshw.2018.05.003.

Orellana, N., Sánchez, E., Benavente, D., Prieto, P., Enrione, J., & Acevedo, C. A. (2020). A new edible film to produce in vitro meat. *Foods, 9*(2), 185. https://doi.org/10.3390/FOODS9020185.

Orkusz, A. (2021). Edible insects versus meat—Nutritional comparison: Knowledge of their composition is the key to good health. *Nutrients, 13*, 1207. https://doi.org/10.3390/NU13041207.

Palmer, L. K., Marsh, J. T., Lu, M., Goodman, R. E., Zeece, M. G., & Johnson, P. E. (2020). Shellfish tropomyosin IgE cross-reactivity differs among edible insect species. *Molecular Nutrition & Food Research, 64*(8), 1900923. https://doi.org/10.1002/MNFR.201900923.

Pietsch, V. L., Bühler, J. M., Karbstein, H. P., & Emin, M. A. (2019). High moisture extrusion of soy protein concentrate: Influence of thermomechanical treatment on protein-protein interactions and rheological properties. *Journal of Food Engineering, 251*, 11–18. https://doi.org/10.1016/j.jfoodeng.2019.01.001.

Pietsch, V. L., Schöffel, F., Rädle, M., Karbstein, H. P., & Emin, M. A. (2019). High moisture extrusion of wheat gluten: Modeling of the polymerization behavior in the screw section of the extrusion process. *Journal of Food Engineering, 246*, 67–74. https://doi.org/10.1016/j.jfoodeng.2018.10.031.

Pojić, M., Mišan, A., & Tiwari, B. (2018). Eco-innovative technologies for extraction of proteins for human consumption from renewable protein sources of plant origin. *Trends in Food Science and Technology, 75*, 93–104. Elsevier Ltd. https://doi.org/10.1016/j.tifs.2018.03.010.

Preece, K. E., Hooshyar, N., Krijgsman, A. J., Fryer, P. J., & Zuidam, N. J. (2017). Intensification of protein extraction from soybean processing materials using hydrodynamic cavitation. *Innovative Food Science and Emerging Technologies*, *41*, 47–55. https://doi.org/10.1016/j.ifset.2017.01.002.

Ramani, S., Ko, D., Kim, B., Cho, C., Kim, W., Jo, C., et al. (2021). Technical requirements for cultured meat production: A review. *Journal of Animal Science and Technology*, *63*(4), 681. https://doi.org/10.5187/JAST.2021.E45.

Ritchie, H., Laird, J., & Ritchie, D. (2017). 3f bio: Halving the cost of mycoprotein through integrated fermentation processes. *Industrial Biotechnology*, *13*, 29–31.

Robal, M., Brenner, T., Matsukawa, S., Ogawa, H., Truus, K., Rudolph, B., et al. (2017). Monocationic salts of carrageenans: Preparation and physico-chemical properties. *Food Hydrocolloids*, *63*, 656–667. https://doi.org/10.1016/j.foodhyd.2016.09.032.

Satari, K. K. (2018). Mucoralean fungi for sustainable production of bioethanol and biologically active molecules. *Applied Microbiology & Biotechnology*, *102*(3), 1097–1117. https://doi.org/10.1007/s00253-017-8691-9.

Scholliers, J., Steen, L., & Fraeye, I. (2020). Partial replacement of meat by superworm (Zophobas morio larvae) in cooked sausages: Effect of heating temperature and insect: Meat ratio on structure and physical stability. *Innovative Food Science and Emerging Technologies*, *66*. https://doi.org/10.1016/j.ifset.2020.102535.

Schreuders, F. K. G., Dekkers, B. L., Bodnár, I., Erni, P., Boom, R. M., & van der Goot, A. J. (2019). Comparing structuring potential of pea and soy protein with gluten for meat analogue preparation. *Journal of Food Engineering*, *261*, 32–39. https://doi.org/10.1016/j.jfoodeng.2019.04.022.

Seah, J. S. H., Singh, S., Tan, L. P., & Choudhury, D. (2021). Scaffolds for the manufacture of cultured meat. *Critical Reviews in Biotechnology*. https://doi.org/10.1080/07388551.2021.1931803.

Sha, L., & Xiong, Y. L. (2020). Plant protein-based alternatives of reconstructed meat: Science, technology, and challenges. *Trends in Food Science and Technology*, *102*, 51–61. Elsevier Ltd. https://doi.org/10.1016/j.tifs.2020.05.022.

Sharma, S., Thind, S. S., & Kaur, A. (2015). In vitro meat production system: Why and how? *Journal of Food Science and Technology*, *52*(12), 7599–7607. https://doi.org/10.1007/S13197-015-1972-3.

Shewry, P. R., Tatham, A. S., Forde, J., Kreis, M., & Miflin, B. J. (1986). The classification and nomenclature of wheat gluten proteins: A reassessment. *Journal of Cereal Science*, *4*(2), 97–106. https://doi.org/10.1016/S0733-5210(86)80012-1.

Shriver, S. K., & Yang, W. W. (2011). Thermal and nonthermal methods for food allergen control. *Food Engineering Reviews*, *3*(1), 26–43. https://doi.org/10.1007/s12393-011-9033-9.

Silva, J. V. C., Jacquette, B., Amagliani, L., Schmitt, C., Nicolai, T., & Chassenieux, C. (2019). Heat-induced gelation of micellar casein/plant protein oil-in-water emulsions. *Colloids and Surfaces A: Physicochemical and Engineering Aspects*, *569*, 85–92. https://doi.org/10.1016/j.colsurfa.2019.01.065.

Smetana, S., Mathys, A., Knoch, A., & Heinz, V. (2015). Meat alternatives: Life cycle assessment of most known meat substitutes. *International Journal of Life Cycle Assessment*, *20*(9), 1254–1267. https://doi.org/10.1007/s11367-015-0931-6.

Sogari, G., Bogueva, D., & Marinova, D. (2019). Australian consumers' response to insects as food. *Agriculture (Switzerland)*, *9*(5). https://doi.org/10.3390/agriculture9050108.

Song, M., Fung, T. T., Hu, F. B., Willett, W. C., Longo, V. D., Chan, A. T., et al. (2016). Association of animal and plant protein intake with all-cause and cause-specific mortality. *JAMA Internal Medicine*, *176*(10), 1453–1463. https://doi.org/10.1001/jamainternmed.2016.4182.

Souza Filho, P. F., Andersson, D., Ferreira, J. A., & Taherzadeh, M. J. (2019). Mycoprotein: Environmental impact and health aspects. *World Journal of Microbiology and Biotechnology, 35*(10). Springer Netherlands https://doi.org/10.1007/s11274-019-2723-9.

Souza Filho, P. F., Nair, R. B., Andersson, D., Lennartsson, P. R., & Taherzadeh, M. J. (2018). Vegan-mycoprotein concentrate from pea-processing industry byproduct using edible filamentous fungi. *Fungal Biology and Biotechnology, 5*(1). https://doi.org/10.1186/s40694-018-0050-9.

Srutee, R., Sowmya, R. S., Uday, S., & A. (2021). Clean meat: Techniques for meat production and its upcoming challenges. *Animal Biotechnology*. https://doi.org/10.1080/10495398.2021.1911810.

Stephens, N., Di Silvio, L., Dunsford, I., Ellis, M., Glencross, A., & Sexton, A. (2018). Bringing cultured meat to market: Technical, socio-political, and regulatory challenges in cellular agriculture. *Trends in Food Science and Technology, 78*, 155–166. Elsevier Ltd. https://doi.org/10.1016/j.tifs.2018.04.010.

Stoffel, F., Santana, W.d. O., Gregolon, J. G. N., Kist, T. B. L., Fontana, R. C., & Camassola, M. (2019). Production of edible mycoprotein using agroindustrial wastes: Influence on nutritional, chemical and biological properties. *Innovative Food Science and Emerging Technologies, 58*. https://doi.org/10.1016/j.ifset.2019.102227.

Stone, A. K., Karalash, A., Tyler, R. T., Warkentin, T. D., & Nickerson, M. T. (2015). Functional attributes of pea protein isolates prepared using different extraction methods and cultivars. *Food Research International, 76*(P1), 31–38. https://doi.org/10.1016/j.foodres.2014.11.017.

Sun, C., Ge, J., He, J., Gan, R., & Fang, Y. (2021). Processing, quality, safety, and acceptance of meat analogue products. *Engineering*, 7(5), 674–678. https://doi.org/10.1016/J.ENG.2020.10.011.

Taylor, S. L., Remington, B. C., Panda, R., Goodman, R. E., & Baumert, J. L. (2015). Detection and control of soybeans as a food allergen. In *Handbook of food allergen detection and control* (pp. 341–366). Elsevier Ltd. https://doi.org/10.1533/9781782420217.3.341.

Teuling, E., Schrama, J. W., Gruppen, H., & Wierenga, P. A. (2019). Characterizing emulsion properties of microalgal and cyanobacterial protein isolates. *Algal Research, 39*. https://doi.org/10.1016/j.algal.2019.101471.

Thavamani, A., Sferra, T. J., & Sankararaman, S. (2020). Meet the meat alternatives: The value of alternative protein sources. *Current Nutrition Reports, 9*(4), 346–355. https://doi.org/10.1007/S13668-020-00341-1/TABLES/1.

Tso, R., Lim, A. J., & Forde, C. G. (2021). A critical appraisal of the evidence supporting consumer motivations for alternative proteins. *Foods, 10*(1). MDPI AG https://doi.org/10.3390/foods10010024.

van der Weele, C., Feindt, P., Jan van der Goot, A., van Mierlo, B., & van Boekel, M. (2019). Meat alternatives: An integrative comparison. *Trends in Food Science and Technology, 88*, 505–512. Elsevier Ltd. https://doi.org/10.1016/j.tifs.2019.04.018.

van Huis, A., & Oonincx, D. G. A. B. (2017). The environmental sustainability of insects as food and feed. A review. *Agronomy for Sustainable Development, 37*(5). https://doi.org/10.1007/S13593-017-0452-8.

WHO. (2002). *Protein and amino acid requirements in human nutrition report of a joint WHO/FAO/UNU expert consultation protein and amino acid requirements in human nutrition report of a joint WHO/FAO/UNU expert consultation protein and amino acid requirements in human.* www.who.int/bookorders.

Xu, J., Chen, Z., Han, D., Li, Y., Sun, X., Wang, Z., et al. (2017). Structural and functional properties changes of β-conglycinin exposed to hydroxyl radical-generating systems. *Molecules, 22*(11). https://doi.org/10.3390/molecules22111893.

Younes, M., Aggett, P., Aguilar, F., Crebelli, R., Filipič, M., Frutos, M. J., et al. (2018). Re-evaluation of carrageenan (E 407) and processed Eucheuma seaweed (E 407a) as food additives. *EFSA Journal, 16*(4). https://doi.org/10.2903/j.efsa.2018.5238.

Yuan, B. (2007). *Algae meat sausage and producing technique thereof.* http://v3.espacenet.com/textdoc?DB=EPODOC&IDX=CN101263917.

Yücetepe, A., Saroğlu, Ö., & Özçelik, B. (2019). Response surface optimization of ultrasound-assisted protein extraction from Spirulina platensis: Investigation of the effect of extraction conditions on techno-functional properties of protein concentrates. *Journal of Food Science and Technology, 56*(7), 3282–3292. https://doi.org/10.1007/s13197-019-03796-5.

Zhong, V., Van Horn, L., & Greenland, P. (2020). Associations of processed meat, unprocessed red meat, poultry, or fish intake with incident cardiovascular disease and all-cause mortality. *JAMA Internal Medicine, 180*(4), 503–512. https://doi.org/10.1001/jamainternmed.2019.6969.

CHAPTER EIGHT

Fabrication of protein nanomaterials as delivery systems

Weijuan Huang[a,b] and Lingyun Chen[b,*]
[a]College of Food Science, South China Agricultural University, Guangzhou, China
[b]Department of Agricultural, Food & Nutritional Science, University of Alberta, Edmonton, AB, Canada
*Corresponding author: e-mail address: lingyun.chen@ualberta.ca

Contents

Abstract

Bioactive compounds in foods, nutraceuticals and pharmaceutical have been gaining interest due to health benefits, which can help to reduce the risk of certain chronic diseases. Recently, nanoencapsulation have attract attention because it is an efficient and promising approach for protection of bioactive compounds, and delivery them to the target physiological sites for controlled release and improvement absorption. Food proteins are promising materials to be fabricated into a variety of nanostructured delivery systems because of their high nutritional value, good functional properties, and health-benefiting effects. Various techniques and approaches are utilized to prepare nanostructured food protein. This chapter introduces the major techniques for the fabrication of nanoparticles and nanoemulsions from food proteins. The basic principles, advantages, and limitations of the techniques are discussed. The encapsulation and release of bioactive compounds in different nanostructured food proteins are illustrated in specific case studies. Due to the fast growing interest of bioactive encapsulation in various sectors, this chapter is of importance for guiding the development of nanostructured food protein loaded with bioactive ingredients for food, nutraceutical and pharmaceutical applications.

1. Introduction

With the emergence of nanotechnology, researchers become more interested in studying the unique properties of nanoscale materials.

Advances in Food and Nutrition Research, Volume 101
ISSN 1043-4526
https://doi.org/10.1016/bs.afnr.2022.05.003

Nanomaterials refer to products that have independent, self-standing structures with at least one dimension below 100 nm. For example, nanoparticles are zero-dimensional (0D) and nanofibers are one-dimensional (1D) nanostructures. Nanomaterials in food, nutraceuticals and pharmaceutical products usually showed novel physicochemical, sensory, and nutritional, such as appearance, texture, stability, flavor, and gastrointestinal fate (Jafari & McClements, 2017). Many bioactive compounds cannot be directly added into foods and pharmaceuticals in their pure form due to poor solubility, sensitivity to harsh conditions during processing such as heating and pressure, undesired sensory attributes, and poor bioavailability, which hamper the applicability and diminish the efficacy of these compounds (Livney, 2015). Nanostructured materials are expected to be more biologically active than larger sized bulk materials of the same chemical composition due to the much larger specific surface area. By reducing particle size, nanotechnology can contribute to improved properties of bioactive compounds, such as solubility, stability, target delivery, controlled release, and prolonged residence time in the gastrointestinal tract for efficient absorption (Ezhilarasi, Karthik, Chhanwal, & Anandharamakrishnan, 2013; Khalil et al., 2019; Mohammadian et al., 2020; Tarhini, Greige-Gerges, & Elaissari, 2017).

Proteins are micronutrients and play an important role to provide food structure and texture due to their good functional properties including emulsifying, foaming and gelling capacity. Proteins are deemed as ideal materials to fabricate nanomaterials suitable for encapsulation of bioactive compounds in pharmaceutical, food, cosmetic, and other products because of their excellent biocompatibility and biodegradability. The protein amphiphilic nature also allows them to interact with both hydrophilic and hydrophobic molecules (Tarhini et al., 2017). Moreover, the amino acid residues in protein molecular chain contain hydroxyl, amine, sulfhydryl, and carboxylic acid functional groups that can be applied to modulate interactions with bioactive molecules for their encapsulation and controlled release (Mohammadian et al., 2020). For example, whey protein hydrolysate could bind with calcium through carboxyl groups and carbonyl groups. The chelation of calcium ions to whey protein hydrolysate caused peptide chain folding and aggregation, leading to the formation of nanoparticles with good stability and improved calcium absorption (Cai, Zhao, Wang, & Rao, 2015). The functional groups also allow protein nanomaterials to be chemically modified to improve their stability and modulate their rate of degradation in gut. For example, the stability of barley protein based nanoparticles in the gastric environment was improved by the succinylation modification, which also slowed down pancreatic digestion of the protein shell, leading to a sustainable release of the bioactive compounds in the small intestine (Liu, Yang, Wang, Liu, & Chen, 2019). In addition, the hydrolysis of food proteins by digestive enzymes results in the formation of biologically active peptides (e.g., antioxidant peptides), making them substrates of choice for functional food development. Based on these superiorities, both animal proteins (e.g., college, gelatin, whey protein) and plant proteins (e.g., soy protein, pea protein, zein) have been widely used to construct nano-vehicles for the encapsulation and delivery of hydrophobic and hydrophilic bioactive molecules such as antibacterial (e.g., curcumin, essential oil) and antioxidant compounds (e.g., carotenoids, polyphenols), unsaturated fatty acids (e.g., ω-3 and ω-6), vitamins and minerals for food and nutraceutical applications.

Given the tremendous progress made in the developing of nanostructured food proteins, in this review, we focused on the major techniques for the fabrication of nanoparticles and nanoemulsions from various food proteins. Accordingly, the approaches to modulate particle size, bioactive molecules encapsulation and release were introduced for specific fabrication technique. The advantages and disadvantages of different techniques were also discussed.

2. Fabrication of protein nanocarriers

2.1 Protein nanoparticles

Various techniques have been utilized for the preparation of protein nanoparticles, including desolvation, nanoprecipitation, self-assembly, polyelectrolyte complexation/complex coacervation, salting out, emulsion/solvent extraction, and electrospray methods. The commonly used protein nanoparticle preparation methods are based on macromolecule precipitation process. The decrease of solubility of the macromolecules in a given solvent induces the precipitation of proteins alone or in combination with other biomolecules, which can be realized either by adding a non-solvent, or by changing the physico-chemical parameters such as pH, salinity or temperature of the protein solution (Miladi et al., 2014). During nanoparticle formation, the protein undergoes conformational changes. The increase of protein unfolding and the decrease of intramolecular hydrophobic interactions in proteins facilitate the formation of nanoparticles. The size of protein nanoparticles fabricated using these methods commonly ranges from 100 to 1000 nm (Jain, Singh, Arya, Kundu, & Kapoor, 2018).

Choice of the preparation method depends on the characteristics of the protein such as solubility, isoelectric point, molecular weight and surface properties, as well as its final use for bioactive molecule encapsulation, delivery and therapy (Tarhini et al., 2017).

2.1.1 Desolvation

Desolvation is the most commonly used method to fabrication protein-based nanoparticles due to its simplicity and advantage of obtaining smaller-sized nanoparticles (Hong et al., 2020). The principle of this method is to reduce the solubility of an aqueous protein solution by adding a desolvating agent such as ethanol or acetone, thereby leading to the formation of protein precipitation in the form of protein nanoparticles (Fig. 1). The particle size can be modulated by the amount of desolvating agent added. For example, the average particle size of β-lactoglobulin nanoparticles increased from a few nm to about 80 nm with increasing the volume ratio of ethanol:water (from 0 to 5 times) (Gülseren, Fang, & Corredig, 2012). Ethanol-water desolvation ratio, temperature, and ionic strength of the medium also significantly affect protein nanoparticle formation. Moreover, desolvation is usually conducted under basic pH to increase electrostatic repulsion between the protein particles, thus, to prevent them from

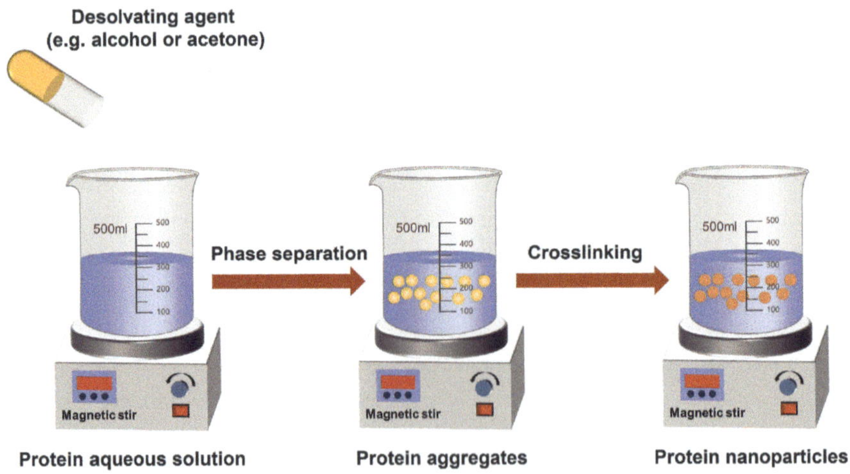

Fig. 1 Preparation of protein nanoparticles by desolvation. In the desolvation process, protein nanoparticles are obtained when a desolvating agent (e.g., ethanol or acetone) is added to an aqueous solution of protein under stirring in order to dehydrate the protein resulting in conformational change. Next, coacervates could be hardened by cross-linking the amino moieties in the protein.

excessive aggregation (Doan & Ghosh, 2019). The prepared protein nanoparticles could be further cross-linked to improve their stability. For example, *N*-(3-dimethylaminopropyl)-*N*-ethylcarbodiimide hydrochloride (EDC) was used as a cross-linker to stabilize egg albumin nanoparticles obtained after desolvation process using ethanol (Prajapati & Srivastava, 2019). The size of soy protein nanoparticles after cross-linking by glutaraldehyde (200–500 nm) were higher than those observed immediately after the desolvation without cross-linking (Teng, Luo, & Wang, 2012).

The desolvation method is widely used in the generation of nanoparticles using β-lactoglobulin (β-Lg) for bioactive compound encapsulation. β-Lg-nanoparticles were fabricated via heat treatment at 85 °C for 30 min before initiating the desolvation method at pH 7 (Fig. 2) (Salah, Mansour, Zogona, & Xu, 2020). Ethanol was added into the protein solution with

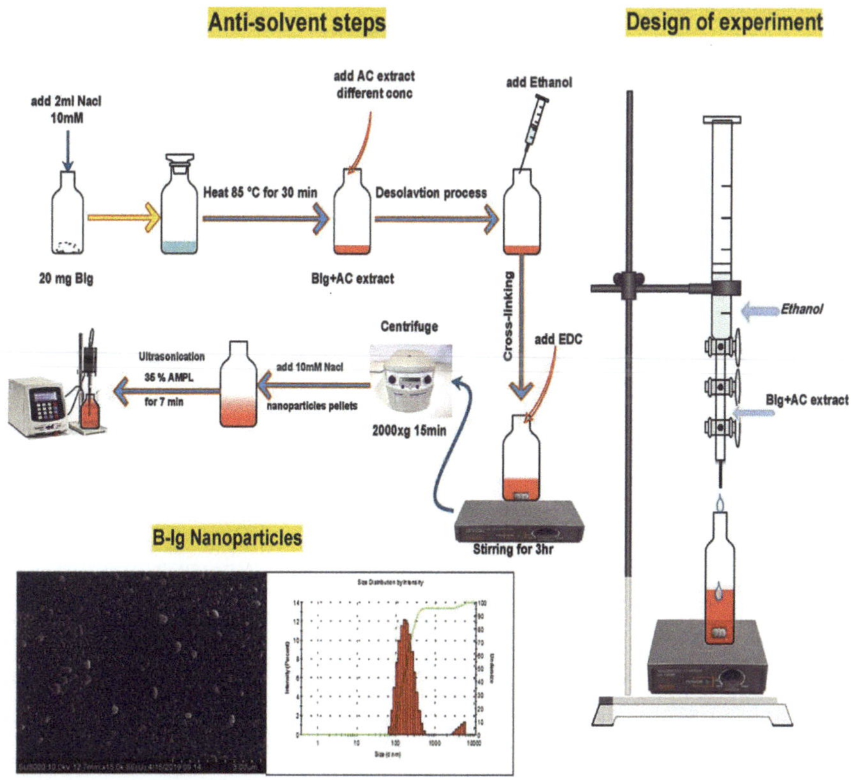

Fig. 2 β-Lg-nanoparticles preparation process by desolvation method. *Reprinted with permission from Salah, M., Mansour, M., Zogona, D., & Xu, X. (2020). Nanoencapsulation of anthocyanins-loaded β-lactoglobulin nanoparticles: Characterization, stability, and bioavailability in vitro. Food Research International, 137, 109635. Copyright (2020) Elsevier Ltd.*

continuous stirring until turbidity developed. *N*-(3-dimethylaminopropyl)-*N*-ethylcarbodiimide hydrochloride (EDC) was used to crosslink the prepared particles. The anthocyanins (AC) extract from red raspberry pomace was encapsulated successfully into β-Lg-nanoparticles with encapsulation efficiency (EE %) of ~77%. This method generated monodisperse particles with the size ranging from 129.13 to 351.85 nm with square morphology. Addition of anthocyanins gradually increased the antioxidant activity of β-Lg-nanoparticles, and improved the nanoparticle stability in mouth, simulated gastric, and simulated intestinal. This work showed that β-Lg nanoparticles prepared by desolvation method is a promising carrier for further application of anthocyanins in food formulations.

Desolvation has been proven as one of the promising methods for the synthesis of protein nanoparticles from β-lactoglobulin, whey protein, and bovine serum albumin (BSA). However, desolvation method has two major drawbacks: the use of organic solvents (acetone and ethanol) and the use of toxic cross-linkers (e.g., glutaraldehyde, EDC). The desolvation process consumes a large amount of organic solvents (e.g., 5 times the protein solution volume) and the desolvating agent usually is added dropwise, which limits the scale-up production of protein nanoparticles. Heat treatment and enzymatic crosslinking of protein nanoparticles are good alternatives for chemical crosslinking.

2.1.2 Non-solvent nanoprecipitation

Non-solvent nanoprecipitation, also called solvent displacement method, is a rapid and easy technique enabling the production of small nanoparticles with narrow or unimodal distribution. Nanoprecipitation requires two miscible solvents, where both protein and bioactive molecules must be soluble in one solvent but insoluble in the other one. Nanoprecipitation occurs when the protein solution is added to the non-solvent. Indeed, as soon as the protein containing solvent diffuses into the dispersing medium, the polymer precipitates involving immediate entrapment of bioactive molecules (Fig. 3) (Elzoghby, Elgohary, & Kamel, 2015; Tarhini et al., 2018). Nanoprecipitation is suitable for preparation of loaded nanoparticles from hydrophobic proteins, such as zein and gliadin because they are soluble in alcohol solutions, whereas precipitate in water (Tarhini et al., 2017).

Anacardic acid is the major constituent of the resin oil, also popularly known as the cashew nut shell liquid, extracted from the mesocarp of bark shells of *Anacardium occidentale* L (Hemshekhar, Sebastin Santhosh, Kemparaju, & Girish, 2012). Anacardic acid (AA) has gained interest

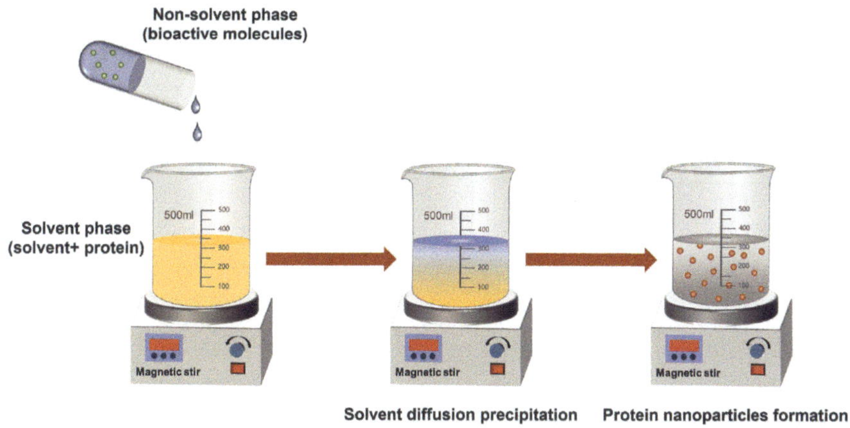

Fig. 3 Preparation of protein nanoparticles by nanoprecipitation. The addition of the non-solvent phase leads to a state of supersaturation which allows the beginning of the nanoprecipitation process and the formation of nanoparticles. The non-solvent nanoprecipitation process includes three steps: generation of supersaturation, nucleation, and growth.

regarding its extraordinary antimicrobial activity. Anacardic acid loaded-zein nanoparticles were prepared by nanoprecipitation (de Araujo, Martin-Pastor, Pérez, Pinazo, & de Sousa, 2021). Zein was dissolved in 70% (v/v) ethanol, which was thereafter gently diluted with ultra-pure water to promote the nanoparticles formation. Dimethylglyoxime was used as a stabilizer agent, which contributed to the particles' long-lasting stabilization and anacardic acid encapsulation. Anacardic acid loaded-zein nanoparticles were spherical, stable, with an average size of 381.6 nm (Fig. 4) and negative surface charge. The OH groups of dimethylglyoxime formed hydrogen bonding with non-bonding electrons, such as N in the primary structure of zein or OH group(s) in anacardic acid, contributed to the stability of the anacardic acid loaded-zein nanoparticles. Moreover, anacardic acid encapsulated in zein nanoparticles was more efficient in reducing *Staphylococcus aureus* and *Candida albicans* biofilms viability than anacardic acid solution.

Since water is usually used as a non-solvent to precipitate alcohol soluble proteins, this method is more economical than desolvation method, and feasible for mass production of protein nanoparticles. Nanoprecipitation is a good strategy for industrial scale production of nanoparticles (Rivas et al., 2017). It should be mentioned that surfactants usually are needed to prevent protein nanoparticles from aggregation and the surfactant nature and concentration affect the particle size.

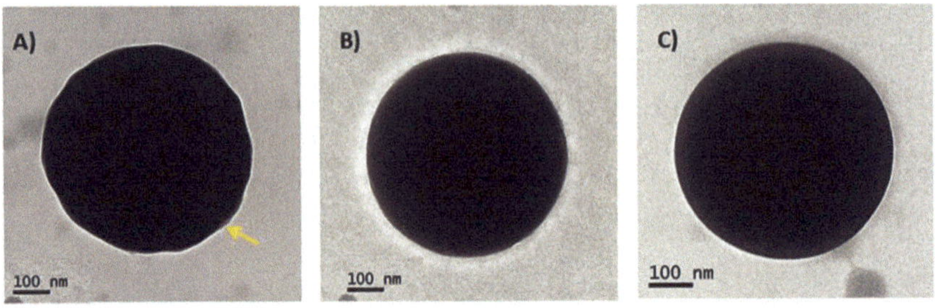

Fig. 4 Transmission electronic microscopic (TEM) images of the nanoparticle formulations. (A) ZA3, Anacardic acid-loaded zein nanoparticles (381.6 nm); (B) ZD, Blank zein-dimethylglyoxime nanoparticles; (C) ZE, Blank zein nanoparticles (392.5 nm). *Reprinted with permission from de Araujo, J. T. C., Martin-Pastor, M., Pérez, L., Pinazo, A., & de Sousa, F. F. O. (2021). Development of anacardic acid-loaded zein nanoparticles: Physical chemical characterization, stability and antimicrobial improvement. Journal of Molecular Liquids, 332, 115808. Copyright (2021) Elsevier Ltd.*

2.1.3 Salting out

The salting-out approach is based on preparing protein coacervates by inducing protein hydrophobic interactions in the presence of a high salt concentration (Zhao, Li, & Xie, 2015). In this method, proteins and bioactive molecules are dissolved into the organic solutions (e.g., acetone and tetrahydrofuran) which are usually water-miscible. The aqueous phase consists of the stabilizer and saturated solution of electrolyte (electrolytes such as magnesium chloride, calcium chloride, magnesium acetate or nonelectrolytes such as sucrose). The organic phase is emulsified in the aqueous phase, under strong shearing force by a mechanical stirring. After emulsification, in order to decrease the ionic strength in the electrolyte, a large amount of distilled water is quickly added into the formed oil/water emulsion under stirring to dilute the emulsion to induce the diffusion of organic solvent into the aqueous phase. At the same time, the hydrophilic organic solvents migrate from the organic phase to aqueous phase, which results in the formation of protein nanoparticles through hydrophobic interactions. Finally, the salting out agent is eliminated by centrifugation or filtration (Fig. 5) (Jain et al., 2018; Tarhini et al., 2017; Wang, Li, Truong-Dinh Tran, Zhang, & Kong, 2016).

Doxorubicin-loaded magnetic silk fibroin nanoparticles (DMSs) were prepared using a one-step potassium phosphate salting-out strategy (Tian, Jiang, Chen, Shao, & Yang, 2014). Briefly, doxorubicin and magnetic Fe_3O_4 nanoparticles (MNPs) are first dispersed in potassium phosphate solution (1.25 M, pH 8) followed by the addition of silk fibroin solution.

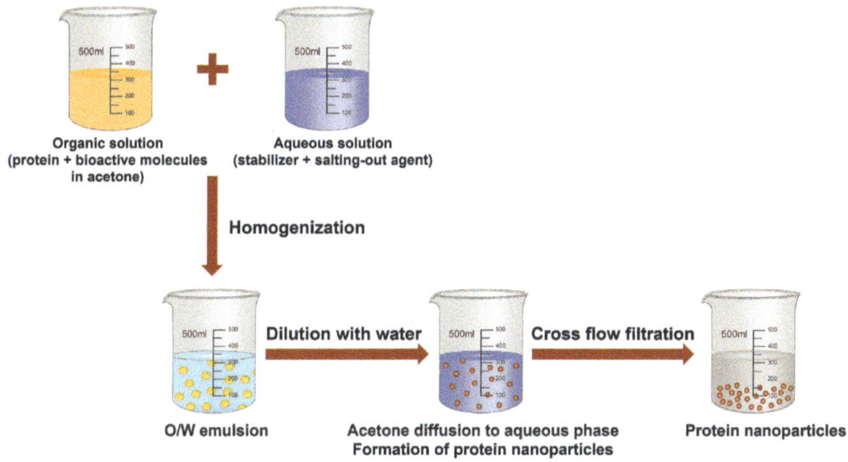

Fig. 5 Schematic diagram of the salting out method for preparing protein nanoparticles. Protein and drug were dissolved in the same solvent, and then emulsified in an aqueous solution containing salting out agents and colloidal stabilizers. Later, water or aqueous solution was added to dilute the emulsion to enhance the diffusion of acetone into the aqueous phase, leading to the formation of protein nanoparticles. Finally, solvent and salting out agent were removed through cross flow filtration.

The samples were than incubated in a refrigerator at −20 °C for 2h before centrifugation at 6000 rpm for 15 min. Subsequently, particles were redispersed in purified water and washed three times. At higher doxorubicin concentration, larger amount of MNPs is required to form DMSs, otherwise particles aggregated into non-dispersible clusters in the salting-out process. The nanoparticles were served as a nanometer-scale delivery system in the chemotherapy of multidrug-resistant cancer under the guidance of a magnetic field.

The simplicity of this approach and relatively high encapsulation efficiency made it attractive. Moreover, the process does not involve heating, thus is suitable for encapsulation of the heat sensitive bioactive molecules. The protein nanoparticles size could be controlled through the appropriated selection of agitation conditions during the emulsification step. However, this technique is restricted to lipophilic bioactive molecules as well as the costly nanoparticles washing step.

2.1.4 Nanoparticle albumin-bound (nab) technology

Albumin is an acidic, very soluble protein (up to 40% *w*/*v*) at pH 7.4. It is stable in the pH range of 4 to 9, soluble in 40% ethanol and can be heated at 60 °C for up to 10 h without any deleterious effects (Loureiro, Azoia, Gomes, & Cavaco-Paulo, 2016). Albumin is a versatile drug carrier in anti-cancer drug delivery system and it also has an actively targeting capacity to tumors (Fu et al., 2009). The readily available albumin includes ovalbumin from egg white, bovine serum albumin (BSA), human serum albumin (HSA). They are biocompatible, biodegradable, and lack of toxicity and immunogenicity, which make them ideal candidates for drug delivery (Elzoghby, Samy, & Elgindy, 2012). Nanoparticle albumin bound technology (nab) is usually applied for the encapsulation of bioactive compounds or drugs with poor water-solubility. In this method, the core ingredient and protein (e.g., albumin) have to be mixed in an aqueous solvent, and then the drug-albumin nanoparticles were formed after the solution passing through a high pressure homogenization for size reduction (Fig. 6). Albumin possesses many chemical groups for drug or ligand attachments on its surface such as amines, carboxylates, and thiols. The size of obtained nanoparticles was in the range 100–200 nm (Desai, 2007). The nab technology was used to encapsulate poorly water-soluble drugs such as paclitaxel and curcumin. Since nanoparticle albumin-bound (nab™) paclitaxel (nab-paclitaxel; Abraxane®) has been approved by FDA for use in patients with metastatic breast cancer who have failed in the combination chemotherapy, the nab-technology has attracted much interest in the anti-cancer drug delivery system (Fu et al., 2009). It is considered as a safe and suitable technique when it comes to intravenously administrated nanoparticles.

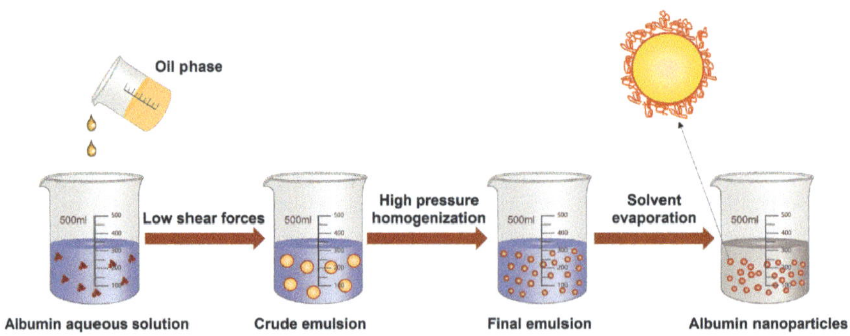

Fig. 6 Schematic representation of nanoparticle albumin bound (nab) technology used to prepare albumin-based nanoparticles. The oil phase containing the drug is added drop-by-drop to the aqueous phase, albumin solution pre-saturated with chloroform. A crude emulsion is formed when the mixture subjected to a low shear forces at low rotating speed, and a final emulsion is produced by homogenizing the crude emulsion using the high pressure homogenizer. The albumin nanosuspensions are yielded by removing the solvent.

An albumin nanoparticle formulation for co-loading paclitaxel (PTX) and curcumin (CCM), both of which have prominent anticancer efficacy, was fabricated via nanoparticle albumin-bound technology (Kim, Lee, Lee, Shin, & Youn, 2016). Specially, bovine serum albumin was dissolved in deionized water; PTX, CCM or PTX/CCM were dissolved individually in 200 μL chloroform: ethanol mixture (9:1). The organic and aqueous phase solutions were gently mixed together and high-pressure homogenized at 20,000 psi for nine cycles. The resulting dispersion was rotary evaporated to remove chloroform at 40 °C under reduced pressure. After centrifugation, the supernatant was lyophilized to recover the PTX/CCM co-bound albumin nanoparticles (PTX/CCM Alb-NPs) (Fig. 7). The particles had size of ~250 nm, and 97.7 ± 1.7% of PTX and 76.2 ± 0.5% of CCM were released from the albumin nanoparticles over the 1-day release test. Furthermore, PTX/CCM Alb-NPs appeared to be efficiently internalized into Mia Paca-2 cells and exhibited a 71% increased IC50 vs PTX Alb-NPs in terms of cytotoxicity to Mia Paca-2 cells. These results suggest that PTX/CCM Alb-NPs are a new potential anticancer agent for combination therapy.

2.1.5 Self-assembly

Protein micelles usually formed from the self-assembly of amphiphilic protein which consist of both hydrophilic and hydrophobic units in an aqueous environment. This self-assembly of amphiphilic proteins is entropically favored and can spontaneously occur above their critical micelle concentration (CMC), resulting in the formation of micelles with a core-shell structure (Hong et al., 2020; Jhaveri & Torchilin, 2014; Tarhini et al., 2017) (Fig. 8). By crosslinking the protein chains, micelles can be stabilized via the solidification process. Protein nanoparticles prepared by self-assembly method possess many advantages, such as high encapsulation efficiency, small nanoparticles, and high stability. However, it is difficult to control the size and shape of nanoparticles (Hong et al., 2020). Native and modified proteins with amphiphilicity and self-assembly behaviors have been designed as artificial building blocks for the construction of self-assembled protein micelles. For example, hydrophilic proteins (e.g., albumin) can become amphiphilicity through hydrophobic modifications. Hydrophobically modified protein molecules can be self-assembled into micelle-like nanospheres upon adding to aqueous solution where the hydrophobic core of micelles acts as a microreservoir for poorly water-soluble bioactive molecules.

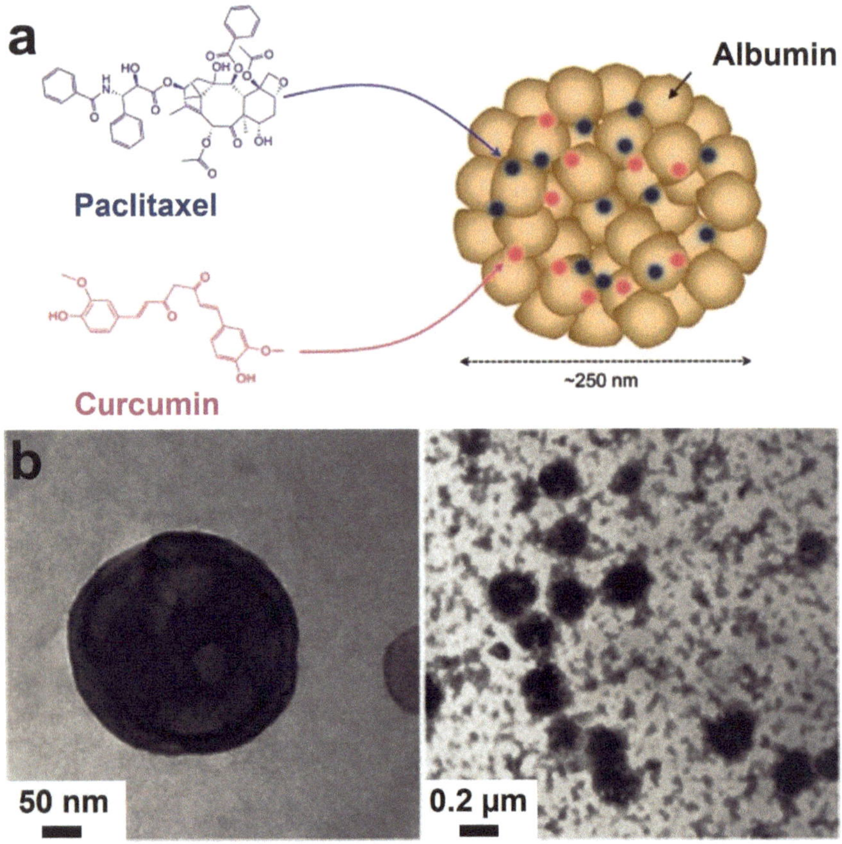

Fig. 7 (A) Chemical structures of paclitaxel and curcumin and an illustration of paclitaxel/curcumin co-bound albumin nanoparticles. (B) Transmission electron microscopic (TEM) images of paclitaxel/curcumin albumin-nanoparticles (PTX/CCM Alb-NPs) with different magnification times. *Reprinted with permission from Kim, B., Lee, C., Lee, E. S., Shin, B. S., & Youn, Y. S. (2016). Paclitaxel and curcumin co-bound albumin nanoparticles having antitumor potential to pancreatic cancer. Asian Journal of Pharmaceutical Sciences, 11(6), 708–714. Copyright (2016) Elsevier Ltd.*

Li et al. report a method of site-selective in situ growth-induced self-assembly (SIGS) to synthesize site-specific human serum albumin-poly(2-(diisopropylamino)ethyl methacrylate) (HSA-PDPA) conjugates that can in situ self-assemble into pH-responsive micelles with tunable morphologies (Fig. 9) (Li et al., 2018). Indocyanine green (ICG), a Food and Drug Administration (FDA) approved near-infrared (NIR) fluorophore, was loaded into the core of sphere-like HSA-PDPA micelles to form

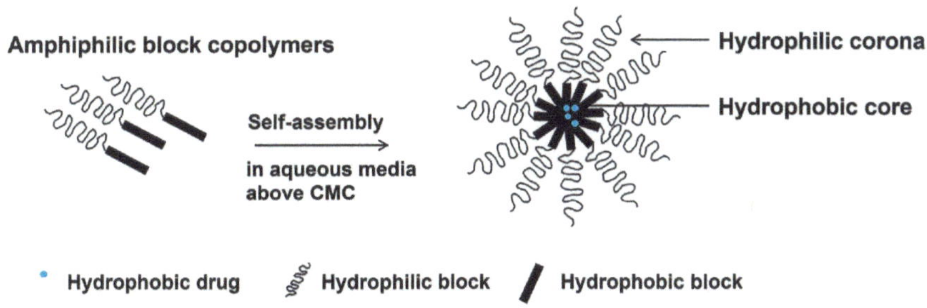

Fig. 8 Protein micelles formed through self-assembly method. Individual protein chains are dissolved in an aqueous solution exceeding the critical micelle concentration (CMC) to spontaneously produce protein micelles during the formation of nanosized aggregates. *Reprinted with permission from Jhaveri, A. M. & Torchilin, V. P. (2014). Multifunctional polymeric micelles for delivery of drugs and siRNA. Frontiers in Pharmacology, 5, 77. Copyright (2014) Frontiers Media S.A.*

pH-responsive fluorescence nanoprobes. The nanoprobes rapidly dissociated into protonated individual unimers at a transition pH of around 6.5, that is the extracellular pH of tumors, which resulted in a sharp fluorescence increase and markedly enhanced cellular uptake. In a tumor-bearing mouse model, they exhibited greatly enhanced tumor fluorescence imaging as compared to indocyanine green alone and pH-nonresponsive nanoprobes. These findings suggest that pH-responsive and site-selective protein – polymer conjugate micelles synthesized by SIGS are promising as a new class of tumor microenvironment-responsive nanocarriers for enhanced tumor imaging and therapy.

2.1.6 Complex coacervation method

Since proteins are amphoteric with multiple charged functional groups, they can be made cationic or anionic by adjusting environmental conditions such as the pH. The charged protein can interact electrostatically with other polymeric electrolytes. The pH-dependent electrostatic interaction between proteins and other polymers can be used to design stable biocompatible nanoparticles for the controlled delivery of bioactive molecules. Complex coacervation is coacervation caused by the interaction of two oppositely charged colloids due to electrostatic forces under aqueous conditions. In coacervation process, liquid–liquid phase separation occurred, giving rise to a polymer-rich dense phase at the bottom and a transparent solution at the top resulting in the desired nanoparticles (Fig. 10). Cationic protein are commonly employed to entrap long stringlike DNA/RNA into nanoparticles by the formation of complexes with anionic oxygen atoms

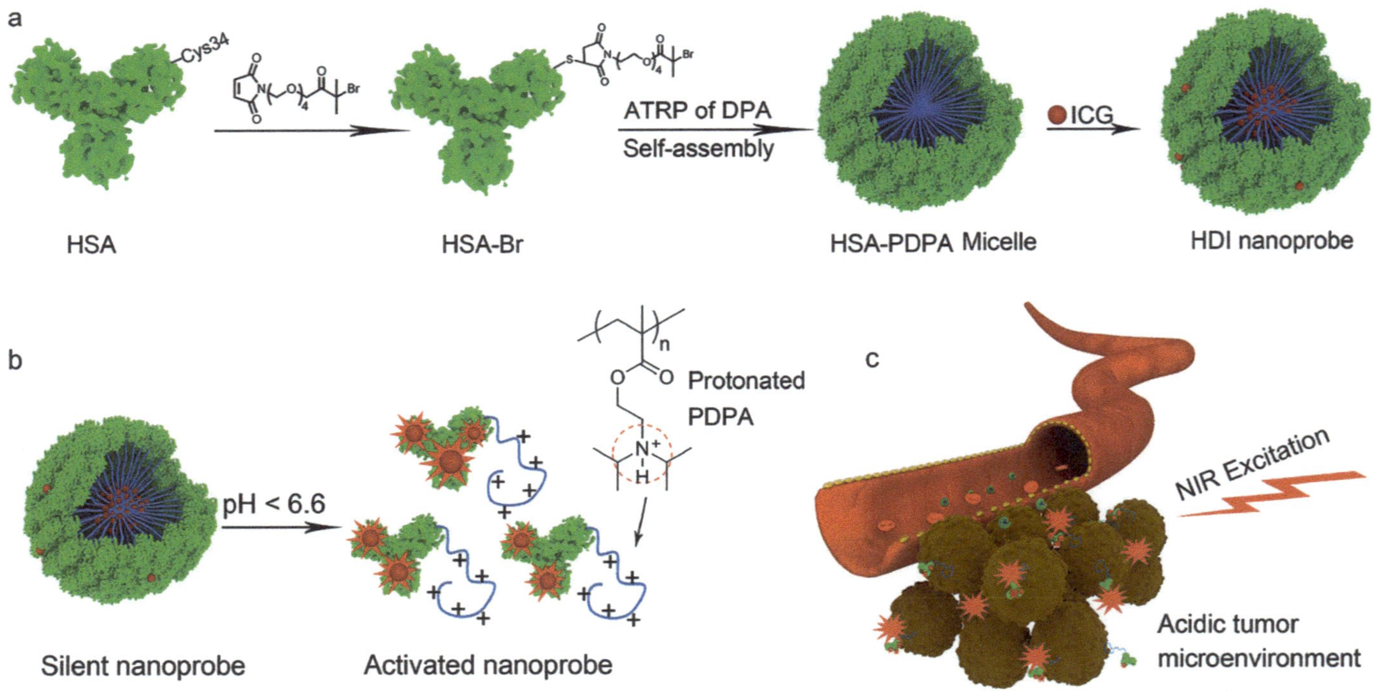

Fig. 9 (A) Schematic illustration of synthesis of HSA-PDPA/ICG (HDI) nanoprobes (B) with pH-Responsiveness (C) for Acidic Tumor Microenvironment-Triggered Fluorescence Imaging. *Reprinted with permission from Li, P., Sun, M., Xu, Z., Liu, X., Zhao, W., & Gao, W. (2018). Site-selective in situ growth-induced self-assembly of protein–polymer conjugates into pH-responsive micelles for tumor microenvironment triggered fluorescence imaging. Biomacromolecules, 19(11), 4472-4479. Copyright (2018) American Chemical Society.*

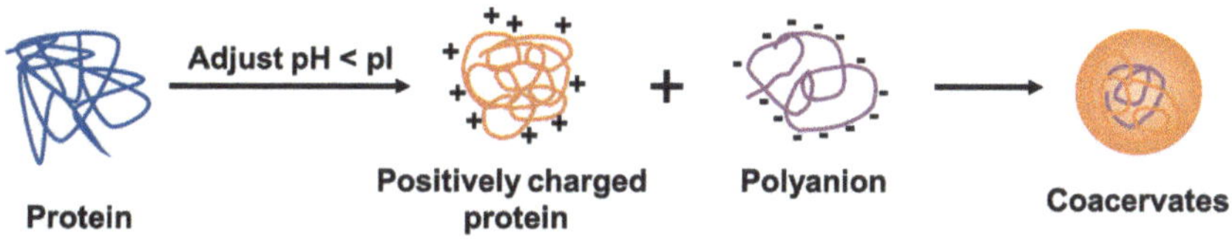

Fig. 10 Fabrication of protein nanoparticles by the polyelectrolyte complexation method. The formation of coacervates between the proteins and polysaccharides due to the initiation of electrostatic interactions between protein and anionic polysaccharides.

in the phosphodiester backbone of the oligonucleotide (Lohcharoenkal, Wang, Chen, & Rojanasakul, 2014). Currently, coacervation of proteins and anionic polysaccharides for encapsulation of active ingredients is of theoretical and practical interest because of the versatility and numerous potential applications of these systems essentially in the fields of food, pharmaceutical, cosmetics and agriculture (De Kruif, Weinbreck, & de Vries, 2004). The formation of coacervates mainly depends on the pH of solvent, which controls charges on the biopolymers surface. During the complex coacervation between protein and anionic polysaccharides, the phase changes occurred between the pI of the protein and the p*Ka* of the polysaccharides.

Ovalbumin (OVA) and propylene glycol alginate (PGA) was used to form coacervate (Zou et al., 2020). The turbidity curve indicated that the soluble complexes and insoluble coacervate occurred at pH 4.78 (pH_c) and 4.08 ($pH_{\phi 1}$), respectively (Fig. 11). The OVA-PGA complexes were formed through the electrostatic interactions between the $-NH_3^+$ groups of OVA and $-COO^-$ groups of PGA.

Ovalbumin (OVA) and high methoxyl pectin (PEC) were applied to fabricate OVA-PEC nanocomplexes to encapsulate and protect vitamin D_3 (VD_3) (Xiang et al., 2020). Vitamin D_3 is a nutrient substance very important for human health and the lack of VD3 will lead to osteoporosis, fragile bones, growth retardation and muscle weakness. Vitamin D_3 is a fat-soluble ingredient, unstable to light, and easily oxidatively decomposed in air. In this work, the formation of OVA-PEC-VD_3 nanocomplexes is ascribed to the electrostatic interactions, hydrogen bonding and hydrophobic interactions among OVA, PEC and VD_3. The VD_3-loaded OVA-PEC nanocomplexes appears as spherical nanoparticles with regular shapes observed by a transmission electron microscope (TEM). The nanocomplexes have a small particle size of about 250 nm and the particle size distribution in the system is relatively uniform (Fig. 12). The encapsulation

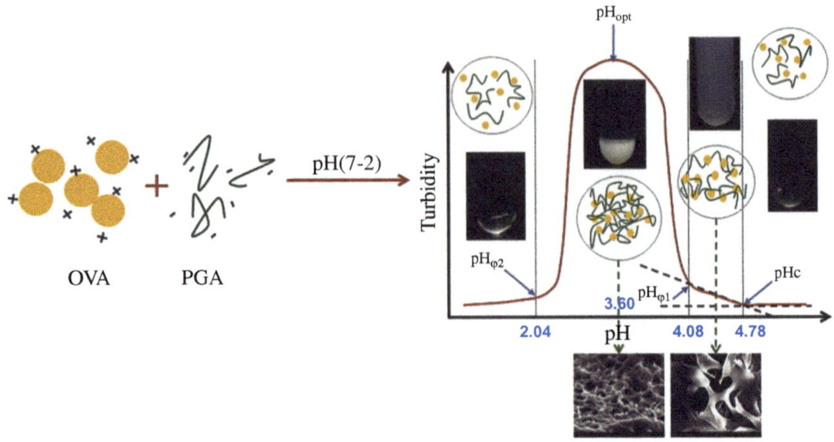

Fig. 11 The formation of coacervates between the ovalbumin (OVA) and propylene glycol alginate (PGA): Phase separation behavior and structure of OVA and PGA mixture. *Reprinted with permission from Zou, W., Mourad, F. K., Zhang, X., Ahn, D. U., Cai, Z., & Jin, Y. (2020). Phase separation behavior and characterization of ovalbumin and propylene glycol alginate complex coacervates. Food Hydrocolloids, 108, 105978. Copyright (2020) Elsevier Ltd.*

efficiency and loading capacity of vitamin D_3 by OVA-PEC complexes could reach to 96.37% and 1.93%, respectively. Moreover, the storage stability of VD_3 was obviously improved after encapsulation. The in vitro release experiment indicated that only a small amount of vitamin D_3 was released from OVA-PEC-VD3 nanocomplexes in the simulated gastric fluid (11.25%) after 60 min, while 98.39% vitamin D_3 was released in the simulated intestinal fluid within 30 min. Therefore the protein–polysaccharide complex coacervates can effectively encapsulate vitamin D_3 and deliver it to the small intestine for release, which have potential applications in food and medicine.

2.1.7 Emulsion/solvent extraction

In a typical emulsion/solvent extraction method, protein aqueous solution is emulsified in oil by using a high-speed homogenizer or ultrasonic shear and the nanoparticles are formed at the W/O interface (Jain et al., 2018) (Fig. 13). Surfactants (e.g., Span 80) are usually added as stabilizers to generate nanoparticles. Protein nanoparticles formed by this method can be chemically stabilized by addition of crosslinkers or thermally stabilized by addition of the W/O emulsion to preheated oil having a temperature of nearly or over 100 °C. The oil phase is then removed using an organic

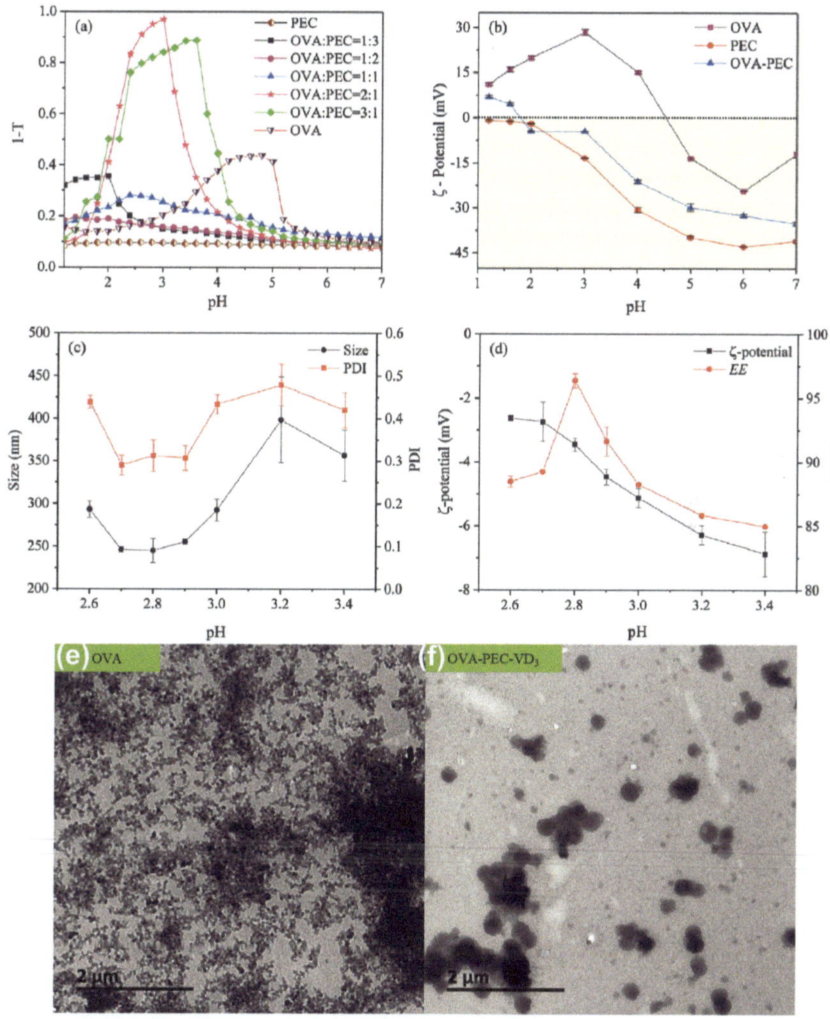

Fig. 12 Optimization of preparation conditions for OVA-PEC-VD_3 nanocomplexes at different pH. (A) Turbidity (1-T) curves of OVA, PEC, OVA-PEC complex solutions at different mass ratios (0.05 wt%). (B) ζ-Potential of OVA (0.025 wt%), PEC (0.025 wt%) and OVA-PEC (2:1, 0.05 wt%) complex solutions at different pH. (C) Size and PDI. (D) ζ-Potential and encapsulation efficiency (EE) for vitamin D_3 (OVA:PEC = 2:1, the concentration of VD_3 = 10 mg/L). TEM images of (E) OVA and (F) OVA-PEC-VD_3 nanocomplexes. *Reprinted with permission from Xiang, C., Gao, J., Ye, H., Ren, G., Ma, X., Xie, H., Fang, S., Lei, Q., & Fang, W. (2020). Development of ovalbumin-pectin nanocomplexes for vitamin D3 encapsulation: Enhanced storage stability and sustained release in simulated gastrointestinal digestion. Food Hydrocolloids, 106, 105926. Copyright (2020) Elsevier Ltd.*

Fig. 13 Preparation of protein nanoparticles by emulsion/solvent extraction method. A protein solution (in aqueous buffer [W]) is dispersed using high speed homogenization or ultrasonication to form an emulsion system (W/O), followed by the solvent being removed to form nanoparticles.

solvent, thus forming protein nanoparticles where the size of the internal phase determines the ultimate size of particles. Afterward, the solidified nanoparticles can be collected by ultracentrifugation. Other than single emulsion (W/O emulsion), double emulsion (W/O/W emulsion) also could be used in this method to develop protein nanoparticles for greater efficiency of encapsulation (Arun, Malrautu, Laha, Luo, & Ramakrishna, 2021). The protein concentration and relative volumes of the water and oil phases (W:O) during emulsification significantly affect the fabrication of the protein nanoparticles (Jain et al., 2018; Mishra et al., 2006). Small-sized nanoparticles were realized by increasing concentrations of BSA and the volume of the water phase.

Whey protein isolate nanoparticles (WPI NPs) were produced by the emulsion/solvent extraction technique. The whey protein solution (10 wt %) was slowly added into the oil phase (corn oil containing hydrophobic emulsifier polyglycerol polyricinoleate) and mixed with a high speed homogenizer to form a water-in-oil (W/O) emulsion. Three different mixing speeds of 7000, 11,000 and 20,000 rpm were conducted to form the W/O emulsion to produce WPI NPs with different sizes. The W/O emulsion was further emulsified using a magnetic stirrer at 500 rpm and heated at 80 °C for about 15 min to induce cross-linking. Then, the emulsions were centrifuged and wash to remove oil and recover protein nanoparticles. The individual WPI NP particles show a wrinkled surface. The particle size distribution became more narrow with increasing mixing speed. The particle size ranges from 200 to 500 nm for mixing at 7000 rpm, 200 to 300 nm at 11000 rpm, and 100 to 200 nm at 20000 rpm (Fig. 14). However, the yield of the protein nanoparticles decreased dramatically with increasing mixing speed.

This emulsion/solvent extraction method is suitable for hydrophobic drug. The major drawbacks of this method include harsh processing

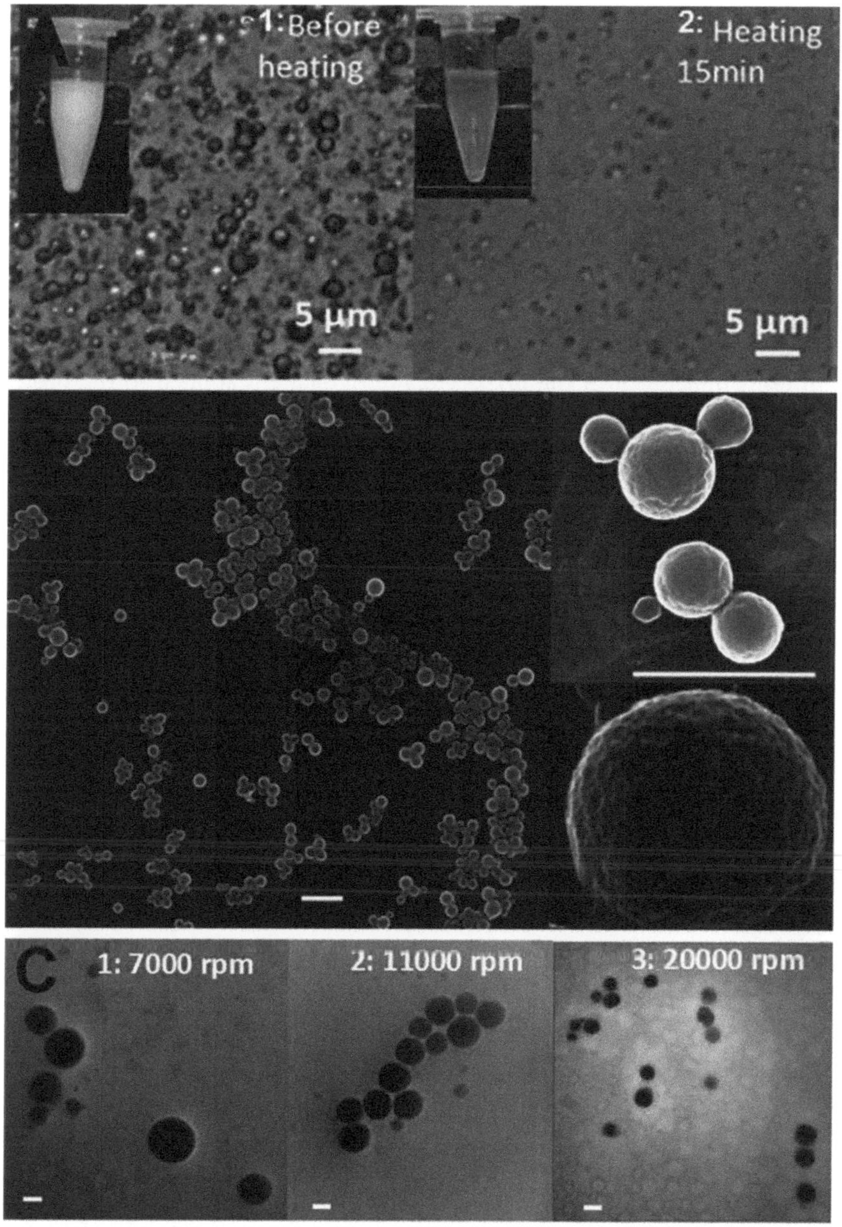

Fig. 14 (A) Heating at 80 °C of a W/O emulsion where the water droplets contain whey protein. (1) The emulsion after homogenization and before heating. (2) The emulsion after heating for 15 min at 80 °C. (B) SEM photograph of heat-induced WPI NPs prepared at a mixing speed of 7000 rpm. The scale bar represents 1 µm. (C) TEM photographs and size distribution (volume-average diameter) of WPI NPs prepared at different mixing speeds: (1) 7000 rpm, (2) 11000 rpm, (3) 20000 rpm. Scale bar represents 200 nm. *Reprinted with permission from Wu, J., Shi, M., Li, W., Zhao, L., Wang, Z., Yan, X., Norde, W., & Li, Y. (2015). Pickering emulsions stabilized by whey protein nanoparticles prepared by thermal cross-linking. Colloids and Surfaces B: Biointerfaces, 127, 96–104. Copyright (2015) Elsevier Ltd.*

conditions (e.g., the use of organic solvent). Compared with other protein nanoparticles preparation method, the emulsion/solvent extraction produce larger particles (Jain et al., 2018).

2.1.8 Electrospray technique

Electrospray technique or electrohydrodynamic atomization is a physical method to fabricate protein nanoparticles. Electrospray is a variant of the method of electrospinning (Tapia-Hernández et al., 2015). In a typical electrospraying process, protein and bioactive ingredients are dispersed in a suitable solvent and loaded in a syringe with a capillary nozzle. Then, the protein solution is pulled out of the nozzle. A high voltage is applied to the protein solution and the charged droplets were sprayed from the tip of the nozzle to form a liquid jet stream, which results in the formation of an aerosolized droplet consisting of protein nanoparticles in the colloidal size. Meanwhile, bioactive molecules were encapsulated in the protein nanoparticles (Fig. 15).

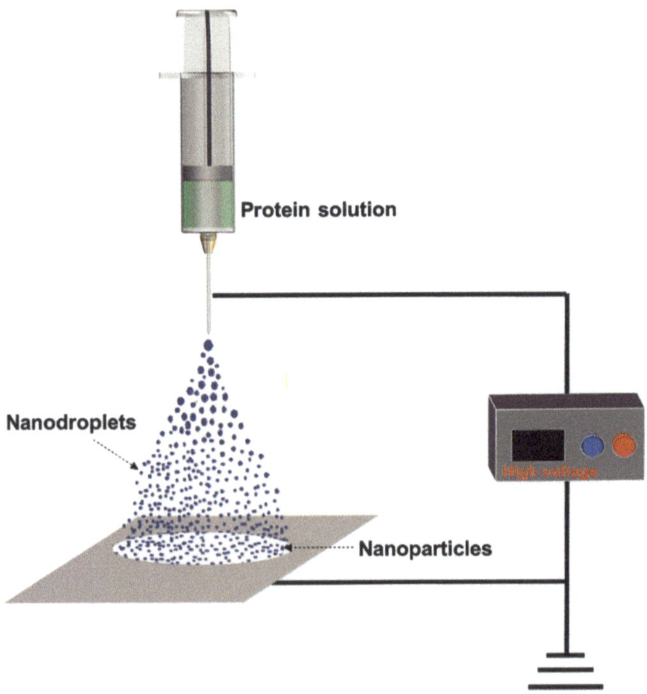

Fig. 15 Production of protein nanoparticles by the electrospraying method. The electrospray technique generates nanoparticles by ejecting a liquid jet stream through a nozzle that forms an aerosolized droplet by applying a high voltage to a protein solution supplied through a nebulizer.

The evaporation of solvent takes place when the electrosprayed jet travels from the capillary to the collector, so the distance must be long enough in order to completely evaporate the solvent, otherwise, protein particles will fuse together (Arun et al., 2021; Hong et al., 2020; Jain et al., 2018). Protein nanoparticles fabricated with this electrospray technique have relatively low hydrodynamic diameters, as the surface tension of the droplets is effectively countered by the electric force. Moreover, protein nanoparticles have high drug loading efficiency. However, this technique may induce some macromolecule degradation due to the stress involved in the operation parameters (e.g., thermal stress in drying, shear stress in the nozzle).

Walnut protein isolate nanoparticles was produced by electrospray technique for curcumin encapsulation (Asadi et al., 2021). The researchers found that the optimized condition for the production of the nanoparticles was protein concentration of 3.5%, a solvent mixture of deionized water: 2-propanol of 2:1 (*v*/v), electrospraying at a fixed distance of 17.5 cm, voltage of 20 kV, and flow rate of 0.09 mL/h. Walnut protein isolate particles had a relatively round shape with an average diameter of 142.3 ± 38.4 nm and the average diameter of curcumin-walnut protein particles increased to 292.7 ± 97.1 nm (Fig. 16). The curcumin-walnut protein particles were still spherical but their structure was less regular and smooth due to formation of spherical agglomerates. When the ratio of curcumin to walnut protein was 1:100 (*w*/w), and the encapsulation efficiency was around 61.5%. The in vitro release of curcumin from curcumin-walnut protein nanoparticles in the simulated gastric fluid (SGF) and simulated intestine fluid (SIF) indicated that the encapsulation of curcumin by electrospray technique could slow down curcumin release in the gastric fluid ($15.49 \pm 2.88\%$) around, while in the simulated small intestine fluid, the curcumin-walnut protein nanoparticles were hydrolyzed, leading to a rapid release of curcumin. In addition, the antioxidant activity of curcumin-walnut protein nanoparticles ($ABTS.^{+}$ radical scavenging activity$= 61.96 \pm 3.21\%$) after in vitro digestion was significantly higher than walnut protein nanoparticles ($ABTS.^{+}$ radical scavenging activity$= 46.96 \pm 1.61\%$) due to the bioactive peptides released from walnut protein nanoparticles which have kept curcumin soluble. Thus, the walnut protein-based nanoparticles could act as a unique food-grade carrier to boost the solubility of curcumin and they are useful for application in pharmaceuticals, functional foods, or beverages with therapeutic prospects.

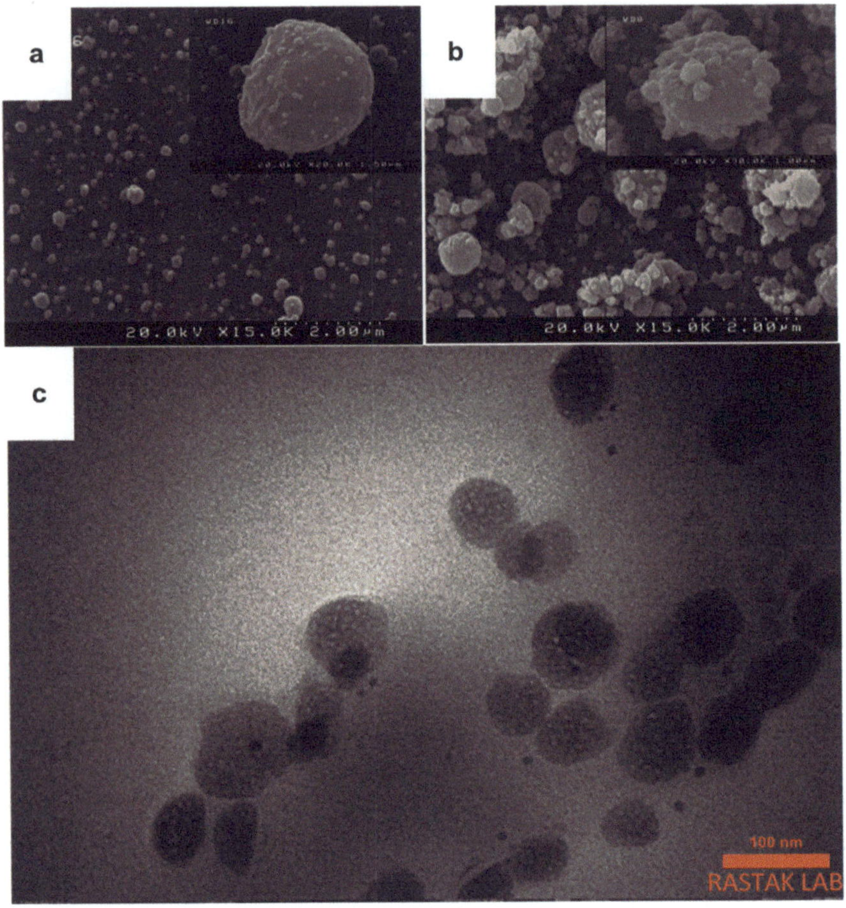

Fig. 16 Field emission scanning electron microscopy (FE-SEM) images of (A) walnut protein nanoparticles and (B) curcumin-walnut protein (curcumin:protein, 1:100 *w*/w) nanoparticles; (C) High resolution transmission electron microscopy (HR-TEM) image of curcumin-walnut protein (curcumin:protein, 1:100 w/w) nanoparticles fabricated via electrospraying at optimal conditions: 20 kV, 17.5 cm and 0.09 mL/h. *Reprinted with permission from Asadi, M., Salami, M., Hajikhani, M., Emam-Djomeh, Z., Aghakhani, A., & Ghasemi, A. (2021). Electrospray production of curcumin-walnut protein nanoparticles. Food Biophysics, 16(1), 15–26. Copyright (2020) Springer Nature.*

2.2 Protein nanoemulsion

Nanoemulsion is a colloidal dispersion of two immiscible liquids that is thermodynamically unstable (Lee, Choi, Li, Decker, & McClements, 2011; McClements, 2012). In a typical nanoemulsion, one of the liquids forms the dispersed phase and other liquid forms the dispersing medium. The general types of nanoemulsions include biphasic nanoemulsions (oil-in-water (O/W), water-in-oil (W/O)) and multiple nanoemulsions (W/O/W) depending on constituents and relative distribution of internal dispersed phase(s) and the continuous phase. The droplets diameter of nanoemulsion ranges from 10 to 200 nm (Kumar, Bishnoi, Shukla, & Jain, 2019). An emulsion is formed by mixing oil, water, and surfactants together. Protein molecules can be used as surfactants to stabilize nanoemulsion. Compared to conventional coarse emulsions, nanoemulsions have attracted growing interest in foods, pharmaceuticals, cosmetics, personal care products, and chemicals because of their extremely small droplet size, higher kinetic stability, and better bioavailability than those of conventional emulsions. The appearance of coarse emulsion usually is cloudy, but nanoemulsions can be optically transparent because the droplets in nanoemulsions are so small that they only scatter light weakly. Thus, nanoemulsions can be incorporated into transparent products (e.g., fortified soft drinks and waters) without adversely affecting their clarity. Moreover, nanoemulsions can increase the bioavailability of lipophilic substances encapsulated inside, which is useful in pharmaceutical and nutraceutical products.

The methods used to fabricate nanoemulsions can be divided into two categories: high energy and low energy approaches. High energy method includes high-pressure homogenization, microfluidization, and ultrasonication whereas low energy methods include the phase inversion emulsification method and the self-nanoemulsification method (Lee et al., 2011).

2.2.1 High-energy approaches

Nanoemulsions generally require large energy input by applying extremely intense disruptive forces with mechanical devices that helps to break up the dispersed phase into droplets within the continuous phase (Espitia, Fuenmayor, & Otoni, 2019). Techniques such as high-shear mixing, high-pressure microfluidics, and ultrasonication are used to provide the mechanical energy necessary to overcome the significant interfacial energy barrier for producing dispersions of nanodroplets (Sheth, Seshadri, Prileszky, & Helgeson, 2020). Their versatility lies in the fact that almost any oil can be subjected to nanoemulsification. However, high-energy approaches need special instruments and they produce high operational temperatures which sometimes rules out thermosensitive drugs (Singh et al., 2017).

2.2.1.1 High-pressure valve homogenization

High-pressure valve homogenization (HPVH) is one of the common methods used for emulsification in the food industry due to the ease of operation, scalability, reproducibility, and high throughput (Golfomitsou, Mitsou, Xenakis, & Papadimitriou, 2018). Usually, a coarse emulsion was preliminarily prepared through high-speed homogenization. Then the coarse emulsion is passed through a small valve (<500 μm) with high pressure ranging from 500 to 5000 psi into a homogenization chamber to produce nanoemulsions with very small droplet sizes (<500 nm) (Fig. 17). A high intensity of shear forces, turbulence, and cavitation are generated in this process that cause the breakdown of emulsion droplets, resulting in smaller particle sizes (Tan & Nakajima, 2005). The number of passes through the valve, the homogenization pressure, and the viscosity ratio of the two phases (usually oil and water) being homogenized affect the size of droplets produced through high-pressure homogenization. The droplet size decreases with an increase in the number of passes and homogenization pressure. High-pressure valve homogenization is suitable for nanoemulsion formation in low-viscosity formulations (Håkansson, 2018).

Pea protein-stabilized nanoemulsions were prepared to encapsulate vitamin D with the aim to develop novel non-dairy functional foods for vitamin D fortifications (Walia & Chen, 2020). Nanoemulsions were prepared using pea protein solution as aqueous phase and canola oil as lipid phase. Vitamin D_3 was dissolved in the lipid phase. First, a coarse emulsion was prepared by breaking down the particles using a high-speed homogenizer, followed by processing through a high pressure homogenizer. Nanoemulsions stabilized by pea protein showed the spherical shape and smooth surface. The researchers investigated the effect of number of

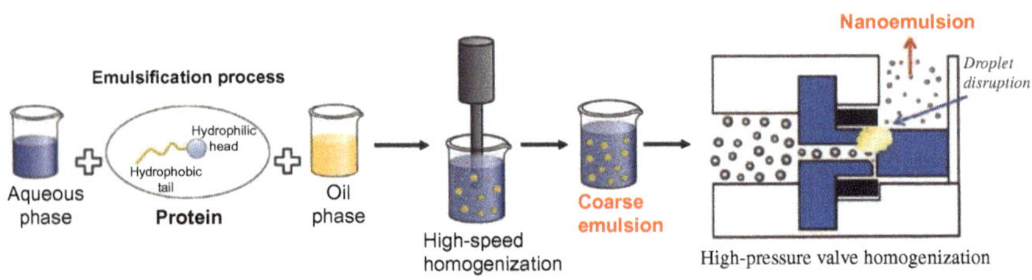

Fig. 17 Schematic illustration of the nanoemulsification process through high-speed homogenization and high-pressure valve homogenization. *Adapted with permission from Santamaria-Echart, A., Fernandes, I. P., Silva, S. C., Rezende, S. C., Colucci, G., Dias, M. M., & Barreiro, M. F. (2021). New trends in natural emulsifiers and emulsion Technology for the Food Industry. In M. Á. Á. P. Lage & P. Otero (Eds.), Food additives London: IntechOpen. Copyright (2021) IntechOpen and McClements, D. J., & Rao, J. (2011). Food-grade nanoemulsions: Formulation, fabrication, properties, performance, biological fate, and potential toxicity. Critical Reviews in Food Science and Nutrition, 51(4), 285–330. Copyright (2011) Taylor & Francis.*

homogenization cycles of 1–5 for nanoemulsions with different oil content (0.5% and 5% *w/v*) on droplet size. The droplet size of nanoemulsion prepared with 0.5% oil was reduced from 257 to ~230 nm from single to three homogenization cycles. Similarly, the sample with 5% oil composition showed a decrease in size from 377, 350 to 294 nm on passing the emulsion through homogenizer from 1 to 3 homogenizing cycles, respectively. Homogenization conditions of 20 kpsi and two homogenization cycles were identified as optimal conditions for producing stable nanoemulsions. The nanoemulsions exhibited controllable sizes (170–350 nm), good stability with zeta-potential of −25 mV, and high vitamin encapsulation efficiency (94–96%). Cellular uptake efficiency of small sized nanoemulsions (233 nm) was ~2.5 times higher than large sized nanoemulsions (350 nm). The vitamin D transport efficiency across Caco-2 cells for small sized nanoemulsions (233 nm) was ~5.3 times greater than free vitamin D suspension. This research demonstrated that pea protein can be used as an effective emulsifier for preparing food nanoemulsions.

2.2.1.2 Microfluidization

Microfluidization is an emerging technique to develop smaller droplet sizes. A microfluidizer is quite similar in design to a high-pressure homogenizer because it involves the use of high pressure to force the emulsion to premix through narrow orifices to promote the breakdown of droplets. It is highly used in the pharmaceutical industry to produce emulsions, and in the food and beverage industry to produce flavored emulsions, nutritional emulsions and homogeneous milk (Augustin & Udabage, 2007; Mccrae, 2010; Yi, Yue, Fang, & Yokoyama, 2014).

The working principle of microfluidization is to divide the emulsion flowing through a channel into two streams, let each stream pass through a separate fine channel, and then guide the two streams to contact each other in the interaction chamber (Fig. 18A). Then, in the interaction chamber, two streams of macroemulsions strike each other at high velocity, generating intense destructive power and breaking down the droplets to form fine emulsions. When the mixture flows through these microchannels to the collision zone, very fine droplets can be prepared (Kumar et al., 2019). In the microfluidization technology, the droplet size is positively correlated with the homogeneous pressure, number of pass, and emulsifier concentration, but negatively correlated with the viscosity ratio of the dispersed phase to the continuous phase. The main advantages of microfluidizers are the combination of higher peak shear rate, large volume of nano-droplet flux and

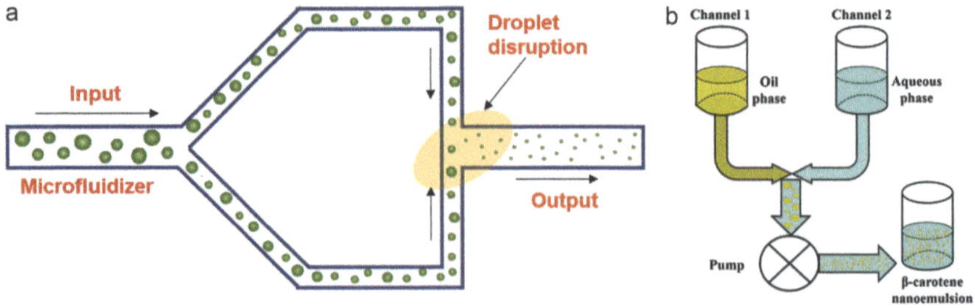

Fig. 18 Schematic illustration of (A) a typical microfluidization process and (B) novel dual-channel microfluidization. *(Panel (B) reprinted with permission from Luo, X., Zhou, Y., Bai, L., Liu, F., Deng, Y., & McClements, D. J. (2017). Fabrication of β-carotene nanoemulsion-based delivery systems using dual-channel microfluidization: Physical and chemical stability. Journal of Colloid and Interface Science, 490, 328–335. Copyright (2016) Elsevier Inc.*

narrower and smaller nanoemulsion particle size distributions. However, microfluidizers are more expensive, and their high wear rate reduces the process efficiency (Gulzar & Benjakul, 2020).

The nanoemulsions were fabricated using a modified high-pressure dual-channel microfluidization method (Bai, Huan, Gu, & McClements, 2016; Luo et al., 2017), which can be used to create nanoemulsions in a single step. In this method, nanoemulsions can rapidly be formed using a continuous process in which an oil phase and an aqueous phase are fed through separate input channels that are made to impinge upon each other, and so an initial pre-mixing stage is not needed (Fig. 18B). Dual channel microfluidization is therefore capable of saving time, energy, and costs, as well as reducing waste. Luo et al. encapsulated β-carotene in O/W nanoemulsions that fabricated through high-pressure dual channel microfluidization (Luo et al., 2017). Whey protein isolate (WPI) solution and β-carotene in corn oil served as aqueous phase and oil phase, respectively. The nanoemulsions were prepared by homogenizing 10% (*w*/w) oil phase with 90% (w/w) aqueous phase. Simultaneously, the oil phase and the aqueous phase were fed into the dual-channel microfluidizer through two separate glass reservoirs, and then made to impinge upon each other under high pressure. The droplet diameter was <500 nm when whey protein isolate concentration was larger than 0.2%. Increasing the pressure from 9 to 19 kpsi, the mean droplet diameter in the whey protein emulsions (1 wt%) decreased from 170 to 136 nm. The physical and chemical stability of these nanoemulsions were characterized during storage at neutral pH conditions at refrigeration (4 °C), ambient (25 °C), and elevated (55 °C) temperatures. At 4 °C and 25 °C, the nanoemulsions remained physically stable throughout 14 days storage, with little change in particle size or evidence of creaming. There was a pronounced decrease in the amount of β-carotene remaining in the nanoemulsions during storage due to chemical degradation, which led to color fading. The rate of chemical degradation and color fading increased with increasing temperature, with the most rapid change occurring between 25 °C and 55 °C. This study showed that dual-channel microfluidization is an efficient method of continuously producing nanoemulsions from proteins.

2.2.1.3 Ultrasonication

Ultrasonication method uses high-frequency sound waves from 20 kHz and above. It can be used to form a nanoemulsion in situ or to reduce the size of a pre-formed emulsion. Ultrasonicators consist of a piezoelectric probe which generates intense disruptive force at its tip (Fig. 19). When dipped in a sample, ultrasonic waves produce cavitation bubbles which continue to grow until they implode (Singh et al., 2017). Ultrasonicators converts electrical waves into pressure waves, leading to size reduction of droplets with increasing sonication time, power level, and emulsifier concentration (McClements, 2011). The formation and collapse of microbubbles are caused by cavitation induced by the pressure fluctuations of the sound waves, physical shear, and the extreme turbulence caused by the ultrasonicator, which ultimately

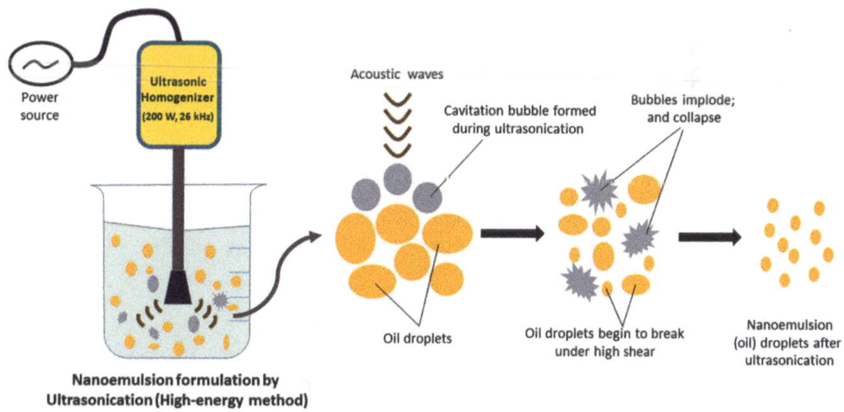

Fig. 19 Schematic illustration of nanoemulsion preparation through ultrasonication using the sonic probe. Piezoelectric probe supplies sound energy which gives rise to unstable growth of cavitation bubbles until they implode and collapse, generating highly localized shear stresses to break emulsion droplets into nano-sized droplets. *Reprinted with permission from Pal, N., Kumar, N., & Mandal, A. (2019). Stabilization of dispersed oil droplets in nanoemulsions by synergistic effects of the gemini surfactant, PHPA polymer, and silica nanoparticle. Langmuir, 35(7), 2655–2667. Copyright (2019) American Chemical Society.*

results in nanodroplets (Kumar et al., 2019). The energy expenditure in ultrasonication is the least in comparison to other high-energy methods (Singh et al., 2017). This method also requires a low emulsifier concentration compared to other mechanical methods and is one of the most economic methods (Espitia et al., 2019). Ultrasonicators are common lab scale instruments for nanoemulsion development. Ultrasonication method is always combined with other methods (e.g. pH shifting, high shear homogenization) to prepare nanoemulsion.

Phosphatidylcholine-soybean protein isolate (PC-SPI) nanoemulsions were prepared by ultrasonication (Teng et al., 2020). Soy protein as an emulsifier plays a vital role in the formation of small oil droplets. The introduction of external emulsifiers phosphatidylcholine can enhance the emulsification properties of SPI in producing nanoemulsions. Sunflower oil containing β-carotene was used as oil phase. The aqueous phase (PC-SPI solution) and oil phase were homogenized with an Ultra-Turrax T18 homogenizer at room temperature at 20,000 rpm for 5 min to form a coarse emulsion. Different ultrasonic treatments at a frequency of 20 kHz and various output powers (200, 300, 400, 500 or 600 W), various treatment times (6, 7, 8, 9 or 10 min), pulse duration 4 s, off time 2 s were studied. The researchers found that the optimal conditions to produce nanoemulsions were 1.5% SPI, 0.20% PC, 500 W ultrasonic power and 9 min ultrasonic time. Particle size decreased from 594 to 292 nm with ultrasonic power increasing from 200 to 500 w. Particle size of the nanoemulsions dropped from 590 to 288 nm when the ultrasound time increased from 6 to 9 min. After degradation of the nanoemulsions during storage, β-carotene was released. The β-carotene retention rate of the nanoemulsion was maintained above 86% at different temperatures (4 °C, 25 °C, and 55 °C) in the absence of light for up to 30 days, exhibiting excellent resistance to droplet coalescence during storage. This may be attribute to the antioxidant activities of SPI and PC.

2.2.2 Low-energy approaches

Low-energy methods use the energy input from the chemical potential of the components to form nanoemulsions. Nanoemulsions develop spontaneously at the oil-water interface by gently mixing the components (Aswathanarayan & Vittal, 2019). The low energy methods involved in the nanoemulsion production includes membrane emulsification, spontaneous emulsification, solvent displacement and phase inversion temperature methods. Low energy methods are sometimes limited by oil type and emulsifiers that can be used (Singh et al., 2017).

2.2.2.1 Membrane emulsification

Membrane emulsification technology is a drop-by-drop emulsification method through porous membranes, which was introduced in Japan in 1988. As an alternative method of preparing emulsion, membrane emulsification technology has received more and more attention in the past 25 years. The principle of this method is to pass the dispersed phase through the pores of a microporous membrane to a continuous phase under the effect of force (Joscelyne & Tragardh, 1999) (Fig. 20). The formation and separation of emulsified droplets at the end of the pores occurs through the drop-by-drop mechanism. Nanoemulsions were produced using different membrane devices (e.g., Shirasu porous glass membranes, track-etched polymeric membranes, and microengineered membranes) (Vladisavljević, 2018).The dispersed phase in the form of droplets can be a pure liquid or an emulsion (Laouini, Fessi, & Charcosset, 2012). Both immiscible liquids can be used as the continuous phase to generate simple emulsions, such as oil-in-water (O/W) or water-in-oil (W/O) droplets (Dong et al., 2011; Laouini et al., 2012). Multiple emulsions or double emulsions such as water-in-oil-in-water (W/O/W) and oil-in water-in oil (O/W/O) emulsions can also be prepared by using this method. The membrane emulsification technology used for the preparation of nanoemulsions has the characteristics of low energy consumption, adjustable droplet size, unimodal distribution, high encapsulation rate, etc., and does not require shearing

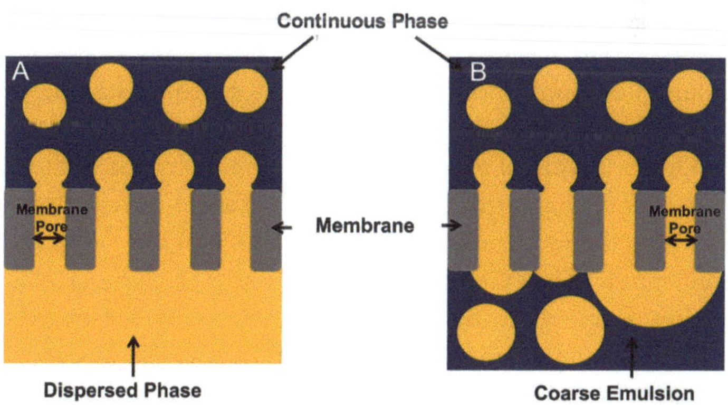

Fig. 20 Schematic illustration of emulsion production by direct and premix membrane emulsification. (A) Direct membrane emulsification and (B) Premix membrane emulsification. *Reprinted with permission from Piacentini, E., Drioli, E., & Giorno, L. (2014). Membrane emulsification technology: Twenty-five years of inventions and research through patent survey. Journal of Membrane Science, 468, 410–422. Copyright (2014) Elsevier B.V.*

or thermal degradation (Schadler & Windhab, 2006). Compared to high-energy processes, it uses less amount of surfactants and has a narrower particle size distribution. The main disadvantage of this method is that the flux of the dispersed phase through the membrane is very low, making it difficult to increase the yield (Laouini et al., 2012; Océane, Eyad, Georgio, Hatem, & Catherine, & Charcosset., 2019).

Protein-stabilized O/W emulsions containing β-carotene were produced by premix membrane emulsification using polymeric microfiltration membranes (Trentin, De Lamo, Güell, López, & Ferrando, 2011). Bovine serum albumin (BSA) and whey protein concentrate (WPC) were used as protein emulsifiers while a nonionic small-molecule surfactant, Tween 20, was used both as a control and coemulsifier. O/W emulsions were prepared in a two step emulsification system. The first step consisted of dissolving the β-carotene in the sunflower oil and then preparing a coarse emulsion. In the second step, the coarse emulsion was loaded into the premix reservoir and forced with nitrogen (500 or 900 kPa) through the membrane to obtain fine emulsions. BSA and WPC as coemulsifiers with Tween 20 increased the physical and chemical stability compared to that of emulsions stabilized with single proteins. Furthermore, during storage at 22 °C and 35 °C, the chemical degradation rate of emulsions with WPC/Tween 20 was slower than those with BSA/Tween 20 whereas Tween 20-stabilized emulsions led to the highest rate of β-carotene reduction during storage at 35 °C.

2.2.2.2 Spontaneous emulsification

In this technique, spontaneous emulsions are formed on mixing water and oil together with an emulsifier by gentle stirring at a particular temperature (Fig. 21). The formation of nanoemulsions by the migration of the surfactant and/or solvent molecules from the dispersed phase to the continuous phase without causing the spontaneous curvature change of the surfactant is called self-emulsification (Silva, Cerqueira, & Vicente, 2011). The excess oil phase is removed by evaporation under reduced pressure. In the pharmaceutical industry, the spontaneous emulsification method is widely used for the packaging and transport of fat-soluble drug molecules. Emulsions prepared by this method are generally called self-emulsifying drug delivery systems. This method can be carried out in many different ways: the composition of the two phases can be changed; environmental conditions such as temperature, pH and ionic strength can be changed; and/or mixing conditions

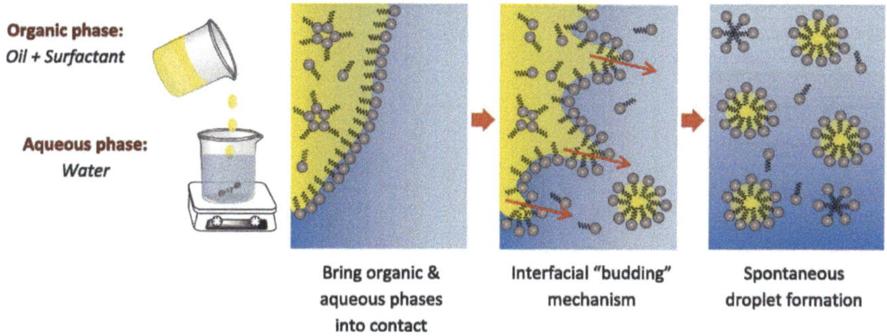

Fig. 21 Schematic diagram of nanoemulsions preparation by the spontaneous emulsification method: Fine oil droplets are spontaneously formed when an organic phase containing a surfactant is mixed with an aqueous phase. The surfactant moves from the organic phase to the water phase (red arrows), leading to interfacial turbulence and spontaneous oil droplet formation. *Reprinted with permission from McClements, D. J., & Rao, J. (2011). Food-grade nanoemulsions: Formulation, fabrication, properties, performance, biological fate, and potential toxicity. Critical Reviews in Food Science and Nutrition, 51(4), 285–330. Copyright (2011) Taylor & Francis.*

such as stirring speed, addition speed and order of addition can be changed (Choi & McClements, 2020; Singh et al., 2017). In the emulsification process, a relatively high concentration of synthetic surfactant is required to form a nanoemulsion. Therefore, due to solvent removal, regulation, cost, or sensory reasons, the spontaneous emulsification method is more suitable for drug delivery, and cannot be widely used in the food industry.

Eugenol-encapsulated nanoparticles were produced using a low-energy self-emulsification process with naturally occurring emulsifiers NaCas and zein (Fig. 22) (Wang & Zhang, 2017). Eugenol, the major constituent of clove oil, possesses significant antioxidant, anti-inflammatory and cardiovascular properties, in addition to analgesic and local anesthetic activity (Pramod, Ansari, & Ali, 2010). The self-emulsification was performed by deprotonation and neutralization procedures without organic solvents and specific equipment. The deprotonated eugenol in hot alkaline was added to NaCas/zein mixtures with different mass ratios at pH 11.5 and then neutralized to pH 7.0. All nanoemulsions were dispersed in water in a homogeneous manner, with a uniform milky appearance. The nanoemulsions showed a well-defined diameter (around 109–139 nm) and a negative surface potential (from −28.5 to −35.8 mV) with spherical morphology. The entrapment efficiency (EE) of 1% (*v*/v) eugenol reached 84.24% by 2% (m/v) NaCas/zein at a mass ratio of 1:1. This formulation also showed

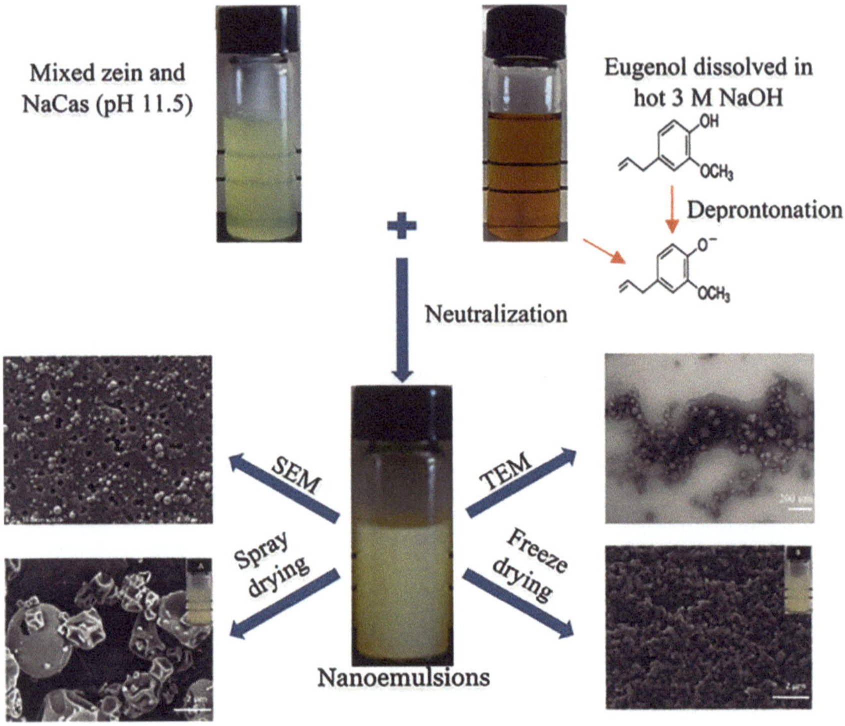

Fig. 22 Eugenol-loaded nanoemulsion prepared from zein and sodium caseinate (NaCas). *Reprinted with permission from Wang, L., & Zhang, Y. (2017). Eugenol nanoemulsion stabilized with zein and sodium caseinate by self-assembly. Journal of Agricultural and Food Chemistry, 65(14), 2990–2998. Copyright (2017) American Chemical Society.*

the narrowest size distribution and extraordinary stability during ambient storage (22 °C) up to 30 days and retained good redispersibility after spray- or freeze-drying.

2.2.2.3 Emulsion phase inversion

Emulsion phase inversion (EPI) includes phase inversion composition (PIC), and phase inversion temperature (PIT). The PIC method is based on changing the water/oil ratio of the system through progressively adding one of the components (water or oil) over a mixture of the other components at a constant temperature during the procedure to obtain nanoemulsions (Solans & Solé, 2012). The preparation of O/W nanoemulsions by the phase inversion composition method can be divided into two steps: (i) mixing of the organic phase (oil + surfactant) to obtain a fixed composition and (ii) adding water resulting in the formation of O/W nanoemulsions (Fig. 23). The mechanism is similar to the temperature-induced curvature change in the PIT method. Water addition leads to a progressively increase in the hydration grade of the surfactant, and the spontaneous curvature of surfactant changes from

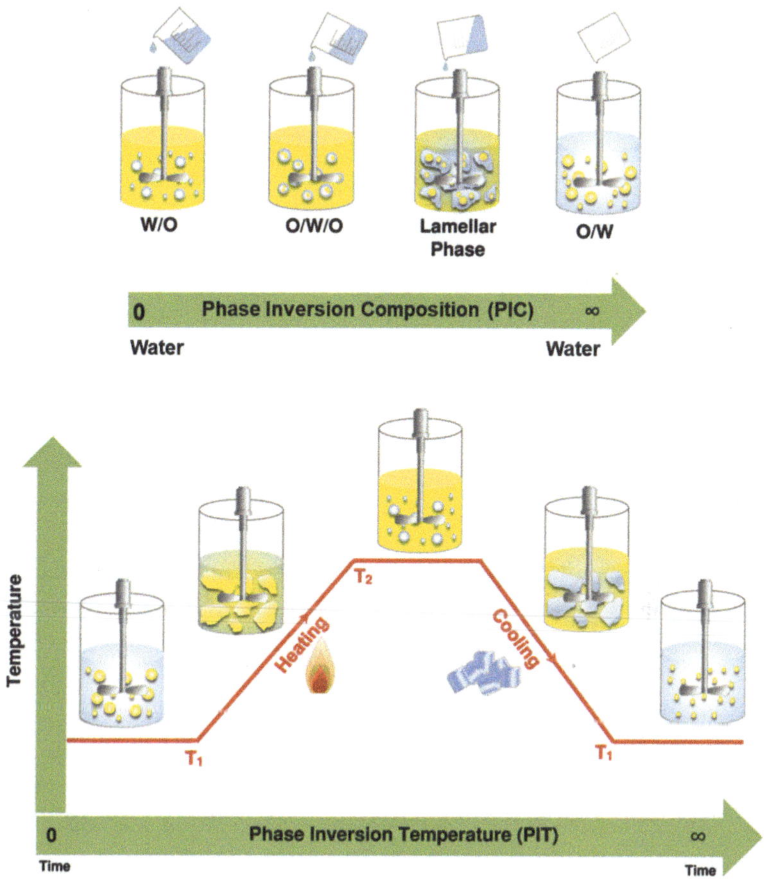

Fig. 23 Schematic illustration of emulsion production by phase inversion composition and phase inversion temperature method. In the phase inversion composition (PIC), titrating water into a mixture containing oil and surfactant initially leads to the formation of a water-in-oil emulsion that then inverts into an oil-in-water emulsion. In the phase inversion temperature (PIT) method, cooling or decrease in temperature induces a phase inversion. *Reprinted with permission from Saffarionpour, S. (2019). Preparation of food flavor nanoemulsions by high-and low-energy emulsification approaches. Food Engineering Reviews, 11(4), 259–289. Copyright (2019) Springer Nature.*

negative to positive (Roger, & Kevin., 2016). The phase inversion composition method has low cost, does not require the use of organic solvents and it has high thermodynamic stability. However, this method is not suitable for highly hydrophobic compounds (Aswathanarayan & Vittal, 2019). The phase inversion composition method was shown to be unsuitable for producing nanoemulsions from label-friendly surfactants, such as whey protein, casein, Quillaja saponin, and sucrose monoesters (Mayer, Weiss, & McClements, 2013).

Phase inversion temperature (PIT) methods was applied to form nanoemulsions by exploiting changes in aqueous/oil solubility of surfactants in response to temperature fluctuation (Fig. 23). At low temperatures, the temperature-sensitive surfactants are hydrophilic with positive spontaneous curvature, but as the temperature rises, they become more hydrophobic with negative spontaneous curvature (Jiang, Liao, & Charcosset, 2020). Oil-in-water nanoemulsion is favored at low temperatures, and the water-in-oil emulsion is favored at high temperatures. Owing to dehydration of the polymer chain, a nonionic surfactant like polyoxyethylene becomes lipophilic at high temperatures and hydrophilic at low temperatures as a result of the continuous process of rapid cooling and heating (Divya, Divyasree, Vamsidhar, Bhavani, & Panigrahi, 2021). This is a low-cost method, i.e., easy to scale up and provides better control over temperature (can be adjusted and readjusted several times) that results in required droplet size. This is a bottom-up process and nascent droplets remain stable over a considerable period of time due to rich surfactant coverage. Since input of heat is necessary, phase inversion temperature methods may rule out utilization of thermosensitive drugs (Singh et al., 2017). This method is limited to non-ionic surfactants (Yukuyama, Ghisleni, Pinto, & Bou-Chacra, 2016) and may require a high quantity of emulsifiers (Espitia et al., 2019).

Su et al. fabricated nanoemulsions by the phase inversion temperature method (Su & Zhong, 2016). The objective of this study was to investigate the phase inversion temperature formation of lemon oil nanoemulsions by combinations of Tween 20 and sodium caseinate (NaCas). The emulsions prepared with combinations of sodium caseinate and Tween 20 had lower turbidity and smaller droplets than those prepared with one surfactant. The nanoemulsions prepared with 2% sodium caseinate, 0.4–1.2% Tween 20 and 1.5% lemon oil had a volume-area mean diameter of around 100 nm and were stable during 15-day storage. Therefore, sodium caseinate can be used to partially replace synthetic surfactants to prepare flavor oils nanoemulsions using the phase inversion temperature method.

3. Conclusions and future perspectives

This chapter gives an overview on methods and techniques used for nanostructured food protein development with respect to the basic processing and mechanism behind these techniques, advantages, and limitations, as well as applications for encapsulation and controlled release of bioactive compounds. The first part summarized physical, chemical and self-assembly methods to construct protein nanoparticles. The second part introduced high-energy (high-pressure homogenization, microfluidization, and ultrasonication) and low-energy (membrane emulsification, spontaneous emulsification, and emulsion phase inversion) approaches to fabricate nanoemulsion. The technology advancement has allowed more capacity to fabricate food proteins into various nanostructures with techno-functional properties, leading to improved stability and bioavailability of bioactive compounds for food, nutraceutical and pharmaceutical applications.

Although there are many different choices of techniques to develop nanostructured food protein, it is still difficult to precisely control the structure and size of nano-vehicles. It is also difficult to develop complex structures for protein based nano-materials for multifunctional properties based on the current techniques. In addition, most of techniques to fabricate nanostructured food protein are laboratory scale preparation method, but it is difficult to scale-up these techniques in a cost-effective manner, which hampers their commercial application. Therefore, easily scale-up-able techniques to produce food protein-based nanocarriers should be developed in the future studies. Moreover, the current studies focused on prepare food protein based nanocarriers, but a few of them have been used in real food, nutraceuticals, and pharmaceutical products formulations. The real products matrices and processing conditions may affect the properties of protein based nanocarriers. For example, the long-term storage stability of protein nanocarriers with bioactive ingredients in liquid products, especially acidic products, is a big issue. Therefore, the applicability of food protein nanocarriers should be systematically investigated in future studies.

References

Arun, A., Malrautu, P., Laha, A., Luo, H., & Ramakrishna, S. (2021). Collagen nanoparticles in drug delivery systems and tissue engineering. *Applied Sciences*, *11*(23), 11369.

Asadi, M., Salami, M., Hajikhani, M., Emam-Djomeh, Z., Aghakhani, A., & Ghasemi, A. (2021). Electrospray production of curcumin-walnut protein nanoparticles. *Food Biophysics*, *16*(1), 15–26.

Aswathanarayan, J. B., & Vittal, R. R. (2019). Nanoemulsions and their potential applications in food industry. *Frontiers in Sustainable Food Systems*, *3*, 95.
Augustin, M. A., & Udabage, P. (2007). Influence of processing on functionality of milk and dairy proteins. *Advances in Food and Nutrition Research*, *53*, 1–38.
Bai, L., Huan, S., Gu, J., & McClements, D. J. (2016). Fabrication of oil-in-water nanoemulsions by dual-channel microfluidization using natural emulsifiers: Saponins, phospholipids, proteins, and polysaccharides. *Food Hydrocolloids*, *61*, 703–711.
Cai, X., Zhao, L., Wang, S., & Rao, P. (2015). Fabrication and characterization of the nano-composite of whey protein hydrolysate chelated with calcium. *Food & Function*, *6*(3), 816–823.
Choi, S. J., & McClements, D. J. (2020). Nanoemulsions as delivery systems for lipophilic nutraceuticals: Strategies for improving their formulation, stability, functionality and bioavailability. *Food Science and Biotechnology*, *29*(2), 149–168.
de Araujo, J. T. C., Martin-Pastor, M., Pérez, L., Pinazo, A., & de Sousa, F. F. O. (2021). Development of anacardic acid-loaded zein nanoparticles: Physical chemical characterization, stability and antimicrobial improvement. *Journal of Molecular Liquids*, *332*, 115808.
De Kruif, C. G., Weinbreck, F., & de Vries, R. (2004). Complex coacervation of proteins and anionic polysaccharides. *Current Opinion in Colloid & Interface Science*, *9*(5), 340–349.
Desai, N. (2007). Nanoparticle albumin bound (nab) technology: Targeting tumors through the endothelial gp60 receptor and SPARC. *Nanomedicine: Nanotechnology, Biology, and Medicine*, *4*(3), 339.
Divya, K., Divyasree, R., Vamsidhar, M., Bhavani, K., & Panigrahi, B. B. (2021). Emulsion, micro emulsion and nano emulsion: A review. *World Journal of Pharmacy and Pharmaceutical Sciences*, *10*(4), 641–660.
Doan, C. D., & Ghosh, S. (2019). Formation and stability of pea proteins nanoparticles using ethanol-induced desolvation. *Nanomaterials*, *9*(7), 949.
Dong, H. O., Balakrishnan, P., Oh, Y. K., Kim, D. D., Yong, C. S., & Choi, H. G. (2011). Effect of process parameters on nanoemulsion droplet size and distribution in SPG membrane emulsification. *International Journal of Pharmaceutics*, *404*(1–2), 191–197.
Elzoghby, A. O., Elgohary, M. M., & Kamel, N. M. (2015). Implications of protein-and peptide-based nanoparticles as potential vehicles for anticancer drugs. *Advances in Protein Chemistry and Structural Biology*, *98*, 169–221.
Elzoghby, A. O., Samy, W. M., & Elgindy, N. A. (2012). Albumin-based nanoparticles as potential controlled release drug delivery systems. *Journal of Controlled Release*, *157*(2), 168–182.
Espitia, P. J. P., Fuenmayor, C. A., & Otoni, C. G. (2019). Nanoemulsions: Synthesis, characterization, and application in bio-based active food packaging. *Comprehensive Reviews in Food Science and Food Safety*, *18*(1), 264–285.
Ezhilarasi, P., Karthik, P., Chhanwal, N., & Anandharamakrishnan, C. (2013). Nanoencapsulation techniques for food bioactive components: A review. *Food and Bioprocess Technology*, *6*(3), 628–647.
Fu, Q., Sun, J., Zhang, W., Sui, X., Yan, Z., & He, Z. (2009). Nanoparticle albumin-bound (NAB) technology is a promising method for anti-cancer drug delivery. *Recent Patents on Anti-Cancer Drug Discovery*, *4*(3), 262–272.
Golfomitsou, I., Mitsou, E., Xenakis, A., & Papadimitriou, V. (2018). Development of food grade O/W nanoemulsions as carriers of vitamin D for the fortification of emulsion based food matrices: A structural and activity study. *Journal of Molecular Liquids*, *268*, 734–742.
Gülseren, İ., Fang, Y., & Corredig, M. (2012). Whey protein nanoparticles prepared with desolvation with ethanol: Characterization, thermal stability and interfacial behavior. *Food Hydrocolloids*, *29*(2), 258–264.

Gulzar, S., & Benjakul, S. (2020). Characteristics and storage stability of nanoliposomes loaded with shrimp oil as affected by ultrasonication and microfluidization. *Food Chemistry*, *310*, 125916.

Håkansson, A. (2018). Fabrication of nanoemulsions by high-pressure valve homogenization. In *Nanoemulsions* (pp. 175–206). Elsevier.

Hemshekhar, M., Sebastin Santhosh, M., Kemparaju, K., & Girish, K. S. (2012). Emerging roles of anacardic acid and its derivatives: A pharmacological overview. *Basic & Clinical Pharmacology & Toxicology*, *110*(2), 122–132.

Hong, S., Choi, D. W., Kim, H. N., Park, C. G., Lee, W., & Park, H. H. (2020). Protein-based nanoparticles as drug delivery systems. *Pharmaceutics*, *12*(7), 604.

Jafari, S. M., & McClements, D. J. (2017). Nanotechnology approaches for increasing nutrient bioavailability. *Advances in Food and Nutrition Research*, *81*, 1–30.

Jain, A., Singh, S. K., Arya, S. K., Kundu, S. C., & Kapoor, S. (2018). Protein nanoparticles: Promising platforms for drug delivery applications. *ACS Biomaterials Science & Engineering*, *4*(12), 3939–3961.

Jhaveri, A. M., & Torchilin, V. P. (2014). Multifunctional polymeric micelles for delivery of drugs and siRNA. *Frontiers in Pharmacology*, *5*, 77.

Jiang, T., Liao, W., & Charcosset, C. (2020). Recent advances in encapsulation of curcumin in nanoemulsions: A review of encapsulation technologies, bioaccessibility and applications. *Food Research International*, *132*, 109035.

Joscelyne, S. M., & Tragardh, G. (1999). Food emulsions using membrane emulsification: Conditions for producing small droplets. *Journal of Food Engineering*, *39*(1), 59–64.

Khalil, I., Yehye, W. A., Etxeberria, A. E., Alhadi, A. A., Dezfooli, S. M., Julkapli, N. B. M., et al. (2019). Nanoantioxidants: Recent trends in antioxidant delivery applications. *Antioxidants*, *9*(1), 24.

Kim, B., Lee, C., Lee, E. S., Shin, B. S., & Youn, Y. S. (2016). Paclitaxel and curcumin co-bound albumin nanoparticles having antitumor potential to pancreatic cancer. *Asian Journal of Pharmaceutical Sciences*, *11*(6), 708–714.

Kumar, M., Bishnoi, R. S., Shukla, A. K., & Jain, C. P. (2019). Techniques for formulation of nanoemulsion drug delivery system: A review. *Preventive Nutrition and Food Science*, *24*(3), 225–234.

Laouini, A., Fessi, H., & Charcosset, C. (2012). Membrane emulsification: A promising alternative for vitamin E encapsulation within nano-emulsion. *Journal of Membrane Science*, *423-424*, 85–96.

Lee, S. J., Choi, S. J., Li, Y., Decker, E. A., & McClements, D. J. (2011). Protein-stabilized nanoemulsions and emulsions: Comparison of physicochemical stability, lipid oxidation, and lipase digestibility. *Journal of Agricultural and Food Chemistry*, *59*(1), 415–427.

Li, P., Sun, M., Xu, Z., Liu, X., Zhao, W., & Gao, W. (2018). Site-selective in situ growth-induced self-assembly of protein–polymer conjugates into pH-responsive micelles for tumor microenvironment triggered fluorescence imaging. *Biomacromolecules*, *19*(11), 4472–4479.

Liu, G., Yang, J., Wang, Y., Liu, X., & Chen, L. (2019). Protein-lipid composite nanoparticles for the oral delivery of vitamin B12: Impact of protein succinylation on nanoparticle physicochemical and biological properties. *Food Hydrocolloids*, *92*, 189–197.

Livney, Y. D. (2015). Nanostructured delivery systems in food: Latest developments and potential future directions. *Current Opinion in Food Science*, *3*, 125–135.

Lohcharoenkal, W., Wang, L., Chen, Y. C., & Rojanasakul, Y. (2014). Protein nanoparticles as drug delivery carriers for cancer therapy. *BioMed Research International*, *2014*.

Loureiro, A., Azoia, N. G., Gomes, A. C., & Cavaco-Paulo, A. (2016). Albumin-based nanodevices as drug carriers. *Current Pharmaceutical Design*, *22*(10), 1371–1390.

Luo, X., Zhou, Y., Bai, L., Liu, F., Deng, Y., & McClements, D. J. (2017). Fabrication of β-carotene nanoemulsion-based delivery systems using dual-channel microfluidization: Physical and chemical stability. *Journal of Colloid and Interface Science*, *490*, 328–335.

Mayer, S., Weiss, J., & McClements, D. J. (2013). Vitamin E-enriched nanoemulsions formed by emulsion phase inversion: Factors influencing droplet size and stability. *Journal of Colloid and Interface Science*, *402*, 122–130.

McClements, D. J. (2011). Edible nanoemulsions: Fabrication, properties, and functional performance. *Soft Matter*, *7*(6), 2297–2316.

McClements, D. J. (2012). Nanoemulsions versus microemulsions: Terminology, differences, and similarities. *Soft Matter*, *8*(6), 1719–1729.

Mccrae, C. H. (2010). Homogenization of milk emulsions:Use of microfluidizer. *International Journal of Dairy Technology*, *47*(1), 28–31.

Miladi, K., Ibraheem, D., Iqbal, M., Sfar, S., Fessi, H., & Elaissari, A. (2014). Particles from preformed polymers as carriers for drug delivery. *EXCLI Journal*, *13*, 28.

Mishra, V., Mahor, S., Rawat, A., Gupta, P. N., Dubey, P., Khatri, K., et al. (2006). Targeted brain delivery of AZT via transferrin anchored pegylated albumin nanoparticles. *Journal of Drug Targeting*, *14*(1), 45–53.

Mohammadian, M., Waly, M. I., Moghadam, M., Emam-Djomeh, Z., Salami, M., & Moosavi-Movahedi, A. A. (2020). Nanostructured food proteins as efficient systems for the encapsulation of bioactive compounds. *Food Science and Human Wellness*, *9*(3), 199–213.

Océane, A., Eyad, A., Georgio, N., Hatem, F., & Catherine, & Charcosset. (2019). Comparison of three processes for parenteral Nanoemulsion production: Ultrasounds, microfluidizer, and premix membrane emulsification. *Journal of Pharmaceutical Sciences*, *108*(8), 2708–2717.

Prajapati, A., & Srivastava, A. (2019). Preparation, characterization and encapsulation efficiency of egg albumin nanoparticles using EDC as crosslinker. *Journal of Scientific & Industrial Research*, *78*, 703–705.

Pramod, K., Ansari, S. H., & Ali, J. (2010). Eugenol: A natural compound with versatile pharmacological actions. *Natural Product Communications*, *5*(12). 1934578X1000501236.

Rivas, C. J. M., Tarhini, M., Badri, W., Miladi, K., Greige-Gerges, H., Nazari, Q. A., et al. (2017). Nanoprecipitation process: From encapsulation to drug delivery. *International Journal of Pharmaceutics*, *532*(1), 66–81.

Roger, & Kevin. (2016). Nanoemulsification in the vicinity of phase inversion: Disruption of bicontinuous structures in oil/surfactant/water systems. *Current Opinion in Colloid & Interface Science*, *25*, 120–128.

Salah, M., Mansour, M., Zogona, D., & Xu, X. (2020). Nanoencapsulation of anthocyanins-loaded β-lactoglobulin nanoparticles: Characterization, stability, and bioavailability in vitro. *Food Research International*, *137*, 109635.

Schadler, V., & Windhab, E. J. (2006). Continuous membrane emulsification by using a membrane system with controlled pore distance. *Desalination*, *189*(1–3), 130–135.

Sheth, T., Seshadri, S., Prileszky, T., & Helgeson, M. E. (2020). Multiple nanoemulsions. *Nature Reviews Materials*, *5*(3), 214–228.

Silva, H. D., Cerqueira, M.Â., & Vicente, A. A. (2011). Nanoemulsions for food applications: Development and characterization. *Food and Bioprocess Technology*, *5*(3), 854–867.

Singh, Y., Meher, J. G., Raval, K., Khan, F. A., Chaurasia, M., Jain, N. K., et al. (2017). Nanoemulsion: Concepts, development and applications in drug delivery. *Journal of Controlled Release*, *252*, 28–49.

Solans, C., & Solé, I. (2012). Nano-emulsions: Formation by low-energy methods. *Current Opinion in Colloid & Interface Science*, *17*(5), 246–254.

Su, D., & Zhong, Q. (2016). Lemon oil nanoemulsions fabricated with sodium caseinate and tween 20 using phase inversion temperature method. *Journal of Food Engineering*, *171*, 214–221.

Tan, C. P., & Nakajima, M. (2005). β-Carotene nanodispersions: Preparation, characterization and stability evaluation. *Food Chemistry*, *92*(4), 661–671.

Tapia-Hernández, J. A., Torres-Chávez, P. I., Ramírez-Wong, B., Rascón-Chu, A., Plascencia-Jatomea, M., Barreras-Urbina, C. G., et al. (2015). Micro-and nanoparticles by electrospray: Advances and applications in foods. *Journal of Agricultural and Food Chemistry*, *63*(19), 4699–4707.

Tarhini, M., Benlyamani, I., Hamdani, S., Agusti, G., Fessi, H., Greige-Gerges, H., et al. (2018). Protein-based nanoparticle preparation via nanoprecipitation method. *Materials*, *11*(3), 394.

Tarhini, M., Greige-Gerges, H., & Elaissari, A. (2017). Protein-based nanoparticles: From preparation to encapsulation of active molecules. *International Journal of Pharmaceutics*, *522*(1–2), 172.

Teng, F., He, M., Xu, J., Chen, F., Wu, C., Wang, Z., et al. (2020). Effect of ultrasonication on the stability and storage of a soy protein isolate-phosphatidylcholine nanoemulsions. *Scientific Reports*, *10*(1), 1–9.

Teng, Z., Luo, Y., & Wang, Q. (2012). Nanoparticles synthesized from soy protein: Preparation, characterization, and application for nutraceutical encapsulation. *Journal of Agricultural and Food Chemistry*, *60*(10), 2712–2720.

Tian, Y., Jiang, X., Chen, X., Shao, Z., & Yang, W. (2014). Doxorubicin-loaded magnetic silk fibroin nanoparticles for targeted therapy of multidrug-resistant cancer. *Advanced Materials*, *26*(43), 7393–7398.

Trentin, A., De Lamo, S., Güell, C., López, F., & Ferrando, M. (2011). Protein-stabilized emulsions containing beta-carotene produced by premix membrane emulsification. *Journal of Food Engineering*, *106*(4), 267–274.

Vladisavljević, G. T. (2018). Fabrication of nanoemulsions by membrane emulsification. In *Nanoemulsions* (pp. 287–346). Elsevier.

Walia, N., & Chen, L. (2020). Pea protein based vitamin D nanoemulsions: Fabrication, stability and in vitro study using Caco-2 cells. *Food Chemistry*, *305*, 125475.

Wang, Y., Li, P., Truong-Dinh Tran, T., Zhang, J., & Kong, L. (2016). Manufacturing techniques and surface engineering of polymer based nanoparticles for targeted drug delivery to cancer. *Nanomaterials*, *6*(2), 26.

Wang, L., & Zhang, Y. (2017). Eugenol nanoemulsion stabilized with zein and sodium caseinate by self-assembly. *Journal of Agricultural and Food Chemistry*, *65*(14), 2990–2998.

Xiang, C., Gao, J., Ye, H., Ren, G., Ma, X., Xie, H., et al. (2020). Development of ovalbumin-pectin nanocomplexes for vitamin D3 encapsulation: Enhanced storage stability and sustained release in simulated gastrointestinal digestion. *Food Hydrocolloids*, *106*, 105926.

Yi, J., Yue, L., Fang, Z., & Yokoyama, W. (2014). The physicochemical stability and in vitro bioaccessibility of beta-carotene in oil-in-water sodium caseinate emulsions. *Food Hydrocolloids*, *35*(3), 19–27.

Yukuyama, M. N., Ghisleni, D. D., Pinto, T. J., & Bou-Chacra, N. A. (2016). Nanoemulsion: Process selection and application in cosmetics- -a review. *International Journal of Cosmetic Science*, *38*(1), 13–24.

Zhao, Z., Li, Y., & Xie, M.-B. (2015). Silk fibroin-based nanoparticles for drug delivery. *International Journal of Molecular Sciences*, *16*(3), 4880–4903.

Zou, W., Mourad, F. K., Zhang, X., Ahn, D. U., Cai, Z., & Jin, Y. (2020). Phase separation behavior and characterization of ovalbumin and propylene glycol alginate complex coacervates. *Food Hydrocolloids*, *108*, 105978.

CHAPTER NINE

Food protein-derived bioactive peptides for the management of nutrition related chronic diseases

Xinyi Cao, Wang Liao*, and Shaokang Wang
Key Laboratory of Environmental Medicine and Engineering of Ministry of Education, and Department of Nutrition and Food Hygiene, School of Public Health, Southeast University, Nanjing, Jiangsu, China
*Corresponding author: e-mail address: wangliao@seu.edu.cn

Contents

Abstract

Dietary intervention via modifications of dietary pattern or supplementations of naturally derived bioactive compounds has been considered as an efficient approach in management of nutrition related chronic diseases. Food protein-derived bioactive peptide is representative of natural compounds which show the potential to prevent or mitigate nutrition related chronic diseases. In the past decades, substantial research has been conducted concentrating on the characterization, bioavailability, and activity assessment of bioactive peptides. Although various activities of bioactive peptides have been reported, the activity testes of most peptides were only conducted in cells and animal models. Some clinical trials of bioactive peptides were also reported but only

Advances in Food and Nutrition Research, Volume 101
ISSN 1043-4526
https://doi.org/10.1016/bs.afnr.2022.04.004

limited to antihypertensive peptides, antidiabetic peptides and peptides modulating blood lipid profile. Hereby, clinical evidence of bioactive peptides in management of nutrition-related chronic diseases is summarized in this chapter, which aims at providing implications for the clinical studies of bioactive peptides in the future.

1. Introduction

During the past decades, rapid changes in our diets and lifestyles have occurred with industrialization, urbanization, and globalization. The excessive consumption of energy dense food and ultra-processed food which are high in fat is a feature of the transition of dietary pattern. On the other hand, sedentary lifestyle is present in an increasing number of individual. As a result, the prevalence of some chronic diseases, such as cardiovascular diseases and metabolic diseases, is rising at an alarming rate, which also leads to a significant increase in disability and mortality (Consultation, 2003). Hence, it is warranted to develop strategies of nutritional intervention, which is a cost-effective approach for the management of these diseases.

Naturally derived bioactive compounds have been considered as a promising candidate for the dietary intervention in control of nutrition related chronic diseases. In addition to phytochemicals (Zhang et al., 2015) and dietary fibers (Soliman, 2019), food protein-derived bioactive peptide is also a representative of natural compound with the potential to prevent or mitigate chronic diseases (Cicero, Fogacci, & Colletti, 2017). Since the first bioactive peptide identified from snake venom 5 decades ago, the research of bioactive peptides concentrated on the peptide preparation and characterization, activity evaluation, absorption and transportation, and structure–function relationship of the peptides (Liao, Jahandideh, Fan, Son, & Wu, 2018), and these works have been reviewed substantially. However, clinical trial of bioactive peptides is still at the infant stage, which is a pivotal factor that impedes the commercialization of bioactive peptides. Hereby, we collected the clinical evidences of bioactive peptides in this book chapter, which could be a summary of the most recent progress in clinical trials of bioactive peptides, but also, it could provide some implications for the future research directions of bioactive peptides.

2. An overview of nutrition related chronic diseases

2.1 The prevalence of nutrition related chronic diseases

It has been estimated that the chronic diseases associated with overnutrition, such as hypertension, cardiovascular diseases, type 2 diabetes mellitus (T2DM), and obesity account for 40 million death every year, which is equivalent to 70% of all deaths worldwide (Ribeiro et al., 2019). On the other hand, the high prevalence of these diseases has posed a challenge to the healthcare system as well as the social economy (Branca et al., 2019).

Obesity is considered as a risk factor of most of the nutrition related chronic diseases. According to global survey of obesity done in 2015 conducted in 195 countries, 604 million adults and 108 million children were obese (Afshin et al., 2017). The incidence of chronic diseases often parallels with the incidence of overweight or obesity. According to the data from National Health and Nutrition Examination Survey (NHNES) in the USA, during 1988–2010, the average body mass index (BMI) of adults increased by 0.37% per year. The waist circumference increased by 0.37 and 0.27% per year in women and men, respectively. Consistently, the prevalence of nutrition related chronic diseases in the United States was also increasing during this period (Saklayen, 2018). The same trend was also present in China. The prevalence of overweight and obesity increased from 20% to 29% during the past decades, while the incidence of all of the major types of nutrition-related chronic diseases were increasing substantially (Wang, Mi, Shan, Wang, & Ge, 2007). Hereby, to retard the incidence of overweight and obesity has been considered as an essential guideline in the development of dietary reference intakes in many countries.

Notably, the number of people suffering from more than 1 risk factors of nutrition related chronic diseases is increasing rapidly. The metabolic syndrome (MetS) is defined as meeting three or more of the five criteria including high blood glucose, low levels of high-density lipoprotein cholesterol (HDL-C), high level of triglycerides, large waist circumference, and high blood pressure. People with advanced age was considered to be more susceptible to MetS. However, the prevalence of MetS among young adults (18–30 years) is increasing, which has been estimated to be 1.8–7% (Nolan, Carrick-Ranson, Stinear, Reading, & Dalleck, 2017). In China, the pooled prevalence of MetS among residence over 15 years was 24.5%, and individuals living in urban area had a higher risk than those living in the rural areas (Qin et al., 2020).

Collectively, overnutrition related chronic diseases and the associated risk factors are affecting a large number of people among both old and young population. Notably, some naturally derived bioactive compounds have been used as the supplements for the intervention of some chronic diseases, particularly hypertension, T2DM, and dyslipidemia. The pathophysiology of these diseases will be introduced briefly in the following sections.

2.2 The pathophysiology of nutrition related chronic diseases

2.2.1 The pathophysiology of hypertension

Essential hypertension accounts for about 95% of all hypertensive cases (Carretero & Oparil, 2000). Although the etiology of essential hypertension is complex, which could be affected by both of genetic and environmental factors, the dysregulated renin angiotensin system (RAS) plays a pivotal role in the development of hypertension. In the classic RAS, angiotensin I (Ang I) is cleaved into angiotensin II (Ang II) with the action of angiotensin converting enzyme (ACE). The vasoconstrictor Ang II exerts its detrimental effects through the angiotensin type I receptor (AT1R). On the contrary, the counterbalancing effects against the Ang II/AT1R axis could be exerted via another G-protein-coupled receptor, namely angiotensin type II receptor (AT2R). Hereby, ACE inhibitors or AT1R blockers are used as the first-line therapy for hypertension management (Beevers, Lip, & O'Brien, 2001). Notably, the view on the RAS was evolved 2 decades ago with the identification angiotensin converting enzyme 2 (ACE2), which is a homolog of ACE but with opposite functions. ACE2 degrades Ang II to generate Ang (1–7) to exert effects against the Ang II/AT1R axis (Donoghue et al., 2000), which is mediated by the Mas receptor (MasR) (Santos et al., 2003). The protective roles of the ACE2/Ang (1–7) axis in cardiovascular functions have been reported in the past decade, and this axis has been considered as a new target for cardiovascular therapy (Patel, Zhong, Grant, & Oudit, 2016).

In addition to the direct vasoconstricting effect, persistent high level of Ang II could increase blood pressure indirectly through multiple mechanisms. Chronically elevated angiotensin II may impair the endothelial transport of L-arginine via increasing oxidative stress, which could consequently reduce the availability of vasodilator nitric oxide (NO). In turn, the reduced NO bioavailability activates the sympathetic nervous system, which is the central regulator of blood pressure, following by exacerbating the hypertensive effects of the Ang II/AT1R axis (Forrester et al., 2018). Besides, Ang II-mediated activation of nicotinamide adenine dinucleotide phosphate (NADPH) oxidase in dendritic cells increases the production of superoxide that activates T cells to produce cytokines, such as interleukin-17, interferon-γ, and tumor necrosis factor-α (TNF-α), which further leads to vascular inflammation and systemic vascular resistance (Hoch et al., 2009).

Clinically, obesity is a risk factor of essential hypertension. It has been reported that excessive bodyweight may contribute to 78% and 65% of the risk of hypertension in men and women, respectively. The prevalence of hypertension has been found to be 60% and 74% in overweight and morbidly obese patients, respectively. However, the mechanisms underlying the association between blood pressure and BMI have not been fully elucidated (Le Jemtel, Richardson, Samson, Jaiswal, & Oparil, 2017).

2.2.2 The pathophysiology of T2DM

Insulin and glucagon are two essential hormones that regulate blood glucose reciprocally. When the level of circulating glucose increases, pancreatic β-cells secretes insulin for glucose clearance from blood. In the early stage after the meal, insulin secreted to suppress hepatic gluconeogenesis. While after 1–2 h of a meal ingestion, insulin stimulates skeletal muscle and adipose tissue to uptake glucose (Grayson, Seeley, & Sandoval, 2013). The phosphatidylinositol 3-kinase (PI3K)/protein kinase B (Akt) pathway and the mitogen-activated protein kinase (MAPK) pathway are two major signaling cascades that mediate most of the metabolic effects of insulin. The PI3k/Akt pathway is responsible for glucose transport, glycogen and protein synthesis, lipogenesis, and the suppression of gluconeogenesis. The MAPK pathway controls the mitogenic effects in adipocytes (Huang, Liu, Guo, & Su, 2018). Insulin resistance is considered as a common feature during the development ofT2DM, and to improve insulin sensitivity is a strategy for the management of T2DM (Taylor, 2012).

Glucagon, a major counter-regulatory hormone to insulin actions, is released from pancreatic α-cells and plays pivotal roles in regulating blood glucose and cellular homeostasis (Habegger et al., 2010). During the fasting state, glucagon secretion is stimulated by hypoglycemia, promoting hepatic glucose production via glycogenolysis and gluconeogenesis to maintain a steady level of blood glucose (Jiang & Zhang, 2003). Meanwhile, glucagon facilitates hepatic mitochondrial fatty acid oxidation to supply energy for glucose production and promotes the conversion of fatty acids into ketone bodies for the brain to spare use of glucose (Longuet et al., 2008; von Meyenn et al., 2013). The actions of glucagon are transduced by its binding to the glucagon receptor (Gcgr) on the cellular membrane, which further stimulates the adenylate cyclase for the intracellular elevation of cyclic adenosine monophosphorylate (cAMP). As a result, the protein kinase A (PKA) is activated, and induces phosphorylation of cAMP response element-binding protein (CREB) and forkhead box protein O1 (Foxo1) that increase the transcriptions of downstream gluconeogenic genes (Wu et al., 2018). Therefore, glucagon plays a key role in the fasting state when food is unavailable and insulin secretion decreases. Most of the T2DM patients is accompanied with hyperglucagonemia that has been recognized as a contributor to hyperglycemia in diabetic patients (Færch et al., 2016).

2.2.3 The pathophysiology of dyslipidemia

Dyslipidemia is a heterogeneous disorder with multiple etiologies. Atherogenic dyslipidemia, which is characterized by elevated triglycerides and low level of HDL-C, is a major type of dyslipidemia associated with overnutrition. Atherogenic dyslipidemia is a dependent risk factor of coronary heart disease. In turn, a number of chronic diseases are strongly associated with atherogenic dyslipidemia. Among the patients with coronary heart disease, 33% of them had a high level of triglycerides, and 64% of them had a low level of HDL-C (Bamba & Rader, 2007). In addition, the prevalence of abnormal triglycerides and HDL-C in patients with T2DM is high. As indicated by a cross-sectional study, elevated triglycerides and low HDL-C level were present in T2DM patients with a proportion of 30% and 37%, respectively. Notably, the prevalence of obesity was 30% in the same population (Ford, Giles, & Dietz, 2002). Thus, there is a complex overlap that is present among dyslipidemia, obesity and T2DM.

Normally, dietary lipids could be transported or diffused into intestinal enterocytes and then packed into chylomicrons. Free fatty acids could be delivered to adipose tissue, skeletal muscle and other tissues by circulating chylomicrons through the lipoprotein lipase-mediated lipolysis (Bamba & Rader, 2007). After the delipidation, hepatic clearance of chylomicrons is executed by hepatic low-density lipoprotein (LDL) receptor. Besides, triglycerides could be derived from dietary fat, endogenous fat storage, and de novo lipogenesis, which undergo progressive delipidation and hepatic clearance in the very low-density lipoprotein (VLDL) particle. Under some pathogenic conditions, such as insulin resistance in T2DM patients, abnormal production and clearance of lipoprotein particles occur. Hepatic insulin resistance suppresses the secretions of chylomicrons and VLDL. In addition, insulin resistance also contributes to an increased number of triglyceride-rich lipoproteins, which competes for the clearance of lipoprotein particles (Musunuru, 2010).

2.3 Management of nutrition related chronic diseases by lifestyle modifications

2.3.1 Physical activity and nutrition related chronic diseases

It has been accepted that sedentary lifestyle is a risk factor for the development of overweight or obesity which may ultimately leads to the other chronic diseases. Generally, the physical activity is defined as "any form of physical exercise that is repeated regularly for a certain period of time." The physical training intervention mainly consists of walking, using an exercise bike, jogging, and weight training (Oja et al., 2011). There has been strong evidence showing that physical activity is essential for preventing weight gain in general, as well as for maintaining body weight after weight loss. A meta-analysis including 41 randomized controlled trials with 3476 overweight or obese individuals found that physical activity alone could contribute to significant weight loss. High-intensity physical activity was more effective than moderate activity. In addition, the combination of exercise and diet restriction resulted in a greater weight loss in BMI than diet restriction alone. Without diet, high-intensity physical training (~60% of the maximum oxygen uptake/pulse) led to greater weight loss than low-intensity physical training (Pedersen & Saltin, 2015; Shaw, Gennat, O'Rourke, & Del Mar, 2006). Besides, the enhanced physical activity has beneficial effects to reduce risk factors for nutrition chronic diseases in both children and adults. There is a cross-sectional study reporting that children with low physical activity had 3 times the higher risk of metabolic diseases, as compared with children high level of physical activity (Koga, Kawaguchi, & Aizawa, 2006). As documented by a cohort study, improved physical activity was associated with significant changes in several important metabolic variables including fasting blood glucose, proinsulin and HDL cholesterol. The risk of cardiovascular diseases for men performing moderate, regular and athletic physical activity was 25% ($P=0.127$), 34% ($P=0.022$) and 71% ($P=0.009$) lower, respectively, as compared with sedentary men (Byberg, Zethelius, McKeigue, & Lithell, 2001). Thus, enhancement of physical activity is important for the prevention of nutritional chronic diseases.

2.3.2 Dietary intervention of nutrition related chronic diseases

Since diet plays an essential role in the development of nutrition-related chronic diseases, dietary modifications could be significant to prevent and manage a number of diseases. Restriction of calories, particularly the restriction of calorie from fat, is a straightforward way to manage chronic diseases associated with lipid metabolism disorder. It has been reported that the level of LDL-C could be reduced by 10–15%, and plasma triglycerides could be reduced by 20–40% by the restrictions of total caloric and saturated fat intake (Gau & Wright, 2006).

The association between dietary pattern and chronic diseases has been discussed. It was suggested that plant-based diets could reduce the risk of coronary heart disease events and cerebral vascular disease events by 40% and 29%, respectively (Kahleova, Levin, & Barnard, 2017). In addition, plant-based diet is efficient for the management of nutrition related chronic diseases. Plant-based diets have been shown to be effective for weight loss. A recent study showed that in overweight or obese subjects a mean BMI was reduced by 4.4 kg/m^2 after a 6-month, whole-food, plant-based diet without energy restrictions, while the reduction of the control group was only 0.4 kg/m^2 (Wright, Wilson, Smith, Duncan, & McHugh, 2017). A low-fat, vegetarian diet was reported to mitigate the development of atherosclerotic plaque in clinical trials, when combined with exercise and stress management (Daubenmier et al., 2007). Furthermore, a meta-analysis including 7 randomized controlled trials and 32 observational studies found that vegetarian diets showed a more pronouncing effect on lowering blood pressure, as compared with omnivorous diets (Yokoyama et al., 2014). Collectively, there has been substantial attractions on the beneficial effects of plant-based dietary pattern. Changing from an unhealthy diet to a plant-based diet might be a simple and low-cost intervention for the management of overnutrition related diseases (Kahleova et al., 2017).

In addition to the modification of diet, supplementation of some naturally derived bioactive compounds is also helpful to prevent or intervene nutrition-related chronic diseases. Phytochemical is one of the representatives of the dietary bioactive compounds. Since the word "phytochemical" is originated from appending the word "chemical" to the Greek word "phyto," phytochemical is defined as non-nutrient plant chemicals with potential health benefits to reduce the risk of chronic diseases. Major types of phytochemicals that have been characterized include phenolics, phenolics, and flavonoids (Liu, 2013). The efficacy of these compound in ameliorating several chronic diseases have been demonstrated in animal models and humans. A large amount of studies has shown that phytochemicals exert the health promoting effects via regulating various intracellular signaling pathways (Manandhar, Kabekkodu, & Pai, 2020; Zhu, Du, & Xu, 2018). While a recent study found that micro-RNA and long non-coding RNAs also involved in the actions of phytochemicals, indicating the epigenetic actions of phytochemical (Kang, 2019; Mishra et al., 2019). Dietary fiber is another type of bioactive compounds that has been implicated to mitigate chronic diseases. Dietary fiber is resistant to the gastrointestinal enzymes and could be utilized by bacteria present in the large intestine to produce short-chain fatty acids. The production of short-chain fatty acids further alters the gut microbiota, which contributes to the health effects of dietary fibers. The activities of dietary fibers in controlling blood glucose, reducing blood pressure, retarding the development of atherosclerosis, and decreasing the cardiovascular risk have been demonstrated in various human studies. The recommended amount of dietary fiber intake is available in the Dietary Reference Intake of many countries. For adults, it is recommended to be 25–30 g per day. However, the tolerance of dietary fiber is varied by individuals, overconsumption may result in bloating and abdominal discomfort (Soliman, 2019). Notably, food protein-derived bioactive peptide is an emerging type of bioactive compounds which shows a great potential for the intervention of nutritional chronic diseases. The production, function as well as clinical evidence of bioactive peptides will be discussed in the following sections.

3. The potential of food protein-derived bioactive peptides in improving human health

3.1 An overview of food protein-derived bioactive peptides

In addition to their regular nutritional value, food-derived bioactive peptides are promising ingredients for the development of nutraceuticals and functional food, which could be applied for the prevention and management of chronic diseases due to their various biological activities. Bioactive peptides are inactive within the parent proteins, but can be released through enzymatic hydrolysis, fermentation, and other food processing approach (Vermeirssen, Camp, & Verstraete, 2004). Bioactive peptides with numerous biological activities have been characterized from various food proteins, such as milk, egg, fish, chicken, and legume proteins. The physiologically regulatory activities of peptide that have been identified include antihypertensive, antioxidant, anti-inflammatory, antimicrobial, osteoprotective, immunomodulatory, opioid, and so on (Chakrabarti, Guha, & Majumder, 2018). The production, bioactivity, and absorption will be introduced briefly in the following parts.

3.2 The production of bioactive peptides

3.2.1 Hydrolysis

Enzymatic hydrolysis is a common approach used for the production of bioactive peptides. Both of gastrointestinal enzymes, such as pepsin and trypsin, and non-gastrointestinal enzymes, such as thermolysin and alcalase, are used to release peptides from parent proteins. While some studies found that the non-gastrointestinal enzymes have broader cleavage sites than the gastrointestinal enzymes (Ulug, Jahandideh, & Wu, 2021). Following the hydrolysate production, the activity-guided purification is employed as a conventional strategy to generate purified peptides. However, the conventional approach of hydrolysis has several disadvantages, such as a large quantity of enzyme required, and a low efficiency of the hydrolysis. Besides, it is not suitable for the large-scale production of the peptides. Indeed, enzymatic membrane reactors were developed for the conversion of total proteins to hydrolysates via continuous production. The enzymatic membrane reactor is composed by a series of membrane separation with enzymatic reactions. As compared with the conventional hydrolysis, the enzymatic membrane reactor contributes to an improved efficiency of enzyme-catalyzed bioconversion as we as the production yields, but also it is easily to be scaled up (Lemes et al., 2016). Several studies have demonstrated the feasibility to produce casein-derived bioactive peptides in an ultrafiltration enzymatic membrane reactor (Luo, Morthensen, Meyer, & Pinelo, 2014; Wu et al., 2013; Zhao, Xu, Yang, & Katiyo, 2013). Despite the high efficiency, the purity of the final product was not as expected. In addition, since the overall hydrolytic process occurs via two consecutive reactions, the hydrolysis of the third bond is controlled by the accumulation of the intermediary products. Notably, the enzymatic membrane reactor was modified in some studies to improve the purity of the peptides (Korhonen & Pihlanto, 2003).

Alternative to enzymatic hydrolysis, novel technologies, such as subcritical water hydrolysis, have been developed to generate bioactive peptides from food proteins (Ulug et al., 2021). The supercritical water extraction system is cost-effective as it does not require enzyme or solvent. However, a number of parameters could affect the extraction efficiency which mainly include the system temperature, pressure, the solid to water ratio, particle size and the flow rate. Besides, the subcritical water hydrolysis is not specific to the cleavage bonds in proteins, which makes it difficult to generate peptides with a consistent profile in different batch (Powell, Bowra, & Cooper, 2016; Shitu, Izhar, & Tahir, 2015).

As mentioned above, the activity-guided fractionation is a common approach used for the conventional peptides screening and identifications. Since this approach involves several steps of separation and purification, which is usually via liquid chromatography, it is inefficient. Thus, in silico simulating proteolysis was developed by using bioinformatics software. Peptides generated from the in silico proteolysis system could be subjected to the model based on quantitative structure–activity relationship to predict the peptide activity (Anekthanakul, Hongsthong, Senachak, & Ruengjitchatchawalya, 2018; Gu, Majumder, & Wu, 2011). Notably, the machine learning algorithm has been introduced into the activity prediction of bioactive peptides in recent years (Kumar et al., 2015; Wang et al., 2021, 2020), which raises new opportunities for the activity predicting modeling as well as the peptide screening.

3.2.2 Fermentation

Fermentation is another commonly used approach used for the generation of bioactive peptides, in particular to release peptides from dairy proteins. Bioactive peptides could be generated by both starter and non-starter bacteria. Some peptides and amino acids degraded from milk proteins during fermentation possess typical flavor and aroma, which contribute to a special characteristic of the final product (Chai, Voo, & Chen, 2020). Moreover, the antigenicity of milk protein could be reduced via fermentation (Abd El-Salam & El-Shibiny, 2021). The lactic bacteria, such as *Lactococcus lactis*, *Lactobacillus helveticus* and *Lactobacillus delbrueckii*, are major strains used for the fermentation. The lactic bacteria hydrolyze the dairy protein into peptides and amino acids, which are used as the nitrogen sources for their growth (Brown et al., 2017). At least 16 peptidases have been characterized from lactic acid bacteria. The proteolytic system of lactic acid bacteria provides transport systems specific for amino acids, di- and tripeptides and oligopeptides of up to 18 amino acids. Longer oligopeptides which are not transported into the cells can be a source for the liberation of bioactive peptides in fermented milk products when further degraded by intracellular peptidases after cell lysis (Korhonen & Pihlanto, 2003).

The production of bioactive peptides from soybean proteins via fermentation has also been reported. Soybean protein-derived bioactive peptides could be released by bacterial or fungi mediated fermentation (Sanjukta & Rai, 2016). Apart from peptides with various bioactivities, soybean fermentation also contributes to reducing the antinutritional components such as proteinase-inhibitors, phytic acid, and oxalic acids. Besides, the level of components associated with immunoreactivity in soybean proteins could also be reduced during fermentation, which is due to the degradation of immunoreactive proteins during hydrolysis (Mukherjee, Chakraborty, & Dutta, 2016).

3.3 Bioactivities of food protein-derived peptides

3.3.1 Antioxidant peptides

Peptides with various bioactivities have been characterized from food proteins. Oxidative stress which is resulted from the excessive production of reactive oxygen species or/and the impaired cellular antioxidant system plays a critical role in the pathology of multiple chronic diseases (Sharifi-Rad et al., 2020). While, most of the antioxidant peptides were characterized by in vitro assays based on scavenging of stable free radicals. Peptides that were identified by assays related to the activity of the antioxidant enzymes have been rarely reported (Liu, Xing, Fu, Zhou, & Zhang, 2016; Wen, Zhang, Zhang, Duan, & Ma, 2020). In addition, antioxidant peptides identified from the in vitro assay might not have biological relevance. As reported, a series of antioxidant peptides characterized from egg white ovotransferrin did not show the antioxidant activity in the endothelial cells (Jahandideh, Chakrabarti, Davidge, & Wu, 2016). Thus, it is highly recommended to develop suitable and cost-effective assay, such as cell-based assay, to identify antioxidant peptides.

3.3.2 Antihypertensive peptides

Antihypertensive peptide is another major type of bioactive peptides that has been extensively studied. Since ACE in the RAS plays an essential role in generating the vasoconstricting Ang II, most of the antihypertensive peptides were initially characterized by in vitro ACE-inhibitory assay and thus claimed as ACE-inhibitory peptides. Although the blood pressure-lowering effect of many ACE-inhibitory peptides have been demonstrated hypertensive animal models, the in vivo ACE inhibition was not involved in the mechanisms of the peptides, which raised new opportunities for the mechanistic study of antihypertensive peptides (Wu, Liao, & Udenigwe, 2017). Interestingly, a recent study showed that ACE2, which plays counter-balancing effect against ACE in the RAS, is a potential target of food protein-derived bioactive peptides (Liao & Wu, 2021). It has been reported that a tripeptide IRW derived from egg white ovotransferrin could activate ACE2 in different systems including the human recombinant ACE2 in the cell-free system, the ACE2 in vascular smooth muscle cells, and the circulating ACE2 in spontaneously hypertensive rats (SHRs) (Liao, Bhullar, Chakrabarti, Davidge, & Wu, 2018). More importantly, the blockage of the Mas receptor that is the down-stream receptor of ACE2 could abolish the blood pressure-lowering effect of IRW in SHRs, which indicated that IRW exerts its antihypertensive peptides in SHRs dependent on the ACE2/MasR axis (Liao, Fan, Davidge, & Wu, 2019). However, the ACE2 -activating effect of IRW should be further validated in relevant clinical trials.

3.3.3 Antidiabetic peptides

The intestinal brush border α-glucosidase is a key enzyme responsible for carbohydrate digestion. Thus, α-glucosidase inhibitory peptides were identified from food protein sources with an implication to have antidiabetic activity (Yu, Yin, Zhao, Liu, & Chen, 2012). In addition, gut-derived hormones glucose-dependent insulinotropic polypeptide (GIP) and glucagon-like peptide-1 (GLP-1) stimulate pancreatic insulin secretion to promote glucose uptake by skeletal muscle and adipose tissue. GIP and GLP-1 are endogenous substrates of dipeptidyl peptidase IV (DPP-IV), which could be inactivated by DPP-IV rapidly (Gautier, Fetita, Sobngwi, & Salaün-Martin, 2005). Hereby, DPP-IV inhibitors are another major type of antidiabetic peptides characterized from food proteins (Nongonierma & FitzGerald, 2014). Interestingly, results from mechanistic studies showed that the insulin signaling in the adipose tissue of diabetic mice or rats could be activated by the treatment of those peptides, which further contributes to improving the insulin sensitivity of the animal model (Li, Liu, He, & Wu, 2018). Such findings suggested that the antidiabetic peptide might not only target a single molecule in vivo, instead it could improve the systemic glycemic control via intracellular signaling transductions. On the other hand, hyperglucagonemia which leads to excessive hepatic glucose production is another hallmark of T2DM in addition to insulin resistance (Lund, Bagger, Christensen, Knop, & Vilsbøll, 2014). Thus, future mechanistic studies of antidiabetic peptides may need to consider the effects of the peptides on hepatic glucose metabolism.

3.3.4 Anti-obesity peptides

The energy imbalance resulted from the excessive energy intake could lead to adipocytes dysfunction and the development of obesity. Adipocytes secrete a number of adipocytokines to regulate adipose metabolism (Guilherme, Virbasius, Puri, & Czech, 2008). It has been reported that lactotripeptides IPP and VPP could induce preadipocytes differentiation, and the secretion of adipokine in 3T3-F442A cells (Chakrabarti & Wu, 2015), indicating their potential against obesity. Besides, the treatment of thermoase and pepsin-digested egg white hydrolysate was reported to reduce the size of adipocytes in high-fat-diet-fed rat (Jahandideh et al., 2019). Peptides with the activity of adipogenic differentiation were also characterized from this hydrolysate (Jahandideh, Liu, & Wu, 2018). Moreover, the treatment of black soybean-derived peptides could reduce body weight gain via decreasing appetite, suggesting a different mechanism from modulating adipose biology of the peptides in mitigating obesity (Jang et al., 2008). However, signaling cascades underlying the actions of the peptides have not been fully elucidated. Additionally, the development of obesity is linked to insulin resistance, thus whether there is a cross-talk between the regulation of glucose metabolism and lipid metabolism by the peptide treatment need to be further determined.

3.3.5 *Hypolipidemic peptides*

Peptides that can modulate blood lipid profile have been identified mainly from soy, milk and egg proteins. These peptides are thought to exert the hypolipidemic effect by perturbing the absorption of dietary cholesterol in the intestine as well as suppressing the activities of lipogenic enzymes in hepatocytes and adipocytes (Udenigwe & Rouvinen-Watt, 2015). Interestingly, it was reported that the treatment of salmon protein hydrolysate mitigated the atherosclerosis in the ApoE deficient mice by acting both at vascular and systemic levels without altering the profile of blood lipids. However, detailed molecular mechanisms of such actions have not been elucidated (Parolini et al., 2014). In addition, the hypolipidemic effects of the peptides reported include the reductions of cholesterol level, triglycerides level, and LDL-C level, but the evidence is mainly with animal models. Moreover, whether there is peptide that can improve the level of HDL-C is unknown (Udenigwe & Rouvinen-Watt, 2015).

3.3.6 *Peptides with other bioactivities*

The types of bioactive peptides mentioned above have been extensively studied. While, peptides with other bioactivities which is associated with nutrition or non-nutrition related chronic diseases have also been investigated. There are abundant comprehensive reviews introducing the functions of these peptides, such anti-cancer peptides (Sharma et al., 2021), anti-inflammatory peptides (Guha & Majumder, 2019), antimicrobial peptides (Li et al., 2021), and peptides improving bone health (Zong et al., 2016). Thus, we do not provide a detailed introduction of these peptides in this chapter. It should be mentioned that some peptides have been shown with multiple functions. A typical example is that the lactotripeptides IPP and VPP were reported to possess several functions including reduction of blood pressure (Geleijnse & Engberink, 2010), improvement of insulin sensitivity (Chakrabarti, Jahandideh, Davidge, & Wu, 2018), and modulation of adipocyte functions (Chakrabarti & Wu, 2015). The presence of multiple functions of these peptides may be due to the cross-talk among different physiological functions, but also it may be resulted from the multiple molecular targets of the peptides. Hereby, it would be interesting to further explore the mechanisms of the peptides by "Omics" study, which could provide a comprehensive view on the actions of the peptides.

3.4 Bioavailability of bioactive peptides

Bioactive peptides could be transported via three pathways including active transport via peptide transporter 1 (PepT1), paracellular diffusion via tight junctions, and adsorptive transcytosis via endocytosis (Fan, Liao, & Wu, 2019). It has been estimated that more than 90% of the peptides could be hydrolyzed by the enterocyte peptidase expressed on the apical surface of intestinal epithelial cells. While, it has been generally accepted that peptides compose by <6 amino acids are able to pass through the enterocytes (Bouglé & Bouhallab, 2017). Thus, peptides with a larger size are more susceptible to gastrointestinal degradation than smaller peptides. Besides, other structural features of the peptides such as the charge, and hydrophobicity could also affect the bioavailability. Notably, although some peptides could be detected in the circulation (Foltz et al., 2007; Sánchez-Rivera, Martínez-Maqueda, Cruz-Huerta, Miralles, & Recio, 2014), the structure and bioavailability relationship of the peptides is warranted to be further investigated.

3.5 Safety considerations of bioactive peptides

Although very few adverse effects of bioactive peptides have been reported, safety of the peptides is still a bit of concern. Bioactive peptides are prepared in general under mild conditions using enzymatic hydrolysis or fermentation. Toxic and allergenic peptides could form during the process. Besides, amino acid racemization, peptide modification/derivatization, and the Maillard reaction occurred during peptide production could also lead to the formation of numerous compounds with potential hazards (Liu et al., 2020). Thus, bioactive peptides are generally regarded as safe, but procedures during the production that may degrade the quality and safety of the peptide should be avoided (Chakrabarti, Guha, et al., 2018).

4. Clinical evidence of bioactive peptides in management of nutrition related chronic diseases

4.1 Clinical trials of antihypertensive peptides

The antihypertensive peptides are the most extensively studied bioactive peptides. Clinical evidences of antihypertensive peptides have been synthesized in several meta-analyses. Thus, we only show the summary of these meta-analyses in Table 1. Although positive effect was not reported in all of the clinical trials of antihypertensive peptides, results from several meta-analyses demonstrated the efficacy of antihypertensive peptides in reducing blood pressure in hypertensive patients by quantitative synthesis of evidence (Cicero et al., 2013; Fekete et al., 2015; Qin et al., 2013). Recently, by conducting subgroup analyses in the meta-analysis, we found that the effect of the peptide intervention was greater in participants with a

Table 1 Meta-analyses of clinical trials of antihypertensive peptides.

Publication year	Number of the included study	Main conclusions	References
2011	18	The blood pressure-lowering effect of the peptide intervention was more evident in Asian patients than in Caucasian patients	Cicero, Gerocarni, Laghi, and Borghi (2011)
2012	14	The peptides IPP and VPP are effective in moderately reducing systolic in European subjects	Cicero, Aubin, Azais-Braesco, and Borghi (2013)
2013	19	The intervention of lactotripeptides showed blood pressure-lowering effect in adults with prehypertension or mild hypertension	Turpeinen, Järvenpää, Kautiainen, Korpela, and Vapaatalo (2013)
2013	24	The intervention of lactotripeptides could reduce 24-h ambulatory blood pressure	Qin et al. (2013)
2015	30	The intervention of lactotripeptides is effective in blood pressure reduction, especially in Japanese individuals	Fekete, Givens, and Lovegrove (2015)
2021	12	The intervention of bioactive peptides could reduce blood pressure in prehypertensive and hypertensive patients. The basal blood pressure and gender of the participants appear to have a significant impact on the effect size	Liao et al. (2021)

higher basal blood pressure (>140/85 mm Hg). Besides, the effect was more pronouncing in the trials with a lower ratio of male to female (<0.5), indicating that female might be more sensitive to the peptide intervention, as compared with the male participants (Liao et al., 2021).

The blood pressure-lowering effect of milk protein derived peptides, in particular IPP and VPP, have been investigated in a number of clinical trials (Fekete et al., 2015; Qin et al., 2013). Overall, both types of peptides produced from either hydrolysis or fermentation showed a significant effect in reducing the blood pressure in hypertensive patients. However, the peptide intervention showed a stronger effect in the Asian group, as compared with the non-Asian subjects. While, results from a meta-analysis showed that the effect was not related to age of the participants, or the dosage or the duration of the peptide intervention (Fekete et al., 2015).

The blood pressure-lowering effect of egg protein-derived peptides have also been reported. It was found that supplementation of 2g egg protein hydrolysate for 7days NWT-03 significantly decreased the daytime blood pressure as well as 36-h ambulatory blood pressure in subjects with mild hypertension (systolic blood pressure: 140–159mmHg and diastolic blood pressure: 90–99mmHg). However, the effect was not present in subjects with normal blood pressure (systolic blood pressure <130mmHg and diastolic blood pressure <85mmHg) or high normal blood pressure (systolic blood pressure: 130–139mmHg and diastolic blood pressure: 85–89mmHg) (Plat, Severins, Morrison, & Mensink, 2017). Surprisingly, another study showed that the treatment of an egg ovalbumin hydrolysate with a daily dosage of 3g for 6weeks did not affect the blood pressure in adults aged from 50 to 70years. Besides, neither variables associated with cardiovascular risk nor blood glucose was altered by the treatment (Lucey, Heneghan, Manning, Kroon, & Kiely, 2019). The inconsistence between these two studies suggests that peptides present in egg proteins other than ovalbumin could be bioactive component that exhibit the blood pressure-lowering effect.

Notably, the blood pressure-lowering effect of bioactive peptides derived from plant proteins have also been reported. As documented, supplementation of a rice bran-derived peptide LRL with a dosage of 43μg/day for 12weeks significantly decreased blood pressure in subjects with a basal systolic pressure between 130 and 159mmHg and diastolic pressure between 85 and 99mmHg. There was no serious adverse event observed during and after the intervention period, suggesting that this rice bran peptide could safely mitigate the elevated blood pressure (Ogawa et al., 2019). Another study reported the effect of pea protein hydrolysate. However, only seven volunteers were included in the trial (Li et al., 2011). The effect needs to be further determined.

It is worthwhile to be mentioned that the treatment of fermented goat milk containing bioactive peptides for 10weeks significantly lowered the blood pressure in prehypertensive subjects, which was accompanied by the reduced level of circulating Ang II (Lu et al., 2018). Such a finding suggested that the modulation of the RAS might involve in the blood pressure-lowering effect of bioactive peptides, which is consistent with the findings from previous studies in vascular cells and animal models. However, previous mechanistic studies showed multifaceted roles of bioactive peptides in reducing blood pressure in animals, which also include mitigation of vascular inflammation, activation of the endothelial-dependent vasorelaxation, and the underlying intracellular pathways have also been characterized (Majumder & Wu, 2015). Thus, whether the peptides exert the blood pressure-lowering activity in humans via the same signaling pathways needs to be further explored.

4.2 Clinical trials of peptides in ameliorating T2DM

4.2.1 Chronic studies

The clinical effects of some bioactive peptides have been demonstrated by chronic studies (Table 2) or acute studies. In the chronic studies, the duration and dosage of the intervention varied among studies. The treatment of *Lactobacillus* fermented milk with bioactive peptides for 12 weeks showed a significant effect in controlling fasting blood glucose in T2DM patients. However, the effect on HbA1c and blood lipid profile were insignificant. Besides, the age of the participants in this study ranged from 40 to 70 years (Hove et al., 2015), thus the efficacy of the treatment in younger patients was unknown. Bioactive peptides derived from marine protein sources have been reported with the effect of glycemic control. Peptides from marine collagen showed a great effect in reducing fasting blood glucose in Chinese T2DM patients. The intervention lasted for 3 months but the effect occurred in 1.5 months. In addition, the level of HbA1c was reduced significantly by the treatment and insulin sensitivity of the participants were enhanced, which suggested that the peptides intervention could regulate the blood glucose metabolism. It has to be mentioned that the peptides intervention did not show the effect in the healthy control participants, indicating that the peptides could only modulate abnormal glucose metabolism (Zhu, Li, Peng, Zhang, et al., 2010). Notably, the same peptides could show the activity in improving glucose metabolism in T2DM patients with hypertension as well (Zhu, Li, Peng, Li, et al., 2010), indicating a wide scope of application of the marine collagen peptides. Indeed, some other marine food protein-derived peptide also showed similar effects. It was reported that the *Styela clava* flesh tissue hydrolysate exhibited a significant effect in decreasing the level of HbA1c with a tendency to improve the homeostasis model assessment of insulin resistance after the 4-week intervention in patients with both T2DM and hypertension. The treatment also showed significant effects in

Table 2 Clinical evidence of bioactive peptides in reducing blood glucose.

Protein source	Production approach	Type of the participants	Age of the participants	Number of the participants	Study design	Daily dosage	Treatment duration	Decrease of fasting blood glucose by the treatment	References
Whey	Fermentation	Mild hypertensive patients	30–65 years	54	R, PC, DB	25 mL liquid product	12 weeks	0.9 mg/dL	Lee, Skurk, Hennig, and Hauner (2007)
Black Soy	N/A	Prediabetic and newly diagnosed T2DM patients	18–69 years	55	R, PC, DB	4.5 g peptides in tablets	12 weeks	4.9 mg/dL	Kwak et al. (2010)
Marine collagen	Hydrolysis	Patients with T2DM and healthy subjects	Healthy control mean age: 65.78 years Patient control mean: 61.78 years Treatment mean age: 64.22 years	150	R, PC, DB	6.5 g hydrolysate powder	3 months	1.91 mmol/L	Zhu et al. (2010)
Marine collagen	Hydrolysis	Patients with T2DM and primary hypertension, and healthy subjects	Healthy control mean age 65.78 years Patient control mean: 63.66 years Treatment mean age: 67.64 years	150	R, PC, DB	13 g hydrolysate powder	3 months	1.42 mmol/L	Zhu et al. (2010)
Milk	Fermentation	Patients with T2DM	40–70 years	41	R, PC, DB	300 mL liquid product	12 weeks	1.1 mmol/L	Hove et al. (2015)

Continued

Table 2 Clinical evidence of bioactive peptides in reducing blood glucose.—cont'd

Protein source	Production approach	Type of the participants	Age of the participants	Number of the participants	Study design	Daily dosage	Treatment duration	Decrease of fasting blood glucose by the treatment	References
Styela clava flesh tissue	Hydrolysis	Patients with hypertension and T2DM	30–80 years	80	R, PC, DB	500 mg hydrolysate in capsules	4 weeks	13.6 mg/dL	Ko, Jung, Lee, Lee, and Jeon (2017)
Casein	Hydrolysis	Patients with gestational diabetes	Gestational age of $\geq 20^{+0}$ to 35^{+6}	175	R, PC, DB	17 g hydrolysate in liquid	8 days	0.2 mmol/L	Saleh et al. (2018)
Egg	Hydrolysis	Overweight and obese subjects with impaired glucose-tolerance or T2DM	18–70 years	40	R, PC, DB, CR	50 g hydrolysate in capsules	3 days	1.77 mmol/L	Plat, Severins, and Mensink (2019)
Cod	Hydrolysis	Adults without chronic diseases	60-80 years	31	R, DB, CR	40 mg/kg body weight hydrolysate	1 week	0.28 mmol/L	Jensen et al. (2019)
Salmon	Hydrolysis	Healthy Japanese adults	35–65 years	14	R, PC, DB, CR	500 mg hydrolysate powder	1 week	1.85 mg/dL	Takahashi, Kamata, Nishimura, and Nishihira (2021)

R, Randomized; DB, Double-blinded; PC, Placebo-controlled; CR, Cross-over; N/A, Not available.

decreasing blood pressure as well as modulating the blood lipid profile in the same participants (Ko et al., 2017), suggesting multiple effects of the peptides. The glycemic controlling effect was also reported in cod protein peptides. In T2DM patients, the treatment showed a trend to decrease the level of fasting blood glucose after 7 days with a dosage of 40 mg/kg body weight/day. However, the effect of the peptides needs to be further validated in the trial with longer period. Moreover, glycemic controlling peptides are present in salmon protein hydrolysate, since it was reported that the supplementation of 500 mg salmon protein hydrolysate for 7 days could reduce fasting blood glucose in health Japanese adults (Takahashi et al., 2021).

The glycemic controlling effect is also exhibited by some egg protein-derived peptides. In overweight or obese T2DM patients, although the treatment of egg protein hydrolysate NWT03 did not show a significant effect in reducing fasting blood glucose, which might be due to the short intervention duration (3 days), the level of serum insulin of the patients was decreased (Plat et al., 2019), suggesting that the insulin sensitivity might be improved by the treatment.

In addition to the activity in T2DM patients, which account for about 90% of all the diabetic cases, the effect of peptides intervention was also reported in patients with other types of diabetes. It has been reported that in patients with gestational diabetes, a twice-daily dose of 8.5 g of casein-derived protein hydrolysate for 8 days reduced the daily main plasma glucose concentration (Saleh et al., 2018). While, the effects of the intervention on other glycemic biomarkers are warranted to be determined in further studies.

4.2.2 Acute studies

The elevated postprandial glucose (PPG) level is one of the earliest abnormalities of glucose homeostasis associated with T2DM, which is also exaggerated in diabetic patients with fasting hyperglycemia (Association, 2001). Clinical trials of food protein-derived bioactive peptides in regulating PPG have been investigated, among which, whey protein-derived peptides were the most studied. Whey protein hydrolysate showed a potent effect in glycemic control in a 60-min blood glucose test. However, the effect was only observed in lean adults but not in obese subjects (Hutchison et al., 2015). Surprisingly, another study showed that the premeal consumption of whey protein hydrolysate did not show a significant effect on the PPG in healthy young adults. Instead, ingestion of the intact whey protein showed a potent effect in regulating the postprandial glucose response and insulin response (Akhavan, Luhovyy, Brown, Cho, & Anderson, 2010). The controversial results from different studies might be due to the different hydrolysis procedures, which resulted in different amino acids profiles. While, detailed information about the hydrolysate production was unavailable.

The effect of cod protein hydrolysate on postprandial glucose metabolism was also reported. While, the hydrolysate only affected the insulin response but not glucose response in a 3-h test. Besides, the trial was only conducted in healthy subjects, the effect of the hydrolysate in subjects with impaired glucose metabolism was unknown (Dale et al., 2018).

Effects of hydrolysates derived from different protein sources on postprandial glucose metabolism have been compared in healthy individuals. Among the tested hydrolysate derived from pea, rice, soy, gluten, whey, and egg proteins, gluten protein hydrolysate, pea, soy, gluten, and whey protein hydrolysates showed a significant effect in PPG response. While, there was no statistical difference among these hydrolysates (Tiekou Lorinczova, Deb, Begum, Renshaw, & Zariwala, 2021).

Collectively, the acute effect of food protein hydrolysates has been investigated. However, most of the studies were conducted in healthy individuals. It is recommended to investigate the acute effect in diabetic patients in future studies (Claessens, Calame, Siemensma, van Baak, & Saris, 2009).

4.3 Clinical trials of peptides modulating blood lipid profile

Dyslipidemia is the risk factor for a number of nutrition related chronic diseases. Studies that evaluate the activity of bioactive peptides in ameliorating dyslipidemia in humans are available but limited. It was reported that supplementation of 1 g *Lupinus angustifolius* protein hydrolysate for 28 days significantly decreased the ratio of LDL-C to HDL-C. Besides, the intervention also improved the pro-inflammatory response of peripheral blood mononuclear cells as well as the cellular antioxidant capacity, indicating the immunomodulatory potential of the peptides. It has to be mentioned that this trial was conducted in healthy adults (Cruz-Chamorro et al., 2021). Thus, the results indicate the potential of *Lupinus angustifolius*-derived peptides in preventing dyslipidemia and inflammatory diseases. Another study that assessed the effect of dry-cured ham-derived peptides found that the consumption of this dry-cured ham product significantly reduced the levels of total cholesterol and LDL-C. Additionally, the level of fasting blood glucose was also decreased by the treatment (Montoro-García et al., 2017).

Since this trial was conducted in adults with prehypertension, the results suggested that dry-cured ham-derived peptides might play a protective role in subjective with a higher cardiovascular risk.

Although the report of clinical trials related to the effect of bioactive peptides on modulating blood lipid profile is limited, abundant studies have demonstrated the efficacy of the consumption of soy protein isolate in ameliorating dyslipidemia in various populations (Moradi, Daneshzad, & Azadbakht, 2020; Reynolds et al., 2006). Since the protein isolate is hydrolyzed by the gastrointestinal enzymes into peptides or amino acids for absorption, it is suggested to investigate the effects of soy protein-derived peptides on lipid profile in future clinical trials.

5. Prospective trends

Food protein-derived bioactive peptides could be a promising candidate for the development of nutraceuticals and functional foods management of nutrition-related chronic diseases. Despite that some strong evidences have been reported, clinical trial of bioactive peptides is still at infancy. Future research trends of bioactive peptides are encouraged to be considered from the following aspects, which aims at accelerating the clinical translation and commercialization of bioactive peptides.

(1) The bioavailability of food protein-derived bioactive peptides needs to be further elucidated. Although the presence of bioactive peptides in circulation after ingestion has been reported, the evidence is not strong since the number of research is limited. Future studies are suggested to further explore the transportation, absorption, and bioavailability of bioactive peptides, and analyze the structural features of peptides that could be able to enter into the blood-stream.

(2) Although bioactive peptides are derived from food proteins, which are considered as a natural compound with fewer adverse effects after the long-term consumption, comprehensive toxic assessments of bioactive peptides are still requested.

(3) By now, clinical evidences of bioactive peptides reported were all based on randomized clinical trials with limited number of participants. Future studies are encouraged to further expand the clinical studies into prospective cohort studies, which could also be helpful to delineate the factors that could impact the effect size of the peptide intervention.

References

Abd El-Salam, M. H., & El-Shibiny, S. (2021). Reduction of Milk protein antigenicity by enzymatic hydrolysis and fermentation. A review. *Food Reviews International, 37*(3), 276–295. https://doi.org/10.1080/87559129.2019.1701010.

Afshin, A., Forouzanfar, M. H., Reitsma, M. B., Sur, P., Estep, K., Lee, A., et al. (2017). Health effects of overweight and obesity in 195 countries over 25 years. *New England Journal of Medicine, 377*(1), 13–27. https://doi.org/10.1056/NEJMoa1614362.

Akhavan, T., Luhovyy, B. L., Brown, P. H., Cho, C. E., & Anderson, G. H. (2010). Effect of premeal consumption of whey protein and its hydrolysate on food intake and postmeal glycemia and insulin responses in young adults. *The American Journal of Clinical Nutrition, 91*(4), 966–975.

Anekthanakul, K., Hongsthong, A., Senachak, J., & Ruengjitchatchawalya, M. (2018). SpirPep: An in silico digestion-based platform to assist bioactive peptides discovery from a genome-wide database. *BMC Bioinformatics, 19*(1), 149. https://doi.org/10.1186/s12859-018-2143-0.

Association, A. D. (2001). Postprandial blood glucose. *Diabetes Care, 24*(4), 775–778.

Bamba, V., & Rader, D. J. (2007). Obesity and atherogenic dyslipidemia. *Gastroenterology, 132*(6), 2181–2190. https://doi.org/10.1053/j.gastro.2007.03.056.

Beevers, G., Lip, G. Y. H., & O'Brien, E. (2001). The pathophysiology of hypertension. *BMJ, 322*(7291), 912–916. https://doi.org/10.1136/bmj.322.7291.912.

Bouglé, D., & Bouhallab, S. (2017). Dietary bioactive peptides: Human studies. *Critical Reviews in Food Science and Nutrition, 57*(2), 335–343. https://doi.org/10.1080/10408398.2013.873766.

Branca, F., Lartey, A., Oenema, S., Aguayo, V., Stordalen, G. A., Richardson, R., et al. (2019). Transforming the food system to fight non-communicable diseases. *British Medical Journal, 364*. https://doi.org/10.1136/bmj.l296.

Brown, L., Pingitore, E. V., Mozzi, F., Saavedra, L., Villegas, M., & J., & M Hebert, E. (2017). Lactic acid bacteria as cell factories for the generation of bioactive peptides. *Protein and Peptide Letters, 24*(2), 146–155.

Byberg, L., Zethelius, B., McKeigue, P., & Lithell, H. (2001). Changes in physical activity are associated with changes in metabolic cardiovascular risk factors. *Diabetologia, 44*(12), 2134–2139.

Carretero, O. A., & Oparil, S. (2000). Essential hypertension. *Circulation, 101*(3), 329–335. https://doi.org/10.1161/01.CIR.101.3.329.

Chai, K. F., Voo, A. Y. H., & Chen, W. N. (2020). Bioactive peptides from food fermentation: A comprehensive review of their sources, bioactivities, applications, and future development. *Comprehensive Reviews in Food Science and Food Safety, 19*(6), 3825–3885. https://doi.org/10.1111/1541-4337.12651.

Chakrabarti, S., Guha, S., & Majumder, K. (2018). Food-derived bioactive peptides in human health: Challenges and opportunities. *Nutrients, 10*(11), 1738 (Retrieved from https://www.mdpi.com/2072-6643/10/11/1738).

Chakrabarti, S., Jahandideh, F., Davidge, S. T., & Wu, J. (2018). Milk-derived tripeptides IPP (Ile-pro-pro) and VPP (Val-pro-pro) enhance insulin sensitivity and prevent insulin resistance in 3T3-F442A preadipocytes. *Journal of Agricultural and Food Chemistry, 66*(39), 10179–10187. https://doi.org/10.1021/acs.jafc.8b02051.

Chakrabarti, S., & Wu, J. (2015). Milk-derived tripeptides IPP (Ile-pro-pro) and VPP (Val-pro-pro) promote adipocyte differentiation and inhibit inflammation in 3T3-F442A cells. *PLoS One, 10*(2), e0117492. https://doi.org/10.1371/journal.pone.0117492.

Cicero, A. F., Aubin, F., Azais-Braesco, V., & Borghi, C. (2013). Do the lactotripeptides isoleucine–proline–proline and valine–proline–proline reduce systolic blood pressure in European subjects? A meta-analysis of randomized controlled trials. *American Journal of Hypertension, 26*(3), 442–449.

Cicero, A. F. G., Fogacci, F., & Colletti, A. (2017). Potential role of bioactive peptides in prevention and treatment of chronic diseases: A narrative review. *British Journal of Pharmacology*, *174*(11), 1378–1394. https://doi.org/10.1111/bph.13608.

Cicero, A., Gerocarni, B., Laghi, L., & Borghi, C. (2011). Blood pressure lowering effect of lactotripeptides assumed as functional foods: A meta-analysis of current available clinical trials. *Journal of Human Hypertension*, *25*(7), 425–436.

Claessens, M., Calame, W., Siemensma, A. D., van Baak, M. A., & Saris, W. H. (2009). The effect of different protein hydrolysate/carbohydrate mixtures on postprandial glucagon and insulin responses in healthy subjects. *European Journal of Clinical Nutrition*, *63*(1), 48–56.

Consultation, J. W. F. E. (2003). Diet, nutrition and the prevention of chronic diseases. *World Health Organization Technical Report Series*, *916*(i-viii), 1–149.

Cruz-Chamorro, I., Álvarez-Sánchez, N., Álvarez-Ríos, A. I., Santos-Sánchez, G., Pedroche, J., Millán, F., et al. (2021). Safety and efficacy of a beverage containing lupine protein hydrolysates on the immune, oxidative and lipid status in healthy subjects: An intervention study (the Lupine-1 trial). *Molecular Nutrition & Food Research*, *65*(14), 2100139. https://doi.org/10.1002/mnfr.202100139.

Dale, H. F., Jensen, C., Hausken, T., Lied, E., Hatlebakk, J. G., Brønstad, I., et al. (2018). Effect of a cod protein hydrolysate on postprandial glucose metabolism in healthy subjects: A double-blind cross-over trial. *Journal of Nutritional Science*, 7, e33. https://doi.org/10.1017/jns.2018.23.

Daubenmier, J. J., Weidner, G., Sumner, M. D., Mendell, N., Merritt-Worden, T., Studley, J., et al. (2007). The contribution of changes in diet, exercise, and stress management to changes in coronary risk in women and men in the multisite cardiac lifestyle intervention program. *Annals of Behavioral Medicine*, *33*(1), 57–68. https://doi.org/10.1207/s15324796abm3301_7.

Donoghue, M., Hsieh, F., Baronas, E., Godbout, K., Gosselin, M., Stagliano, N., et al. (2000). A novel angiotensin-converting enzyme–related carboxypeptidase (ACE2) converts angiotensin I to angiotensin 1-9. *Circulation Research*, *87*(5), e1–e9. https://doi.org/10.1161/01.RES.87.5.e1.

Færch, K., Vistisen, D., Pacini, G., Torekov, S. S., Johansen, N. B., Witte, D. R., et al. (2016). Insulin resistance is accompanied by increased fasting glucagon and delayed glucagon suppression in individuals with Normal and impaired glucose regulation. *Diabetes*, *65*(11), 3473–3481. https://doi.org/10.2337/db16-0240.

Fan, H., Liao, W., & Wu, J. (2019). Molecular interactions, bioavailability, and cellular mechanisms of angiotensin-converting enzyme inhibitory peptides. *Journal of Food Biochemistry*, *43*(1), e12572. https://doi.org/10.1111/jfbc.12572.

Fekete, Á. A., Givens, D. I., & Lovegrove, J. A. (2015). Casein-derived lactotripeptides reduce systolic and diastolic blood pressure in a meta-analysis of randomised clinical trials. *Nutrients*, 7(1), 659–681.

Foltz, M., Meynen, E. E., Bianco, V., van Platerink, C., Koning, T. M., & Kloek, J. (2007). Angiotensin converting enzyme inhibitory peptides from a lactotripeptide-enriched milk beverage are absorbed intact into the circulation. *The Journal of Nutrition*, *137*(4), 953–958.

Ford, E. S., Giles, W. H., & Dietz, W. H. (2002). Prevalence of the metabolic syndrome among US AdultsFindings from the third National Health and nutrition examination survey. *JAMA*, *287*(3), 356–359. https://doi.org/10.1001/jama.287.3.356.

Forrester, S. J., Booz, G. W., Sigmund, C. D., Coffman, T. M., Kawai, T., Rizzo, V., et al. (2018). Angiotensin II signal transduction: An update on mechanisms of physiology and pathophysiology. *Physiological Reviews*, *98*(3), 1627–1738. https://doi.org/10.1152/physrev.00038.2017.

Gau, G. T., & Wright, R. S. (2006). Pathophysiology, diagnosis, and Management of Dyslipidemia. *Current Problems in Cardiology*, *31*(7), 445–486. https://doi.org/10.1016/j.cpcardiol.2006.03.001.

Gautier, J. F., Fetita, S., Sobngwi, E., & Salaün-Martin, C. (2005). Biological actions of the incretins GIP and GLP-1 and therapeutic perspectives in patients with type 2 diabetes. *Diabetes & Metabolism, 31*(3), 233–242. https://doi.org/10.1016/S1262-3636(07)70190-8.

Geleijnse, J. M., & Engberink, M. F. (2010). Lactopeptides and human blood pressure. *Current Opinion in Lipidology, 21*(1), 58–63.

Grayson, B. E., Seeley, R. J., & Sandoval, D. A. (2013). Wired on sugar: The role of the CNS in the regulation of glucose homeostasis. *Nature Reviews Neuroscience, 14*(1), 24–37. https://doi.org/10.1038/nrn3409.

Gu, Y., Majumder, K., & Wu, J. (2011). QSAR-aided in silico approach in evaluation of food proteins as precursors of ACE inhibitory peptides. *Food Research International, 44*(8), 2465–2474. https://doi.org/10.1016/j.foodres.2011.01.051.

Guha, S., & Majumder, K. (2019). Structural-features of food-derived bioactive peptides with anti-inflammatory activity: A brief review. *Journal of Food Biochemistry, 43*(1), e12531. https://doi.org/10.1111/jfbc.12531.

Guilherme, A., Virbasius, J. V., Puri, V., & Czech, M. P. (2008). Adipocyte dysfunctions linking obesity to insulin resistance and type 2 diabetes. *Nature Reviews Molecular Cell Biology, 9*(5), 367–377. https://doi.org/10.1038/nrm2391.

Habegger, K. M., Heppner, K. M., Geary, N., Bartness, T. J., DiMarchi, R., & Tschop, M. H. (2010). The metabolic actions of glucagon revisited. *Nature Reviews. Endocrinology, 6*(12), 689–697. https://doi.org/10.1038/nrendo.2010.187.

Hoch, N. E., Guzik, T. J., Chen, W., Deans, T., Maalouf, S. A., Gratze, P., et al. (2009). Regulation of T-cell function by endogenously produced angiotensin II. *American Journal of Physiology. Regulatory, Integrative and Comparative Physiology, 296*(2), R208–R216. https://doi.org/10.1152/ajpregu.90521.2008.

Hove, K., Brøns, C., Færch, K., Lund, S., Rossing, P., & Vaag, A. (2015). Effects of 12 weeks of treatment with fermented milk on blood pressure, glucose metabolism and markers of cardiovascular risk in patients with type 2 diabetes: A randomised double-blind placebo-controlled study. *European Journal of Endocrinology, 172*(1), 11–20.

Huang, X., Liu, G., Guo, J., & Su, Z. (2018). The PI3K/AKT pathway in obesity and type 2 diabetes. *International Journal of Biological Sciences, 14*(11), 1483–1496. https://doi.org/10.7150/ijbs.27173.

Hutchison, A. T., Feinle-Bisset, C., Fitzgerald, P. C., Standfield, S., Horowitz, M., Clifton, P. M., et al. (2015). Comparative effects of intraduodenal whey protein hydrolysate on antropyloroduodenal motility, gut hormones, glycemia, appetite, and energy intake in lean and obese men. *The American Journal of Clinical Nutrition, 102*(6), 1323–1331.

Jahandideh, F., Chakrabarti, S., Davidge, S. T., & Wu, J. (2016). Antioxidant peptides identified from Ovotransferrin by the ORAC method did not show anti-inflammatory and antioxidant activities in endothelial cells. *Journal of Agricultural and Food Chemistry, 64*(1), 113–119. https://doi.org/10.1021/acs.jafc.5b04230.

Jahandideh, F., de Campos Zani, S. C., Son, M., Proctor, S. D., Davidge, S. T., Chan, C. B., et al. (2019). Egg white hydrolysate enhances insulin sensitivity in high-fat diet-induced insulin-resistant rats via Akt activation. *British Journal of Nutrition, 122*(1), 14–24. https://doi.org/10.1017/S0007114519000837.

Jahandideh, F., Liu, P., & Wu, J. (2018). Purification and identification of adipogenic-differentiating peptides from egg white hydrolysate. *Food Chemistry, 259*, 25–30. https://doi.org/10.1016/j.foodchem.2018.03.099.

Jang, E., Moon, J., Ko, J., Ahn, C., Lee, H., Shin, J., et al. (2008). Novel black soy peptides with antiobesity effects: Activation of leptin-like signaling and AMP-activated protein kinase. *International Journal of Obesity, 32*(7), 1161–1170.

Jensen, C., Dale, H. F., Hausken, T., Lied, E., Hatlebakk, J. G., Brønstad, I., et al. (2019). Supplementation with cod protein hydrolysate in older adults: A dose range cross-over study. *Journal of Nutritional Science, 8*.

Jiang, G., & Zhang, B. B. (2003). Glucagon and regulation of glucose metabolism. *American Journal of Physiology. Endocrinology and Metabolism*, *284*(4), E671–E678.

Kahleova, H., Levin, S., & Barnard, N. (2017). Cardio-metabolic benefits of plant-based diets. *Nutrients*, *9*(8), 848 (Retrieved from https://www.mdpi.com/2072-6643/9/8/848).

Kang, H. (2019). MicroRNA-mediated health-promoting effects of phytochemicals. *International Journal of Molecular Sciences*, *20*(10), 2535 (Retrieved from https://www.mdpi.com/1422-0067/20/10/2535).

Ko, S.-C., Jung, W.-K., Lee, S.-H., Lee, D. H., & Jeon, Y.-J. (2017). Antihypertensive effect of an enzymatic hydrolysate from Styela clava flesh tissue in type 2 diabetic patients with hypertension. *NRP*, *11*(5), 396–401. https://doi.org/10.4162/nrp.2017.11.5.396.

Koga, T., Kawaguchi, A., & Aizawa, H. (2006). Physical activity and cardiovascular risk in children. *The Lancet*, *368*(9544), 1326.

Korhonen, H., & Pihlanto, A. (2003). Food-derived bioactive peptides-opportunities for designing future foods. *Current Pharmaceutical Design*, *9*(16), 1297–1308.

Kumar, R., Chaudhary, K., Singh Chauhan, J., Nagpal, G., Kumar, R., Sharma, M., et al. (2015). An in silico platform for predicting, screening and designing of antihypertensive peptides. *Scientific Reports*, *5*(1), 12512. https://doi.org/10.1038/srep12512.

Kwak, J. H., Lee, J. H., Ahn, C.-W., Park, S.-H., Shim, S.-T., Song, Y. D., et al. (2010). Black soy peptide supplementation improves glucose control in subjects with prediabetes and newly diagnosed type 2 diabetes mellitus. *Journal of Medicinal Food*, *13*(6), 1307–1312.

Le Jemtel, T. H., Richardson, W., Samson, R., Jaiswal, A., & Oparil, S. (2017). Pathophysiology and potential non-pharmacologic treatments of obesity or kidney disease associated refractory hypertension. *Current Hypertension Reports*, *19*(2), 18. https://doi.org/10.1007/s11906-017-0713-6.

Lee, Y.-M., Skurk, T., Hennig, M., & Hauner, H. (2007). Effect of a milk drink supplemented with whey peptides on blood pressure in patients with mild hypertension. *European Journal of Nutrition*, *46*(1), 21–27.

Lemes, A. C., Sala, L., Ores, J. D. C., Braga, A. R. C., Egea, M. B., & Fernandes, K. F. (2016). A review of the latest advances in encrypted bioactive peptides from protein-rich waste. *International Journal of Molecular Sciences*, *17*(6), 950 (Retrieved from https://www.mdpi.com/1422-0067/17/6/950).

Li, S., Liu, L., He, G., & Wu, J. (2018). Molecular targets and mechanisms of bioactive peptides against metabolic syndromes. *Food & Function*, *9*(1), 42–52.

Li, H., Prairie, N., Udenigwe, C. C., Adebiyi, A. P., Tappia, P. S., Aukema, H. M., et al. (2011). Blood pressure lowering effect of a pea protein hydrolysate in hypertensive rats and humans. *Journal of Agricultural and Food Chemistry*, *59*(18), 9854–9860.

Li, S., Wang, Y., Xue, Z., Jia, Y., Li, R., He, C., et al. (2021). The structure-mechanism relationship and mode of actions of antimicrobial peptides: A review. *Trends in Food Science & Technology*, *109*, 103–115. https://doi.org/10.1016/j.tifs.2021.01.005.

Liao, W., Bhullar, K. S., Chakrabarti, S., Davidge, S. T., & Wu, J. (2018). Egg white-derived tripeptide IRW (Ile-Arg-Trp) is an activator of angiotensin converting enzyme 2. *Journal of Agricultural and Food Chemistry*, *66*(43), 11330–11336. https://doi.org/10.1021/acs.jafc.8b03501.

Liao, W., Fan, H., Davidge, S. T., & Wu, J. (2019). Egg white–derived antihypertensive peptide IRW (Ile-Arg-Trp) reduces blood pressure in spontaneously hypertensive rats via the ACE2/ang (1-7)/mas receptor Axis. *Molecular Nutrition & Food Research*, *63*(9), 1900063. https://doi.org/10.1002/mnfr.201900063.

Liao, W., Jahandideh, F., Fan, H., Son, M., & Wu, J. (2018). Chapter one - egg protein-derived bioactive peptides: Preparation, efficacy, and absorption. In F. Toldrá (Ed.), *Vol. 85. Advances in food and nutrition research* (pp. 1–58). Academic Press.

Liao, W., Sun, G., Xu, D., Wang, Y., Lu, Y., Sun, J., et al. (2021). The blood-pressure-lowering effect of food-protein-derived peptides: A meta-analysis of recent clinical trials. *Food*, *10*(10), 2316.

Liao, W., & Wu, J. (2021). The ACE2/ang (1–7)/MasR axis as an emerging target for antihypertensive peptides. *Critical Reviews in Food Science and Nutrition, 61*(15), 2572–2586. https://doi.org/10.1080/10408398.2020.1781049.

Liu, R. H. (2013). Dietary bioactive compounds and their health implications. *Journal of Food Science, 78*(s1), A18–A25. https://doi.org/10.1111/1750-3841.12101.

Liu, L., Li, S., Zheng, J., Bu, T., He, G., & Wu, J. (2020). Safety considerations on food protein-derived bioactive peptides. *Trends in Food Science & Technology, 96*, 199–207. https://doi.org/10.1016/j.tifs.2019.12.022.

Liu, R., Xing, L., Fu, Q., Zhou, G.-H., & Zhang, W.-G. (2016). A review of antioxidant peptides derived from meat muscle and by-products. *Antioxidants, 5*(3), 32 (Retrieved from https://www.mdpi.com/2076-3921/5/3/32).

Longuet, C., Sinclair, E. M., Maida, A., Baggio, L. L., Maziarz, M., Charron, M. J., et al. (2008). The glucagon receptor is required for the adaptive metabolic response to fasting. *Cell Metabolism, 8*(5), 359–371.

Lu, T.-M., Chiu, H.-F., Lu, Y.-Y., Han, Y.-C., Shen, Y.-C., Venkatakrishnan, K., et al. (2018). Efficacy of fermented goat milk on blood pressure in prehypertensive adults: A randomized, placebo-controlled, clinical trial. *Journal of Food Biochemistry, 42*(2), e12474. https://doi.org/10.1111/jfbc.12474.

Lucey, A. J., Heneghan, C., Manning, E., Kroon, P. A., & Kiely, M. E. (2019). Effect of an egg ovalbumin-derived protein hydrolysate on blood pressure and cardiovascular risk in adults with a mildly elevated blood pressure: A randomized placebo-controlled crossover trial. *European Journal of Nutrition, 58*(7), 2823–2833. https://doi.org/10.1007/s00394-018-1832-9.

Lund, A., Bagger, J. I., Christensen, M., Knop, F. K., & Vilsbøll, T. (2014). Glucagon and type 2 diabetes: The return of the alpha cell. *Current Diabetes Reports, 14*(12), 555.

Luo, J., Morthensen, S. T., Meyer, A. S., & Pinelo, M. (2014). Filtration behavior of casein glycomacropeptide (CGMP) in an enzymatic membrane reactor: Fouling control by membrane selection and threshold flux operation. *Journal of Membrane Science, 469*, 127–139. . https://doi.org/10.1016/j.memsci.2014.06.024.

Majumder, K., & Wu, J. (2015). Molecular targets of antihypertensive peptides: Understanding the mechanisms of action based on the pathophysiology of hypertension. *International Journal of Molecular Sciences, 16*(1), 256–283.

Manandhar, S., Kabekkodu, S. P., & Pai, K. S. R. (2020). Aberrant canonical Wnt signaling: Phytochemical based modulation. *Phytomedicine, 76*, 153243. https://doi.org/10.1016/j.phymed.2020.153243.

Mishra, S., Verma, S. S., Rai, V., Awasthee, N., Chava, S., Hui, K. M., et al. (2019). Long non-coding RNAs are emerging targets of phytochemicals for cancer and other chronic diseases. *Cellular and Molecular Life Sciences, 76*(10), 1947–1966. https://doi.org/10.1007/s00018-019-03053-0.

Montoro-García, S., Zafrilla-Rentero, M. P., Celdran-de Haro, F. M., Piñero-de Armas, J. J., Toldrá, F., Tejada-Portero, L., et al. (2017). Effects of dry-cured ham rich in bioactive peptides on cardiovascular health: A randomized controlled trial. *Journal of Functional Foods, 38*, 160–167.

Moradi, M., Daneshzad, E., & Azadbakht, L. (2020). The effects of isolated soy protein, isolated soy isoflavones and soy protein containing isoflavones on serum lipids in post-menopausal women: A systematic review and meta-analysis. *Critical Reviews in Food Science and Nutrition, 60*(20), 3414–3428.

Mukherjee, R., Chakraborty, R., & Dutta, A. (2016). Role of fermentation in improving nutritional quality of soybean meal - A review. *Asian-Australasian Journal of Animal Sciences, 29*(11), 1523–1529. https://doi.org/10.5713/ajas.15.0627.

Musunuru, K. (2010). Atherogenic dyslipidemia: Cardiovascular risk and dietary intervention. *Lipids, 45*(10), 907–914. https://doi.org/10.1007/s11745-010-3408-1.

Nolan, P. B., Carrick-Ranson, G., Stinear, J. W., Reading, S. A., & Dalleck, L. C. (2017). Prevalence of metabolic syndrome and metabolic syndrome components in young adults: A pooled analysis. *Preventive Medicine Reports*, *7*, 211–215. https://doi.org/10.1016/j.pmedr.2017.07.004.

Nongonierma, A. B., & FitzGerald, R. J. (2014). An in silico model to predict the potential of dietary proteins as sources of dipeptidyl peptidase IV (DPP-IV) inhibitory peptides. *Food Chemistry*, *165*, 489–498. https://doi.org/10.1016/j.foodchem.2014.05.090.

Ogawa, Y., Shobako, N., Fukuhara, I., Satoh, H., Kobayashi, E., Kusakari, T., et al. (2019). Rice bran supplement containing a functional substance, the novel peptide leu-Arg-ala, has anti-hypertensive effects: A double-blind, randomized, placebo-controlled study. *Nutrients*, *11*(4), 726 (Retrieved from https://www.mdpi.com/2072-6643/11/4/726).

Oja, P., Titze, S., Bauman, A., de Geus, B., Krenn, P., Reger-Nash, B., et al. (2011). Health benefits of cycling: A systematic review. *Scandinavian Journal of Medicine & Science in Sports*, *21*(4), 496–509. https://doi.org/10.1111/j.1600-0838.2011.01299.x.

Parolini, C., Vik, R., Busnelli, M., Bjørndal, B., Holm, S., Brattelid, T., et al. (2014). A Salmon protein hydrolysate exerts lipid-independent anti-atherosclerotic activity in ApoE-deficient mice. *PLoS One*, *9*(5), e97598. https://doi.org/10.1371/journal.pone.0097598.

Patel, V. B., Zhong, J.-C., Grant, M. B., & Oudit, G. Y. (2016). Role of the ACE2/angiotensin 1-7 Axis of the renin-angiotensin system in heart failure. *Circulation Research*, *118*(8), 1313–1326. https://doi.org/10.1161/CIRCRESAHA.116.307708.

Pedersen, B. K., & Saltin, B. (2015). Exercise as medicine - evidence for prescribing exercise as therapy in 26 different chronic diseases. *Scandinavian Journal of Medicine & Science in Sports*, *25*(Suppl 3), 1–72. https://doi.org/10.1111/sms.12581.

Plat, J., Severins, N., & Mensink, R. P. (2019). Improvement of pulse wave velocity and metabolic cardiovascular risk parameters through egg protein hydrolysate intake: A randomized trial in overweight or obese subjects with impaired glucose tolerance or type 2 diabetes. *Journal of Functional Foods*, *52*, 418–423.

Plat, J., Severins, N., Morrison, S., & Mensink, R. P. (2017). Effects of NWT-03, an egg-protein hydrolysate, on blood pressure in normotensive, high-normotensive and mild-hypertensive men and women: A dose-finding study. *British Journal of Nutrition*, *117*(7), 942–950. https://doi.org/10.1017/S0007114517000836.

Powell, T., Bowra, S., & Cooper, H. J. (2016). Subcritical water processing of proteins: An alternative to enzymatic digestion? *Analytical Chemistry*, *88*(12), 6425–6432.

Qin, X., Qiu, L., Tang, G., Tsoi, M.-F., Xu, T., Zhang, L., et al. (2020). Prevalence of metabolic syndrome among ethnic groups in China. *BMC Public Health*, *20*(1), 297. https://doi.org/10.1186/s12889-020-8393-6.

Qin, L.-Q., Xu, J.-Y., Dong, J.-Y., Zhao, Y., van Bladeren, P., & Zhang, W. (2013). Lactotripeptides intake and blood pressure management: A meta-analysis of randomised controlled clinical trials. *Nutrition, Metabolism, and Cardiovascular Diseases*, *23*(5), 395–402.

Reynolds, K., Chin, A., Lees, K. A., Nguyen, A., Bujnowski, D., & He, J. (2006). A meta-analysis of the effect of soy protein supplementation on serum lipids. *The American Journal of Cardiology*, *98*(5), 633–640. https://doi.org/10.1016/j.amjcard.2006.03.042.

Ribeiro, P. V. D., Andrade, P. A., Hermsdorff, H. H. M., dos Santos, C. A., Cotta, R. M. M., Estanislau, J., et al. (2019). Dietary non-nutrients in the prevention of non-communicable diseases: Potentially related mechanisms. *Nutrition*, *66*, 22–28. https://doi.org/10.1016/j.nut.2019.03.016.

Saklayen, M. G. (2018). The global epidemic of the metabolic syndrome. *Current Hypertension Reports*, *20*(2). https://doi.org/10.1007/s11906-018-0812-z.

Saleh, L., Schrier, N. L., Bruins, M. J., Steegers, E. A., van den Meiracker, A. H., & Visser, W. (2018). Effect of oral protein hydrolysate on glucose control in patients with gestational diabetes. *Clinical Nutrition*, *37*(3), 878–883.

Sánchez-Rivera, L., Martínez-Maqueda, D., Cruz-Huerta, E., Miralles, B., & Recio, I. (2014). Peptidomics for discovery, bioavailability and monitoring of dairy bioactive peptides. *Food Research International, 63*, 170–181.
Sanjukta, S., & Rai, A. K. (2016). Production of bioactive peptides during soybean fermentation and their potential health benefits. *Trends in Food Science & Technology, 50*, 1–10. https://doi.org/10.1016/j.tifs.2016.01.010.
Santos, R. A. S., e Silva, A. C. S., Maric, C., Silva, D. M. R., Machado, R. P., de Buhr, I., et al. (2003). Angiotensin-(1–7) is an endogenous ligand for the G protein-coupled receptor Mas. *Proceedings of the National Academy of Sciences, 100*(14), 8258–8263. https://doi.org/10.1073/pnas.1432869100.
Sharifi-Rad, M., Anil Kumar, N. V., Zucca, P., Varoni, E. M., Dini, L., Panzarini, E., et al. (2020). Lifestyle, oxidative stress, and antioxidants: Back and forth in the pathophysiology of chronic diseases. *Frontiers in Physiology, 11*(694). https://doi.org/10.3389/fphys.2020.00694.
Sharma, P., Kaur, H., Kehinde, B. A., Chhikara, N., Sharma, D., & Panghal, A. (2021). Food-derived anticancer peptides: A review. *International Journal of Peptide Research and Therapeutics, 27*(1), 55–70. https://doi.org/10.1007/s10989-020-10063-1.
Shaw, K. A., Gennat, H. C., O'Rourke, P., & Del Mar, C. (2006). Exercise for overweight or obesity. *Cochrane Database of Systematic Reviews, 4*.
Shitu, A., Izhar, S., & Tahir, T. (2015). Sub-critical water as a green solvent for production of valuable materials from agricultural waste biomass: A review of recent work. *Global Journal of Environmental Science and Management, 1*(3), 255–264.
Soliman, G. A. (2019). Dietary fiber, atherosclerosis, and cardiovascular disease. *Nutrients, 11*(5), 1155 (Retrieved from https://www.mdpi.com/2072-6643/11/5/1155).
Takahashi, Y., Kamata, A., Nishimura, M., & Nishihira, J. (2021). Effect of one-week administration of dipeptidyl peptidase-IV inhibitory peptides from chum salmon milt on postprandial blood glucose level: A randomised, placebo-controlled, double-blind, crossover, pilot clinical trial. *Food & Function, 12*(18), 8544–8551.
Taylor, R. (2012). Insulin resistance and type 2 diabetes. *Diabetes, 61*(4), 778–779. https://doi.org/10.2337/db12-0073.
Tiekou Lorinczova, H., Deb, S., Begum, G., Renshaw, D., & Zariwala, M. G. (2021). Comparative assessment of the acute effects of whey, rice and potato protein isolate intake on markers of glycaemic regulation and appetite in healthy males using a randomised study design. *Nutrients, 13*(7), 2157.
Turpeinen, A. M., Järvenpää, S., Kautiainen, H., Korpela, R., & Vapaatalo, H. (2013). Antihypertensive effects of bioactive tripeptides—A random effects meta-analysis. *Annals of Medicine, 45*(1), 51–56.
Udenigwe, C. C., & Rouvinen-Watt, K. (2015). The role of food peptides in lipid metabolism during dyslipidemia and associated health conditions. *International Journal of Molecular Sciences, 16*(5), 9303–9313 (Retrieved from https://www.mdpi.com/1422-0067/16/5/9303).
Ulug, S. K., Jahandideh, F., & Wu, J. (2021). Novel technologies for the production of bioactive peptides. *Trends in Food Science & Technology, 108*, 27–39. https://doi.org/10.1016/j.tifs.2020.12.002.
Vermeirssen, V., Camp, J. V., & Verstraete, W. (2004). Bioavailability of angiotensin I converting enzyme inhibitory peptides. *British Journal of Nutrition, 92*(3), 357–366. https://doi.org/10.1079/BJN20041189.
von Meyenn, F., Porstmann, T., Gasser, E., Selevsek, N., Schmidt, A., Aebersold, R., et al. (2013). Glucagon-induced acetylation of Foxa2 regulates hepatic lipid metabolism. *Cell Metabolism, 17*(3), 436–447. https://doi.org/10.1016/j.cmet.2013.01.014.
Wang, Y., Mi, J., Shan, X. Y., Wang, Q. J., & Ge, K.-Y. (2007). Is China facing an obesity epidemic and the consequences? The trends in obesity and chronic disease in China. *International Journal of Obesity, 31*(1), 177–188. https://doi.org/10.1038/sj.ijo.0803354.

Wang, L., Niu, D., Wang, X., Khan, J., Shen, Q., & Xue, Y. (2021). A novel machine learning strategy for the prediction of antihypertensive peptides derived from food with high efficiency. *Food*, *10*(3), 550 (Retrieved from https://www.mdpi.com/2304-8158/10/3/550).

Wang, Y.-T., Russo, D. P., Liu, C., Zhou, Q., Zhu, H., & Zhang, Y.-H. (2020). Predictive modeling of angiotensin I-converting enzyme inhibitory peptides using various machine learning approaches. *Journal of Agricultural and Food Chemistry*, *68*(43), 12132–12140. https://doi.org/10.1021/acs.jafc.0c04624.

Wen, C., Zhang, J., Zhang, H., Duan, Y., & Ma, H. (2020). Plant protein-derived antioxidant peptides: Isolation, identification, mechanism of action and application in food systems: A review. *Trends in Food Science & Technology*, *105*, 308–322. https://doi.org/10.1016/j.tifs.2020.09.019.

Wright, N., Wilson, L., Smith, M., Duncan, B., & McHugh, P. (2017). The BROAD study: A randomised controlled trial using a whole food plant-based diet in the community for obesity, ischaemic heart disease or diabetes. *Nutrition & Diabetes*, 7(3), e256. https://doi.org/10.1038/nutd.2017.3.

Wu, J., Liao, W., & Udenigwe, C. C. (2017). Revisiting the mechanisms of ACE inhibitory peptides from food proteins. *Trends in Food Science & Technology*, *69*, 214–219. https://doi.org/10.1016/j.tifs.2017.07.011.

Wu, Y., Pan, Q., Yan, H., Zhang, K., Guo, X., Xu, Z., et al. (2018). Novel mechanism of Foxo1 phosphorylation in glucagon signaling in control of glucose homeostasis. *Diabetes*, *67*(11), 2167–2182. https://doi.org/10.2337/db18-0674.

Wu, S., Qi, W., Li, T., Lu, D., Su, R., & He, Z. (2013). Simultaneous production of multi-functional peptides by pancreatic hydrolysis of bovine casein in an enzymatic membrane reactor via combinational chromatography. *Food Chemistry*, *141*(3), 2944–2951. https://doi.org/10.1016/j.foodchem.2013.05.050.

Yokoyama, Y., Nishimura, K., Barnard, N. D., Takegami, M., Watanabe, M., Sekikawa, A., et al. (2014). Vegetarian diets and blood pressure: A meta-analysis. *JAMA Internal Medicine*, *174*(4), 577–587. https://doi.org/10.1001/jamainternmed.2013.14547.

Yu, Z., Yin, Y., Zhao, W., Liu, J., & Chen, F. (2012). Anti-diabetic activity peptides from albumin against α-glucosidase and α-amylase. *Food Chemistry*, *135*(3), 2078–2085. https://doi.org/10.1016/j.foodchem.2012.06.088.

Zhang, Y.-J., Gan, R.-Y., Li, S., Zhou, Y., Li, A.-N., Xu, D.-P., et al. (2015). Antioxidant phytochemicals for the prevention and treatment of chronic diseases. *Molecules*, *20*(12), 21138–21156 (Retrieved from https://www.mdpi.com/1420-3049/20/12/19753).

Zhao, W., Xu, G., Yang, R., & Katiyo, W. (2013). Preparation of casein phosphopeptides using a novel continuous process of combining an enzymatic membrane reactor with anion-exchange chromatography. *Journal of Food Engineering*, *117*(1), 105–112. https://doi.org/10.1016/j.jfoodeng.2013.02.015.

Zhu, F., Du, B., & Xu, B. (2018). Anti-inflammatory effects of phytochemicals from fruits, vegetables, and food legumes: A review. *Critical Reviews in Food Science and Nutrition*, *58*(8), 1260–1270. https://doi.org/10.1080/10408398.2016.1251390.

Zhu, C.-F., Li, G.-Z., Peng, H.-B., Li, Y., Zhang, F., & Chen, Y. (2010). Therapeutic effects of marine collagen peptides on Chinese patients with type 2 diabetes mellitus and primary hypertension. *The American Journal of the Medical Sciences*, *340*(5), 360–366.

Zhu, C.-F., Li, G.-Z., Peng, H.-B., Zhang, F., Chen, Y., & Li, Y. (2010). Treatment with marine collagen peptides modulates glucose and lipid metabolism in Chinese patients with type 2 diabetes mellitus. *Applied Physiology, Nutrition, and Metabolism*, *35*(6), 797–804.

Zong, G., Li, Y., Wanders, A. J., Alssema, M., Zock, P. L., Willett, W. C., et al. (2016). Intake of individual saturated fatty acids and risk of coronary heart disease in US men and women: Two prospective longitudinal cohort studies. *BMJ*, *355*, i5796. https://doi.org/10.1136/bmj.i5796.

FSC
www.fsc.org
MIX
Papier | Fördert gute Waldnutzung
FSC® C083411

Kolibri 360 GmbH
Ferdinand-Jühlke-Straße 7
99095 Erfurt, Deutschland
produktsicherheit@kolibri360.de